T0229698

Microbiological Decomposition of
Chlorinated Aromatic Compounds

MICROBIOLOGY SERIES

Series Editors
ALLEN I. LASKIN
Sommerset, New Jersey

RICHARD I. MATELES
Stauffer Chemical Company
Westport, Connecticut

Additional Volumes in Preparation

Microbiological Decomposition of Chlorinated Aromatic Compounds

Melissa L. Rochkind-Dubinsky
International Technology Corporation
Knoxville, Tennessee

Gary S. Sayler
Department of Microbiology and
Graduate Program in Ecology
The University of Tennessee
Knoxville, Tennessee

James W. Blackburn
Energy, Environment, and Resources Center
The University of Tennessee
Knoxville, Tennessee

MARCEL DEKKER, INC.   New York and Basel

Library of Congress Cataloging-in-Publication Data

Rochkind-Dubinsky, Melissa L.
  Microbiological decomposition of chlorinated aromatic
compounds.

  (Microbiology series ; v. 18)
  Bibliography: p.
  Includes index.
  1. Organochlorine compounds--Biodegradation.
2. Aromatic compounds--Biodegradation. 3. Bacteria,
Heterotropic. 4. Microbial metabolism. I. Sayler,
Gary S.   II. Blackburn, James W.
          III. Title. IV. Series.
QR92.C55R63   1986   576.1'133   86-16715
ISBN 0-8247-7527-9

MARCEL DEKKER, INC.
270 Madison Avenue, New York, New York 10016

Current printing (last digit):
10 9 8 7 6 5 4 3 2 1

PRINTED IN THE UNITED STATES OF AMERICA

Preface

This book was initiated because of a need to bring together a review of the literature pertaining to microbial metabolism of chlorinated aromatic compounds. The information gathered here is extensive although not exhaustive. Most attention has been given to reports of bacterial, fungal, and cyanobacterial pathways of substrate degradation where metabolites or end products have been identified. Studies that report data on metabolites arising from incubation of the substrate with mixed cultures or environmental samples and studies that show disappearance of the compound have also been evaluated and included.

In addition to separate chapters on each class of chlorinated aromatic compound, reviews of microbial physiology, genetics, and methods of biodegradation assessment are included. One chapter reviews biodegradation of these compounds in scaled-up processes. The potential biodegradation pathways for all classes of chloroaromatic compounds have been brought together into an overview diagram.

Many factors are involved in assessing the biodegradability of a compound, including the nature of the molecule, substrate concentration, environmental parameters, availability of nutrients and growth factors, and presence of degradative microorganisms. Not enough information is currently available to permit extrapolation from one environment to another or to utilize data on a similar compound to assess the biodegradability potential of a given substrate.

This book has been funded in part by the United States Environmental Protection Agency under Contract 68-03-3074 to IT Corporation.

The authors thank Dr. P. R. Sferra, the Technical Project Monitor (Alternative Technologies Division, Hazardous Waste Engineering Research Laboratory, USEPA, Cincinnati), for his helpful support and assistance throughout this project. We also acknowledge Mr. David R. Watkins of the same organization for his considerable help in the planning and inception of this project and Dr. John A. Glaser for his advice and assistance as Technical Project Monitor during the final phases of the project. Mrs. Kim Truong of IT Corporation provided invaluable assistance in reviewing the manuscript and assisting in the production of this book. Her contribution is gratefully appreciated. Dr. David T. Gibson, Director of the Center for Applied Microbiology, University of Texas at Austin, reviewed the book and offered his unique technical perspective on its contents. The authors believe these comments improve the book and are indebted to Dr. Gibson for his cooperation. Dr. Dennis D. Focht, Department of Soil and Environmental Sciences, University of California at Riverside, reviewed the chapter on DDT. We are very appreciative of his helpful comments.

Melissa L. Rochkind-Dubinsky
Gary S. Sayler
James W. Blackburn

Introduction

The first synthesized organochloride compound, ethyl chloride, was prepared about 1440, but large-scale synthesis of industrially important chlorinated organics, including chlorinated aromatic compounds, occurred only during the past few decades (214a). In general, the chlorinated aromatics of industrial synthesis or by-products thereof represent one class of xenobiotic recalcitrant compounds. These compounds have few or no naturally occurring structural analogs and are persistent or resistant to both biological and abiotic degradation. Many of the chlorinated aromatics share similar physicochemical properties of low water solubility and high K_{ow} (octanol-water partition coefficients), which suggest lipophilicity or bioaccumulation potential.

Properties such as persistence, bioaccumulation, and demonstrable chronic and acute toxicity to human and nonhuman animal populations cause immediate concern because of their environmental health effects and their potential for ecosystem perturbations on long- and short-term exposure. These concerns cause the frequent appearance of these chemicals on EPA priority pollutant lists and have led to extensive research on their fate in the environment and their potential for microbiological transformation to less hazardous molecules.

The term biodegradation has had many different meanings. The Biodegradation Task Force, Safety of Chemicals Committee, Brussels (299) has defined biodegradation as the molecular degradation of an organic substance resulting from the complex action of living organisms. A substance is said to be biodegraded to an environmentally acceptable extent when environmentally undesirable properties are lost. Loss of some characteristic function or property of the substance by biodegradation may be referred to as biological transformation. In this book, we have attempted to restrict use of the term biodegradation in favor of more specific terminology. In this respect, biotransformation refers to any alteration of an organic molecule by organisms, and mineralization means the transformation of an organic molecule to its inorganic component parts with release of halide, CO_2, and/or methane.

The potential for microbial transformation of chlorinated aromatic compounds is related to the two fundamental roles of heterotrophic microorganisms in the global ecosystem. Both roles relate to the central concept of microbial decomposition of organic matter to release stored energy in the organic molecules (whether natural or anthropogenic in origin) and to return essential nutrients, such as CO_2, to biogeospheric nutrient pools. The first has thermodynamic implications, and the second relates to elemental and nutrient cycling. Although these simplified generalizations apply to most naturally produced organic matter, certain aromatic polymers, such as lignin, are persistent in the environment. Factors contributing to persistence of organic compounds have been discussed previously (92a) and can be summarized as insolubility, large molecular size or polymeric nature, toxicity, and anthropogenic origin. Although most or all of these factors are important for various chlorinated aromatics, the anthropogenic origin of most chlorinated aromatics is critical for prediction of the rate of elimination of the compound from the environment and its eventual fate.

In general, microbial decomposition or biodegradative capabilities have coevolved with the synthesis of organic matter by plants and animals over the millennia. However, in the case of many chlorinated organics of industrial synthesis, 40−50 years is an unlikely time frame to expect evolution of enzyme systems capable of decomposition of such compounds. Yet, evidence has accumulated that some of these compounds can be biologically transformed and extensively biodegraded by a diversity of heterotrophic microorganisms.

Such evidence has arisen primarily from studies using pure and mixed microbial cultures and from lesser-controlled environmental fate experiments. This evidence has promoted much additional research on the molecular mechanism of biodegradation, which in turn has permitted studies leading to increased knowledge of more detailed aspects of biodegradation itself. Currently, major questions exist as to the rate at which biodegradation occurs in various environments and the potential for kinetic prediction of pollutant fate based on laboratory and environmental observations. Research needs directed at this major question have led to a renewed focus on individual microbial populations that may be specifically responsible for biodegradation of a narrow spectrum of chlorinated aromatic substrates, and on physicochemical environmental parameters that may modulate both the population and their catabolic activity. Developments in these areas have led to the relatively recent detection of bacterial strains harboring extrachromosomal DNA (plasmids) that genetically encodes enzymes which mediate the biodegradation of specific groups of aromatic and chlorinated aromatic compounds. Coupled with the availability of new molecular genetic techniques, such as DNA probe technology, it has become feasible to plan research to examine catabolic gene maintenance and transfer in natural populations and to detect specific biodegradative microorganisms in the environment. Such developments will lead to greater predictive capabilities on the long-term persistence of selected chlorinated aromatic pollutants and to insight in the evolution and reassortment of genes responsible for biodegradation. With these genetic techniques there is a potential, therefore, to enhance biodegradative capacity among natural populations.

This book begins with three chapters that provide an overview of microbial physiology, genetic information transfer and processing, technologies for gene manipulation, and methods of biodegradation assessment. These principles are applicable to microorganisms in general and are specifically relevant to metabolism of chlorinated aromatic compounds. Readers wishing review of these basic microbiological concepts should read these chapters first.

Chapter 4 includes a review of pathways of metabolism of non-chlorinated aromatic compounds from which the chlorinated pollutants are derived. The microbial metabolism of most of these compounds has been studied extensively, and these data form the basis for an understanding of the biotransformation pathways of the more complex chlorinated aromatic molecules.

The following nine chapters discuss the microbial metabolism of the chlorobenzoic acids, chlorobenzenes, chlorophenols, pentachlorophenol, chlorophenoxy and chlorophenyl herbicides, phenylamide and miscellaneous herbicides, chlorinated biphenyls, DDT and related compounds, and chlorinated dioxins and dibenzofurans. These compound classes represent all the major classes of chlorinated aromatic molecules with environmental pollution potential.

Each chapter details the information available on metabolism of these compounds by pure cultures or consortia of bacteria, cyanobacteria, and fungi. Emphasis has been placed on studies in which metabolites have been identified and, where possible, complete or partial pathways have been reported. Studies reporting evolution of CO_2 and chloride release have been reviewed and are discussed as well. The disappearance of these pollutants in soils, water, sewage, and other environments has also been noted, although the body of literature relating to herbicide disappearance is so extensive that it has only been summarized and has not been reviewed as completely as the other topics. The fate of these compounds in scaled-up biological treatment processes is the subject of Chapter 14.

Chapter 15 summarizes the information reviewed here and draws together the pathways of metabolism of chlorinated aromatic molecules into an overview diagram indicating the potential for biodegradation of a compound under optimum conditions. This is an idealized representation, as few environmental situations would comprise all the factors necessary for complete biotransformation and/or mineralization of these recalcitrant compounds.

A glossary of many of the scientific terms that appear in the text is appended in order to aid the reader. Also included are a list of cited references and bibliography of additional references including several excellent reviews on the biodegradation of various classes of organic compounds. All of the compounds noted in this book, including metabolites, are registered in an illustrated alphabetical list in the appendix.

This book is intended to be a general reference for environmental decision makers who are interested in the fate of chlorinated aromatic compounds with respect to microbial activity. It is also meant to provide a resource for scientists and engineers involved in environmental predictive fate assessments. In addition, the book is designed to be a continuing resource for environmental microbiologists interested in the areas of metabolic pathways of chlorinated aromatic compound dissimilation and biodegradative fate of these potential pollutants.

The organization of this book into specific chapters and associated review sections is intended to facilitate its use by these diverse groups. This is an extensive, although not exhaustive, review of the literature pertaining to the biodegradative fate of these classes of chlorinated aromatic compounds.

Contents

Microbiological Decomposition of
Chlorinated Aromatic Compounds

1

Overview of Microbial Physiology

A study of the specific pathway of degradation of a compound by an organism necessarily is limited to specific biochemical reactants, products, and reactions occurring within the cell. Much of the content of this document is concerned with just such features. When genetics of the pathways are discussed, the relationship to the total cell is even more remote. Therefore, it is important to begin with a firm understanding of the physiology of the microbial cell, its structure, requirements for growth and survival, and relationship to its environment. While the cell is often described as a microscopic biochemical reactor, the activities of the cell are intimately connected to and shaped by its external environment. Information presented in this chapter is based on several general references, which may be consulted for further details (4,107,276).

The primary physical difference between bacteria and fungi is the presence of a membrane surrounding the DNA material. The enclosed structure is called the nucleus. Cells containing a membrane-bound nucleus are referred to as eukaryotic cells. The DNA of bacteria and cyanobacteria is contained in a diffuse region without a surrounding membrane called the nucleoid and these forms are called prokaryotes (Figure 1). Although it is tempting to consider bacteria as primitive compared to eukaryotes, the complexity of their biochemical reactions, and their regulatory and adaptive mechanisms, preclude such a label.

A.

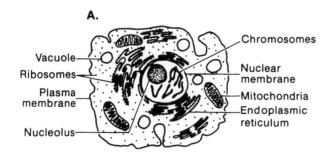

B.

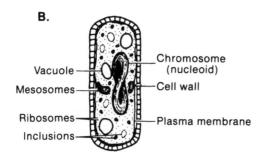

Figure 1   A. Typical eukaryotic animal cell. B. Typical prokaryotic rod-shaped bacterium. (Redrawn from Ref. 4.)

MICROBIAL CELL STRUCTURE

Most bacterial cells average about 1 to 2 microns in length and are rod shaped. Among all bacteria, however, cell size ranges from one-tenth to 100 times the average bacterial size. Although most of the bacteria found in the environment are rod shaped, some water isolates are shaped like commas or as spirals, and many bacteria, especially pathogenic species, are spherical (cocci). Other bacteria take unique shapes and forms as well.

The shape of the cell is conferred by a rigid cell wall composed mostly of peptidoglycan, lipid, lipopolysaccharide, and protein. There are charged polymers within the cell wall which assist in the uptake of ions and some nutrients. The wall also acts as a molecular sieve which prevents entry of some large molecules and prohibits loss of proteins, i.e., enzymes, from within the cell.

Bacteria and cyanobacteria (Figure 2) can be divided into two groups based on cell wall structure and composition. Classically the bacteria have been differentiated into Gram-positive and Gram-negative groups according to a staining procedure called the Gram stain.

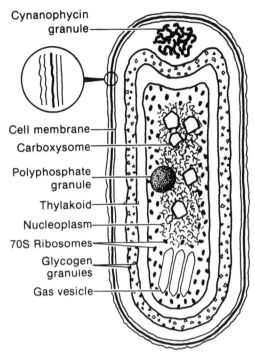

Cynanophycin granule

Cell membrane
Carboxysome

Polyphosphate granule

Thylakoid

Nucleoplasm

70S Ribosomes

Glycogen granules

Gas vesicle

Figure 2 Schematic diagram of cyanobacterial vegetative cell. Insert: Enlarged view of cell envelope, showing outer membrane and peptidoglycan wall layers and cell membrane.

Electron microscopic techniques have shown differences in the form of the cell walls between the two types of bacteria. The Gram-positive cell wall is composed of a single dense layer of peptidoglycan. Embedded in the peptidoglycan matrix are polysaccharides and teichoic acids. The cell wall is closely associated with the cytoplasmic membrane which has a double-track appearance with a central transparent layer. The Gram-negative cell wall is more complex. The outermost layer is a wavy, double-track membrane which differs in chemical composition and in function from the cytoplasmic membrane. This layer is composed of lipopolysaccharides, phospholipids, and proteins. Internal to the outer membrane is a thin rigid layer of peptidoglycan. Between the cytoplasmic membrane and the outer cell wall membrane lies the periplasmic space containing enzymes (Figure 3). In contrast, the fungi have cell walls composed mainly of polysaccharides. The particular types of polysaccharide are characteristic of the taxonomic group of the fungi.

Some bacterial cells are motile, and of these the most common mechanism is by use of flagella, hairlike helical structures several

A.

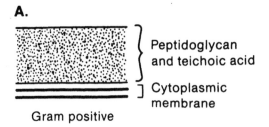

Gram positive

B.

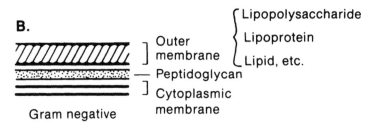

Gram negative

Figure 3 Cell wall structures seen in thin-section electron micro-
scopy. A. Diagrammatic representation of the Gram-positive wall. B.
Diagrammatic representation of the Gram-negative wall. (Redrawn
from Ref. 107).

times the length of the cell. Some genera possess only one or two
flagella, while in other genera the flagella are present over the entire
cell surface. The flagella rotate to propel the cell through the water.
Some bacteria, including some cyanobacteria, move in a characteristic gliding motion by flexing the cell wall against a surface in a manner similar to inchworm movement.

Some bacteria have the ability to attach to solid surfaces. In a
few genera this may be accomplished by hairlike pili or by structures
called holdfasts. In most bacteria attachment occurs by a capsule or
slime layer composed of organic polymers, mostly polysaccharides.
After initial contact with the surface, the cell synthesizes polymers
which bridge the gap and attach firmly to the surface. It then may
become impossible to remove the cell without destroying it. When the
cell divides, the nonattached portion can move, but the new cell
arising from the attached portion of the cell remains in place.

Other functions of the capsule or slime layer include protection
of the cell from such conditions as dessication. In many pathogenic
bacteria, the presence of the capsule affords protection against white
blood cells and antibodies. The capsule seems also to serve as storage
sites for excess nutrients or wastes.

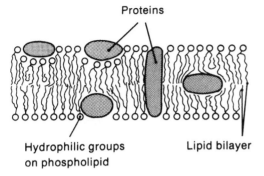

Proteins

Hydrophilic groups   Lipid bilayer
on phospholipid

Figure 4   Structure of membranes, a diagrammatic representation.
(Adapted from Ref. 107.)

In freshwater environments, the cell contains a higher concentration of salts than the surrounding medium. The cell would expand
and lyse without the protection against osmotic shock afforded by
the cell wall together with the cell membrane. The cytoplasmic membrane is internal to the cell wall and has additional functions (Figure
4). It is selectively permeable and often facilitates movement of a
substrate into or out of the cell against a concentration gradient. For
other substrates transport is almost completely prevented. The rate
of transport can be specific for the particular substrate, and two
substrates very closely related structurally can have very different
transport rates. Some substances enter or leave by passive diffusion.
The cytoplasmic membrane also maintains the osmotic gradient, is the
site of enzymes involved with cell wall synthesis, and is the site of
oxidative metabolism and energy conversions. More complex construc-
tions of the cytoplasmic membranes are found in specialized groups
of bacteria such as the cyanobacteria and the methane-utilizing bacteria. The cyanobacteria contain internal membrane structures called
thylakoids which contain the photosynthetic apparatus.
        In fungi (eukaryotes), cell growth occurs only at the tip of the
hypha, and the plasma membrane below the tip contains a large number of membrane-bound vesicles which may hold the enzymes and cell
wall precursors needed for cell growth. In addition, the fungal plasma
membrane is involved with osmotic regulation and nutrient uptake.
Functions such as oxidative metabolism are reserved to certain membrane-bound organelles which are absent in bacteria.
        All cells contain chromosomal DNA; in bacteria it is circular and
double stranded, resembling a helical ladder. Bacteria also contain
extrachromosomal DNA called plasmids which code for auxiliary functions in the cell, such as resistance to antibiotics and heavy metals
and ability to metabolize some organic compounds. Fungi contain a
number of linear chromosomes. The DNA contains the code which

guides the structure and metabolism of the cell. Specific features in
the functioning of DNA will be discussed in a later section.

Eukaryotic cells contain mitochondria, which are organelles bounded by a double membrane. These function in ATP generation and the
oxidative metabolism of substrates, activities carried out in bacteria
at the cytoplasmic membrane. There are other specialized structures
within some bacterial or fungal cells which function in storage of excess nutrients or gaseous products. The Gram-positive genera *Bacillus* and *Clostridium* form spores when exposed to unfavorable conditions. The spores are extremely resistant to heat, dessication, radiation, acids, and chemical disinfectants, yet when exposed to favorable conditions will germinate and form a vegetative cell within hours.

GROWTH REQUIREMENTS

Bacteria of one type or another have been found in all environments
and under all conditions with the possible exception of pure vacuum.
A specific bacterial species may grow under a wide variety of conditions or it may have very exacting requirements for cell growth.

Certain nutrients are required by all cells. Carbon is most important and those cells that obtain it from organic substrates are referred to as heterotrophs. Autotrophs can fix carbon from carbon
dioxide. A few specialized groups can utilize other substrates; methyl-
otrophs, for example, can oxidize methane at aerobic/anaerobic in-
terfaces. Other essential nutrients include phosphorus, usually derived from phosphates, and nitrogen, usually obtained from nitrate
or ammonia. These three elements are the most common nutrients
that limit growth. Other necessary growth factors include sulfur,
magnesium, potassium, calcium, and other metallic elements. While
some bacteria synthesize all their required vitamins and growth factors, other bacteria must obtain some from the environment. Water
is also a specific requirement in cellular metabolism. The bacterial
cell is composed of about 80% water, and water is both the solvent
and a specific cofactor in many biochemical reactions.

An important physical parameter for growth is temperature. Cellular enzyme activity is also governed partially by the ambient temperature. An increase in temperature may inactivate enzymes or may
be lethal to the cells, while a decrease in temperature may simply
inhibit growth. Upon warming, the cells may resume normal cellular
function. An individual microbial species usually has a minimum, an
optimum, and a maximum temperature for growth. Those that grow
best at temperatures below 20°C are called psychrophiles. Mesophiles
grow from about 15°C to about 45°C, and most bacteria are grouped
into this category. Thermophiles grow at temperatures above 50°C.
These names permit categorization of a situation which in reality
represents a gradation of microbial tolerances for temperatures ranging from the arctic environment to thermal springs.

The oxygen requirements of microorganisms vary considerably. Obligate aerobes grow only in the presence of air and use aerobic respiration to obtain energy. Obligate anaerobes grow only in the absence of air. The sensitivity of anaerobes to molecular oxygen is due to lack of enzymes which render the toxic superoxide free radical ion harmless through reduction. Facultative anaerobes will grow in the presence or absence of air using alternate chemical electron acceptors such as O_2 or NO_3. Microaerophiles have a narrow range of tolerance for their gaseous environment and require a reduced oxygen environment or in some cases an increased proportion of carbon dioxide.

Bacteria also respond to changes in pH of the medium. Most bacteria (neutrophiles) grow best at neutral pH (pH 7). However, acid-producing bacteria (acidophiles) grow very well at lower pH values and strains adapted to alkaline environments (basophiles) grow at pH of 8 or 9. Fungi often tolerate extremes of pH better than bacteria. Halophilic bacteria require high salt concentrations, while other species are salt-tolerant although high salt concentrations are not mandatory for survival. Organisms have been isolated from the deep ocean. Some strains survive both at the deep ocean pressure and at atmospheric pressure and are called barotolerant. Other deep ocean strains have only been kept alive by bringing them to the surface in pressurized vessels, and these bacteria are said to be barophilic. Many common strains of bacteria found at atmospheric pressure are killed when placed in a high-pressure environment.

THE CELL GROWTH CYCLE

The bacterial cell normally grows until it reaches a certain size, at which time it divides by binary fission into two identical daughter cells. During this time the bacterial DNA replicates and is partitioned to opposite sides of the growing cell. The cell wall and cytoplasmic membrane divide the cell in one of two ways depending on the particular genus.

In some genera the elongated cell pinches in equatorially until two cells are formed. In other genera a double cytoplasmic membrane is formed in the middle of the cell followed by synthesis of a double cell wall. When wall construction is complete the two daughter cells separate. Cells do not divide in synchrony, so in a culture of cells all stages of the growth cycle are represented.

Fungi grow by elongation at the tip of the hypha. Many move into a yeast stage during which the cell divides by budding. An outgrowth appears at some point on the cell surface and grows until it is almost the size of the mother cell. The cellular organelles including DNA are replicated and a copy is partitioned into the daughter cell which eventually is walled off. The mother cell does not increase greatly in size during cell division.

POPULATION GROWTH

Laboratory studies with pure cultures of bacteria traditionally have demonstrated exponential growth in batch culture, in which all essential nutrients are present in excess and growth parameters are optimal. In this situation, population growth follows a characteristic cycle which begins with the lag phase of growth, during which the cells are adapting to the new environment (Figure 5). Enzyme synthesis induced by contact with a new substrate occurs during this phase. Cells that are preadapted to the substrate or the growth conditions experience a shortened lag phase or no lag phase at all. Following the lag phase is a period of unrestricted multiplication called the log or exponential growth phase, so named because of the binary fission process of cell division. The rate of growth (cell division) depends upon the composition of the growth medium and the environmental parameters, and under optimal conditions a cell may divide every 15 minutes. When a nutrient becomes limiting or when inhibitory or toxic products accumulate, the cell enters the stationary growth phase. The individual cell is still viable although not replicating. This phase is manifested within the total culture by an equivalence between the number of cells produced by cell division and the number of cells dying; thus, there is no net change in the number of cells in the population. When the number of cells dying becomes greater than the number of cells being formed, the death phase ensues. The population eventually stabilizes at a constant low number of surviving cells. Because this cycle is characterized by an abundance of nutrients, it is rarely seen in the natural environment.

In cultures where (1) cells are able to proliferate, (2) there is an absence of inhibitors, (3) there is a homogeneous mixture of cells and nutrients, and (4) the substrate is the limiting factor in growth, cellular growth can be related mathematically with the disappearance of substrate. In this case, Monod kinetics apply as in equation 1:

$$\frac{dX}{dt} = \mu_m \frac{C_a X}{K_s + C_a} \tag{1}$$

The instantanous change in cell concentration over time, dX/dt, is equal to the maximum specific growth rate times a fraction including the substrate concentration in the medium, C_a, the cell concentration, X, and the Monod half-saturation constant, K_s.

The half-saturation constant is equal to the substrate concentration at which the specific growth rate is one-half μ_m (in batch culture this situation occurs at the end of the exponential growth phase). Typically, the constants μ_m and K_s are determined in batch culture tests using linearized graphical plots of the reciprocals of the measured cellular growth and substrate concentrations. Alternatively, they may be determined in a series of continuous culture tests.

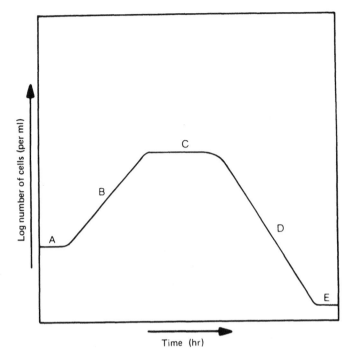

Figure 5 Bacterial growth curve. (A) Lag phase; (B) logarithmic phase; (C) stationary phase; (D) decline phase; (E) surviving population.

CONTINUOUS CULTURE

Techniques for continuous cultures have been developed to provide a constant environment for microbial growth. The physical factors of temperature, pH, O_2 concentration, etc., are well controlled, and nutrients can be supplied at controlled rates coupled with removal of potentially toxic waste materials. Population growth, therefore, occurs at a constant rate which in some systems can be varied by changing the availability of a nutrient. Studies conducted in continuous systems such as the chemostat and the recycling fermentor have helped to establish the energy requirements of cells under growth or maintenance (survival) conditions in addition to exploring the response of cells to various types of nutrient or other growth factor limitations. Maintenance of a cell population under steady-state conditions with selective pressure, such as a nonutilizable carbon source, may enable mutants with capability to metabolize the substrate to be generated and then grow to sufficient population levels to be recovered.

CELL DEATH

The most widespread measure of cell death is loss of reproductive capability. However, the medium used to detect survivors may not be adequate to demonstrate cell division, although the cell may be viable and capable of reproduction in another environment. The problem of defining cell death has not yet been resolved.

PURE- AND MIXED-CULTURE METABOLISM

Populations in which all the cells are of the same species are considered to be a pure culture. Except for the ongoing process of mutation, discussed in a later section, the process of cell reproduction by binary fission with replication of genetic information ensures that the pure culture will express essentially the same properties. The population will be nearly homogeneous in its ability to metabolize a substrate.

In some cases, a substrate may be metabolized only partially by a particular species and a product may accumulate. In a parallel situation, another species may be able to metabolize that product further, although the second species may lack enzymes needed to metabolize the parent substrate. By themselves, neither species could mineralize the substrate of interest. However, a mixed culture of the two organisms might act in concert with one species mineralizing the product resulting from metabolism of the substrate by the other species. A consortium of more than two species may be required to mineralize a substrate and the effective species may be bacteria, fungi, or a mixture of the two.

Cometabolism refers to the fortuitous metabolism of a compound while the cell obtains its carbon and energy from another source (273). Such metabolism may be partial or complete and depends upon enzymes already active in the cell.

SUBSTRATE UPTAKE AND TRANSPORT

Some motile bacteria possess the ability to move along the concentration gradient of a specific compound in its environment. This phenomenon, called chemotaxis, may permit these strains to scavenge some nutrients more efficiently or to move away from toxic or inhibitory compounds.

Some bacteria are able to utilize as carbon and energy source substrates which are too large to enter the cell. These strains secrete hydrolytic enzymes into the culture medium which break the high-molecular-weight compunds (such as proteins, starch, or cellulose) into smaller components which can enter the cell.

The transport of other substrates (lower molecular weight) into cells depends on a number of interrelated factors. The substrate must be able to pass the complex cytoplasmic membrane which is composed of a hydrophobic zone surrounded on both sides by hydrophilic layers. Some lipid-soluble substances can pass across this zone by free diffusion which is dependent on the difference in substrate concentration inside and outside of the membrane. The rate of uptake by this mechanism, called passive transport, depends upon the size and charge of the substrate.

Within the cytoplasmic membranes are proteins which couple substrate transport to an energy-yielding process. Called active transport, this mechanism is the route of entry for most substrates and ions, and the concentration within the cell protoplasm can be much greater than the concentration outside the cell. The proteins involved in active transport can be very specific for a particular substrate to the exclusion of structurally related analogs.

ENZYMES

Enzymes are proteins and are the most efficient known catalysts for biochemical processes. They serve to increase the rate of a reaction, often causing a reaction to occur under physiological conditions which otherwise could only occur under extremes of pH, temperature, or concentration.

Classically, enzymes were grouped into 9 categories based on function. These names are still used frequently. (1) Dehydrogenases mediate the loss of a hydrogen ion from a substrate with the acceptor being other than molecular oxygen. (2) Oxidases catalyze loss of a hydrogen ion with molecular oxygen as the acceptor. (3) Kinases transfer a phosphate group from ATP or other nucleoside triphosphate to the substrate. (4) Phosphatases mediate the hydrolytic cleavage of phosphate esters. (5) Mutases catalyze transfer of a functional group between two positions in the same molecule. (6) Synthetases mediate condensation of two separate molecules coupled with cleavage of ATP. (7) Decarboxylases achieve decarboxylation of the substrate. (8) Thiokinases catalyze the ATP-dependent formation of thiol esters. (9) Carboxylases catalyze the ATP-dependent addition of carbon dioxide to the acceptor substrate.

These categories have been replaced by six classes of enzymes in a formal system developed by the International Enzyme Commission (39). (1) Oxidoreductases act on the CH-OH group of a substrate, requiring NAD^+ or $NADP^+$ as the hydrogen acceptor. This category includes dehydrogenases and oxidases. (2) Transferases catalyze the transfer of an intact group of atoms, such as methyl or phosphorus containing groups, from a donor to an acceptor molecule. Kinases and mutases are included in this group. (3) Hydrolases, including phos-

phatases, mediate the transfer of chemical groups to water. (4)
Lyases, such as decarboxylases, catalyze the addition of groups to
substrates containing double bounds, or the removal of groups from
substrates to yield products with double bonds. (5) Isomerases catalyze a change in the atomic configuration of a molecule without a
change in the number or kind of atoms. (6) Ligases are involved in
the formation of a product resulting from the condensation of two different molecules coupled with the breaking of a pyrophosphate linkage in ATP. This class includes synthetases, thiokinases, and carboxylases.

Most enzymes are notably specific in their actions, catalyzing the
reaction of a particular substrate, but having no activity against a
very closely structurally related substrate. Some enzymes, however,
act on many related compounds. These enzymes act on a specific
structural component of different substrates. Enzyme specificity is
related to two features of the substrate. First is the specific chemical
structure which is attacked by the enzyme. Second, the substrate
must also contain a binding group which binds to the enzyme in such
a way as to permit optimal association of the susceptible structure
with the enzyme. The active site on the enzyme is the area containing both the binding site and the catalytic site, and the three-dimensional configuration of the complex resembles a "lock-and-key"
relationship.

The activity of enzymes can be inhibited either irreversibly or
reversibly. Irreversible inhibitors destroy or bind to a functional
group on an enzyme which is necessary for its catalytic activity. Re-
versible inhibitors may be either competitive or noncompetitive. A
competitive inhibitor has a similar structure to that of the substrate
and therefore can be bound by the enzyme. However, the enzyme has
no activity against the inhibitor. Since the inhibitor and the substrate compete for the binding sites of enzymes, the action of a competitive inhibitor can be partially reversed by increasing the concentration of substrate. Noncompetitive inhibitors bind to the enzyme in
an area other than the binding site, and in so doing alter the catalytic site so as to make it inactive. The affected site is often called the
regulatory site of the enzyme and is reversibly occupied by the inhibitor. Lowered concentrations of the inhibitor increase the activity
of the enzyme. When the inhibitor is a direct product of a series of
reactions involving the enzyme, the regulatory process is called feedback inhibition and is an important cellular mechanism for regulating
metabolic processes such that energy is not wasted on production of
unnecessary metabolites. Enzymes with a regulatory site, called allosteric enzymes, can be stimulated as well as inhibited by specific effector molecules which bind to the regulatory site. The effector molecule may be the substrate itself, signaling the enzyme to initiate the
metabolic pathway. Often only one key enzyme in a pathway is regu-

lated; the activity of the rest of the enzymes is limited by the availability of their specific substrate.

Enzymes may also be controlled at the level of enzyme synthesis. A reduction in the amount of enzyme would reduce the total enzymic activity. The genetic system for synthesis of an enzyme consists of

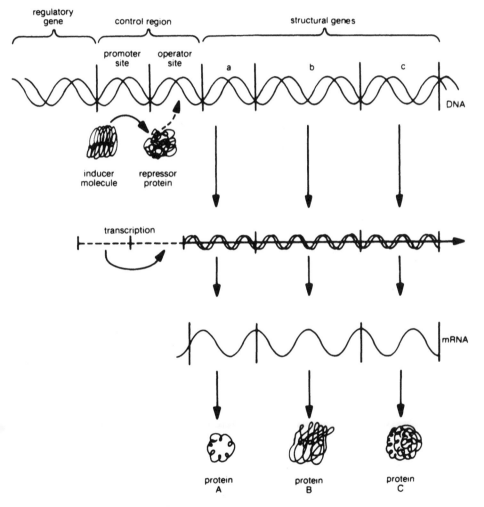

Figure 6 Regulation of enzyme synthesis. (Adapted from Ref. 276).

several parts (Figure 6). This general model shows several structural genes which code for the enzyme proteins. More than one enzyme
may be part of a system. In addition, each system contains one control gene coding for a protein, called the repressor, which binds
specifically to a control site called the operator region. The repressor protein binds to the operator region in the absence of an inducer
molecule, and the entire enzyme system is inactive. When the inducer
is present it binds with the one repressor protein and the complex
has reduced affinity for the operator region. The operator region
then complexes with RNA polymerase at the adjacent promoter region
and transcription of the structural genes is initated. When the concentration of the inducer molecule falls below a critical point, the
operator region is again blocked by the repressor protein and enzyme synthesis ceases.

Very long or branched metabolic pathways, in which an intermediate substrate may be directed to alternative pathways, are regulated by sequential induction, in which sections of the pathway are
under separate regulatory control. The product of one series of steps
acts as the inducer for the next several steps. This prevents the
cell from wasting energy on unnecessary or unproductive metabolic
processes.

Gene systems for a particular function which are grouped in one
place physically on the genome are rare in eukaryotes. Genes for a
particular metabolic pathway are more likely to be scattered over
many chromosomes. However, regulatory genes still function in a similar fashion at separate control sites. Some bacteria (prokaryotes)
also have systems in which the genes are scattered along the chromosome.

A more general type of control is called catabolite repression, in
which the control protein binds to operator sites of many enzyme systems. This permits a favored substrate to be utilized preferentially
before other substrates are metabolized. As long as the substrate of
choice is present, other substrates are not metabolized, even though
they may also be present. When the concentration of the favored substrate is reduced, enzyme systems for metabolism of the other substrates are induced. The favored pathway is more efficient and therefore costs the cell less energy.

Another consideration in the effectiveness of enzyme activity is
the physical location of the enzyme with respect to the substrate. In
eukaryotes the enzyme may be enclosed within membrane-bound organelles. In prokaryotes an enzyme may be enclosed within the periplasmic space between the cell wall and the cytoplasmic membrane,
while the substrate may be extracellular or intracellular.

All of the factors regulating enzyme activity may act in concert.
Control of metabolic processes is finely tuned to the nutritional oppor-
tunities available, so that the cell acts in the most energy-efficient
manner possible.

METABOLIC ENERGY PRODUCTION

All microorganisms need a source of energy for maintenance of cell viability and growth. The manner in which energy is obtained varies, and bacteria can be classified according to the source of their energy requirement. Phototrophs such as cyanobacteria use light directly in a photosynthetic process. Chemotrophs oxidize organic or inorganic compounds. A chemotroph which can derive its carbon requirements from carbon dioxide is a lithotroph, while a chemotroph which utilizes organic carbon is known as an organotroph.

The most common method of gaining energy is through oxidation reactions, which are normally coupled to the formation of ATP and other high-energy molecules. Many different kinds of substrates can be oxidized, but eventually the substrates are modified to metabolites which can enter one of only a few pathways for carbon dissimilation. These pathways can be divided into two categories, fermentation pathways in which organic compounds serve as both the electron donor and electron acceptor, and respiration pathways in which oxygen or an inorganic compound or ion serves as the terminal electron acceptor.

Aerobes utilize a respiratory pathway known as the tricarboxylic acid cycle (Figure 7). For each mole of glucose converted to acetyl-CoA which completes the cycle, 38 moles of ATP are generated. The intermediates in the cycle are precursors to important cell macromolecules and may be utilized to fulfill other needs. Other metabolic reactions act to replace the intermediates in order to maintain functioning of the cycle.

Under anaerobic conditions some facultatively anaerobic bacteria utilize anaerobic respiration. This is an oxidative process utilizing the same pathway for substrate degradation as aerobic respiration, except that nitrate or another inorganic compound is substituted for oxygen as the terminal electron acceptor.

3-KETOADIPATE PATHWAY

One of the major pathways for the degradation of aromatic compounds in bacteria and fungi is the 3-ketoadipate pathway (Figure 8). The primary aromatic substrate is converted to either catechol or protocatechuic acid, each of which undergoes several catabolic reactions in two separate but parallel pathways, until they converge to three common intermediates, 3-ketoadipic acid enol-lactone, 3-ketoadipic acid, and 3-ketoadipyl-CoA, which is cleaved to form succinic acid and acetyl-CoA. These two end products enter the tricarboxylic acid cycle. The pathway is strictly aerobic and regulated exclusively by control of enzyme synthesis. Although the chemistry of the pathway is the same in all bacteria, the pathway is regulated differently in different groups. Details of this pathway are found in the chapter on parent compounds.

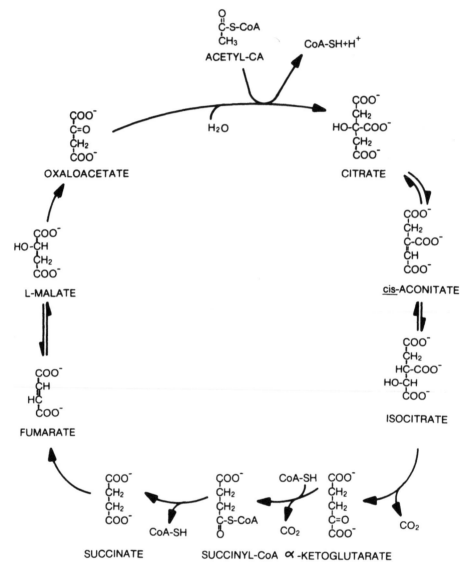

Figure 7   Tricarboxylic acid cycle. (Adapted from Ref. 276.)

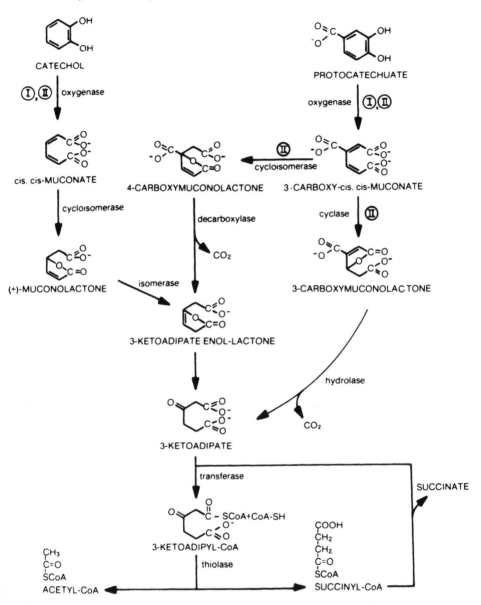

Figure 8 The 3-ketoadipate pathway in bacteria (path I) and fungi (path II). (Adapted from Ref. 66a,93,408.)

The 3-ketoadipate pathway is one of the best studied cellular metabolic processes. The chemistry of the pathway has been elucidated (144,238,289,340-342,409), the enzymes isolated and their amino acid composition and other properties identified (123,307,336,337,345, 348,396,477-480), and microbial regulation of the pathway described (335,338,339). A comprehensive review of the pathway has been published (408).

PLASMIDS

Genes coding for vital functions of the bacterial cell are located on the chromosome and are passed to every daughter cell. However, some metabolic processes, while not essential, confer considerable advantages on cells with those capabilities. The genes for these processes are coded for on plasmids, circular strands of DNA which can replicate autonomously. Plasmids can be passed from cell to cell as well as being replicated in the progeny; thus, an entire population can quickly acquire the specific characteristic. Traits which are often coded for on plasmids include the ability to metabolize unusual substrates including many aromatic compounds, resistance to antibiotics, and ability to survive in the presence of heavy metals. Presence of a specific plasmid is often a guide to the metabolic capability of that cell. Regulation of enzymes coded for on plasmids is similar to that discussed earlier, and catabolic repression may be effective across both plasmid and chromosomal DNA. Thus, a substrate which could be metabolized by two pathways, one on the plasmid and one on the chromosome, may be metabolized by one pathway preferentially while the other is repressed. A pathway for mineralization may involve some steps coded for on the plasmid and others coded for on the chromosome.

Since plasmids may be considered potential vehicles for genetic reassortment and transfer, they may also be viewed as mediators of evolution of biodegradative capabilities within microbial populations. At the population level, the development of DNA probes labeled with 32P or fluorescent reagents can permit detection and monitoring of specific catabolic genes. Such applications would likely utilize colony hybridization techniques to directly probe for complementary target DNA in individual microbial colonies. Information derived from such experiments would allow measuring the selective pressure required to maintain catabolic genes in the natural population. In addition, the survival and transfer of novel catabolic genes originating from recombinant DNA technologies can also be tracked in the environment. Such information can be useful in the finer detailed prediction of the kinetics of biodegradation and the likelihood of utilizing genetically engineered microorganisms to degrade specific chlorinated aromatic pollutants.

2

Cellular Gene Coding and Genetic Technologies

STRUCTURE AND FUNCTION OF DNA

Deoxyribonucleic acid (DNA) consists of four kinds of deoxyribonu-
cleotides linked together in a specific sequence. DNA is usually
double-stranded. Of the four kinds of ribonucleoside bases, there
are two subsets of hydrogen-bonded pairings, adenine with thymine
and guanine with cytosine. RNA (ribonucleic acid) is also composed
of four types of bases, but with uracil substituted for thymine. Each
nucleoside base is joined to the carbon-1 of a pentose sugar (deoxyribose in DNA, ribose in RNA). A phosphate molecule is joined
through an ester linkage to the carbon-5 and the resulting molecule
is known as a nucleotide. The nucleotides are joined by linking a
hydroxyl group on the carbon-3 of a pentose to the phosphate group
on the carbon-5 of another pentose to form a phosphodiester bridge.
The pentose carbons are primed (" ' ") in order to distinguish them
from carbons in the bases; thus the two ends of single strand DNA
are known as the 5'-phosphate end and the 3'-hydroxyl end. DNA
from different species have characteristic relative amounts of the
four nucleotides and this property has served to help identify unknown species of bacteria and establish evolutionary relationships
among the species (Figure 9, Figure 10).

Double-stranded DNA consists of two strands in which the 5'-
end of one strand is paired with the 3'-end of the opposite strand.
The two strands are complementary—an adenine on one strand always
pairs through hydrogen bonding with a thymine on the other strand,
as does guanine with cytosine.

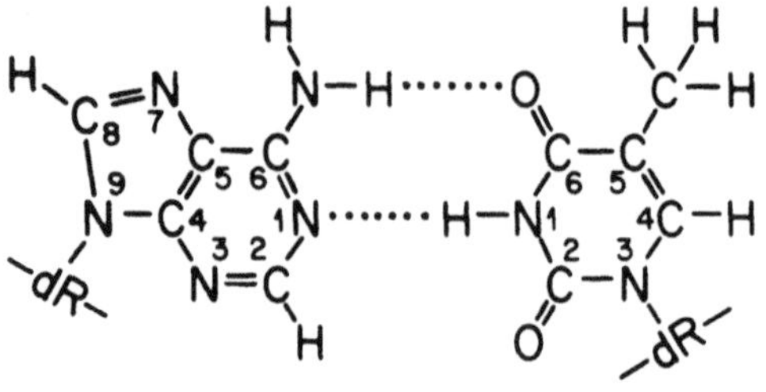

Adenine   Thymine

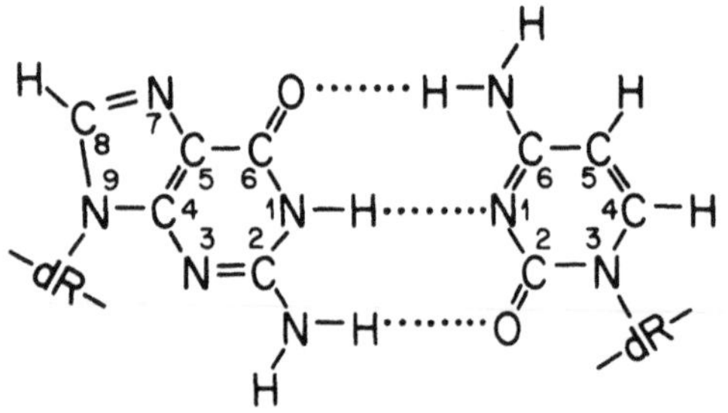

Guanine   Cytosine

Figure 9 The pairing of adenine with thymine and guanine with cytosine by hydrogen bonding. The symbol -dR- represents the deoxyribose moieties of the sugar-phosphate backbones of the double helix. Hydrogen bonds are shown as dotted lines. (Redrawn from Ref. 407A.)

Each set of three bases along a strand is called a codon and codes for a specific message, usually formation of an amino acid. Some triplet sets of bases are stop messages while others are nonsense codons and lead to premature termination of message reading. There is some redundancy in the triplet codes. Several triplets code for the same amino acid, so a change in one base may not cause a functional change in the message. Groups of triplets code for sequences of amino acids which become proteins after some modifications. The DNA segment which codes for a single sequence of amino acids is known as a gene. The products of several genes may combine to form a protein. The total genetic material of a cell is called the genome, consisting of the chromosome and in some cases plasmids. In eukaryotic cells the chromosome includes some proteins also.

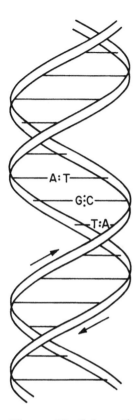

Figure 10 Schematic representation of the DNA double helix. The outer ribbons represent the two deoxyribosephosphate strands. The parallel lines between them represent the pairs of purine and pyrimidine bases held together by hydrogen bonds. Specific examples of such bonding are shown in the center section, each dot between the pairs of bases representing a single hydrogen bond. The direction of the arrows corresponds to the 3' to 5' direction of the phosphodiester bonds between adjacent molecules of 2'-deoxyribose. (Redrawn from Ref. 407A.)

Since bases are read in groups of three, the deletion or addition of one or two bases will cause a shift in the reading frame such that the DNA is no longer read correctly. This alteration is called a frame-shift mutation and may or may not be lethal to the cell.

Hydrogen bonding as well as hydrophobic interactions of the molecules free DNA to take on a highly structured configuration resembling a double helix. Physical and chemical perturbations such as heat or acidity cause denaturation or unwinding of the DNA,

eventually leading to separation of the two strands. The covalent
molecules joining the bases within each strand are not broken. When
the physical or chemical stress is removed, the two strands will anneal spontaneously. When denatured DNA from two different sources
is mixed, a certain percentage of DNA from the separate sources
will anneal to form a DNA-DNA hybrid. The amount of hybridization
depends on the percentage of complementary sections between the
DNA. Strands of RNA can also combine with DNA to form RNA-DNA
hybrids.

The chromosomal DNA of *E. coli* consists of about 4×10^6 base
pairs, has a molecular weight of about 2.6×10^5, and is about 1400
μm long (the *E. coli* cell is about 2 μm long). Eukaryotic cells contain from 10 to 600 times as much DNA as *E. coli* cells. Eukaryotic
DNA is organized into several linear chromosomes, each carrying a
unique set of genes as well as proteins.

Segments of DNA in eukaryotic cells are repeated many times,
while prokaryotes usually lack repetitive sequences. Some regions
of eukaryotic DNA are characterized by inverted repetitions of base
sequences which may be a few to a thousand base pairs long. Eukaryotic genes also contain segments which do not code for an amino
acid and are not translated. These intervening sequence or introns
have been found in all eukaryotic genes yet examined. Their function is unclear. They have been postulated to contain regulatory
signals or to separate the genes into smaller units which can be
readily recombined into new genes.

The definition of what constitutes a gene has modified over the
years as the biochemical exploration of the cell has become more detailed. A gene classically has been taken to mean the genetic material which specifies a single trait. More recently, portions of the DNA
have been classified as structural genes if they code for a single
polypeptide (a portion of a protein) or for a specific type of RNA,
or as regulatory sequences if they function to mark the beginning
and end of structural genes or start or terminate transcription.

There are two functions of the DNA molecule. The first function
is to serve as a template for its own replication. Enzymes separate
the two strands, add new complementary bases to each intact strand,
and ligate the bases. Since the original two strands were complementary, the two new complete DNA molecules are identical, each containing one old strand and one newly synthesized complementary strand.

TRANSCRIPTION

The process of converting the information coded by DNA into RNA
is called transcription and is the second function of DNA (Figure
11). Only portions of the chromosome which code for a specific sequence or sequences of required genes are transcribed at any given

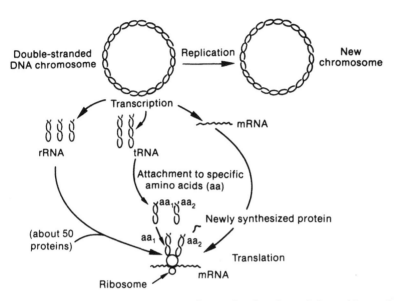

Figure 11 The general plan of synthesis of nucleic acids and proteins. (Redrawn from Ref. 407A.)

time. A single strand of RNA complementary to the DNA strand is generated. Most of the RNA so formed is called messenger RNA (mRNA), which codes for the amino acids that comprise the polypeptides of the proteins. Other sections generated are transfer RNAs (tRNA), ribosomal RNAs (rRNA), and regulatory sequences.

The mRNA serves to carry the genetic message from the DNA in the nucleus or nucleolus to a ribosome, a collection of proteins and RNA units which is the site of protein synthesis in the cytoplasm. A single mRNA molecule contains the message needed to code for one or several polypeptides as well as a leader region and intergenic spacer regions which are not translated.

TRANSLATION

The process of protein biosynthesis according to the code carried by the mRNA is called translation and takes place at the ribosomes in the cytoplasm. There are specific transfer RNAs that recognize each triplet codon which codes for an amino acid. The tRNA molecule has receptor sites for both the mRNA chain and specific amino acids, and serves as an adapter to bring the appropriate amino acid in close proximity to the developing polypeptide. Enzymes then attach the amino acid to the chain and remove the tRNA. When a termination codon is read, biosynthesis stops and the chain is released from the ribosome. The polypeptide finally is subjected to posttranslational modification of some of its amino acids and undergoes folding into its characteristic three-dimensional shape, which renders the molecule biologically active.

Ribosomes are composed of ribosomal RNA and proteins. The rRNAs have a specific three-dimensional structure and serve as a framework for the binding of the polypeptide subunits. The ribosomal proteins are postulated to function in the process of synthesizing the polypeptide chain.

The process of translation is repetitive and a single mRNA molecule can be read simultaneously by several ribosomes spaced closely along the length of the molecule. In bacteria, translation of mRNA begins while the molecule is still being transcribed from the DNA. Thus, in prokaryotes these two processes are closely linked. The prokaryotic mRNA is quickly degraded by nucleases, so that efficient regulatory control over protein synthesis is maintained.

MUTAGENIC EVENTS

Most traits of bacteria are conservative and are reproduced in each generation. However, like all living organisms, bacteria may undergo mutations in which the genetic message is altered. If the alteration is lethal for the cell, the message is not passed on because the cell dies. Other mutations may not result in a change in expression of the message, since there is some redundancy built into DNA codes. In some cases, an alteration in the DNA code may lead to an alteration in the cell's metabolism. Some mutations allow the cell to survive at unfavorable temperatures or in the presence of potentially toxic compounds. In other cases, the cell acquires the ability to use previously unsuitable substrates. Most mutations are either lethal or place the cell at an environmental disadvantage.

Mutations occur randomly, on the order of approximately one in one million cells for a given characteristic. Some agents, including ultraviolet light, some kinds of radiation, and some chemical agents, cause increased mutagenesis. These mutations are characterized by being randomly distributed across the DNA. Mutations can also be selected by applying selective pressure to a population. For instance, in the presence of an unfavorable environment, only those cells which have mutated in such a way as to adapt to the environmental situation will survive.

Cells have powerful mechanisms for excising mutations from DNA. There are enzymes which recognize specific types of mutations and replace them with the correct message. Thus, the number of mutations passed to progeny cells may be a small fraction of the total number of mutagenic events sustained by the cell. In the case of massive mutations, such as radiation damage, the cell sets into motion a complex series of steps designed to foster cell replication at the expense of almost every other cellular function. The resultant cells are usually heavily damaged and do not function normally. The survival rate of these cells is very low.

Classical genetic techniques are based on manipulation of whole cells and environments to select and induce desired mutations. One common induction procedure is to expose a population of cells to a broadly acting mutagen and then place the surviving cells in the desired environment. Some mutations which have occurred in the genetic region of interest may enable those cells to grow or express the desired trait. A disadvantage of this method is that multiple mutations may have occurred in other genetic regions of the cell which may change the properties of the cell in unknown ways. Another method of obtaining mutations is to expose the population directly to the selective property. Only mutant cells which can adapt will grow. However, such adaptation may take weeks, months or longer before the altered population is large enough to be observed.

Some types of genetic alterations in the cell occur as normal cellular events. Exogenous DNA can enter the cell by a variety of mechanisms and once inside can recombine with the chromosomal DNA. This process, called genetic recombination, can result in addition of new genetic information or the substitution of homologous DNA sequences. One method of genetic recombination is the transfer of exogenous DNA into a recipient bacterial cell. This process is called transformation and is the only direct evidence for DNA being the genetic material. Only a minority of recipient cells is competent at any given time to receive the DNA. Once the DNA has entered the cell, it may find its homologous region on the chromosome, recombine, and become a permanent part of the host chromosome, or it may recircularize into an autonomous plasmid which replicates and is passed to daughter cells along with the chromosome.

Conjugation is a method which permits the entry of large segments of DNA from a donor cell into a recipient (Figure 12). Direct cell-to-cell contact is required between the donor cell, which possesses a particular plasmid-encoded mating appendage through which the DNA passes, and the recipient cell which lacks the appendage. Upon contact, the donor cell is stimulated to begin replication of the plasmid and the copy is threaded through the conjugation bridge to the recipient. The donor chromosome itself can be transferred to the recipient cell if the plasmid has integrated into the chromosome. As long as contact can be maintained, transfer of genetic material continues. The transfer of genetic material always begins at the plasmid origin of replication. Therefore, by separating the cells at specific time intervals and noting which traits have been transferred, mapping of the genes along the chromosome can be achieved.

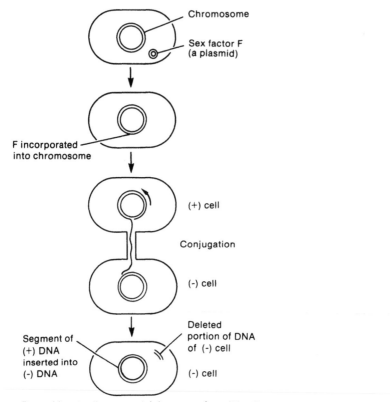

Figure 12   Transfer and recombination of genes during bacterial conjugation. The DNA of the (+) cell is replicated by the rolling-circle
process, and the resulting single strand containing F is introduced
into the (-) cell. (Redrawn from Ref. 276.)

    Transduction is the term given to transfer of genetic material
by bacteriophages which are viruses specific for bacteria (Figure
13). During the packaging of viral DNA into phage heads in the
lytic cycle, some portions of bacterial DNA will be incorporated in-
stead. When these particles are expelled from the cell and infect
another cell, the bacterial DNA is released into the new cell to recombine with homologous host DNA.
    In eukaryotic cells both parental chromosomes contribute genes
to the daughter chromosome. Both parental chromosomes undergo
cleavage at homologous points and segments of the chromosomes are
exchanged. The new combinations of genes are spliced together and
passed to the progeny cells.

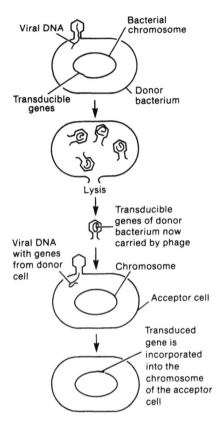

Figure 13 Genetic recombination during viral transduction of bacterial genes into a recipient cell. (Redrawn from Ref. 276.)

Some segments of DNA are highly mobile and can leave their original position in the chromosome to be inserted elsewhere. Each end of these transposable elements or transposons contains short DNA pieces called insertion sequences. The insertion sequences are recognized by specific enzymes which catalyze their insertion into new sites on the chromosome or plasmid.

In recent years techniques have been developed which permit the direct manipulation of specific genes. Many of these techniques are now well established and are being applied to solve specific problems. Descriptions of the fundamental techniques and methods follow. Additional information can be obtained from references which served as the basis for this chapter (276,294,334). In particular, reference 294 contains details of the methods discussed here.

CURRENT BIOCHEMICAL TOOLS FOR GENETIC MANIPULATION

A series of prokaryotic enzymes have been isolated which can be used to cleave DNA and then splice different pieces together to form new strands. These enzymes are now being utilized in the laboratory.

Restriction Endonuclease

Restriction endonucleases are enzymes which specifically recognize certain sequences within double-stranded DNA. These sequences are usually four to six nucleotides long with a twofold axis of symmetry. Examples of restriction endonuclease recognition sequences are shown in Table 1. A particular tetranucleotide recognition site might arise

Table 1   Recognition Sites for Restriction Endonucleases

Enzyme	Recognition sequence	Termini	
	3' 5'		
Bam HI	GGATCC[a]	G	GATCC
	CCTAGG	CCTAG	G
EcoRI	GAATTC	G	AATCC
	CTTAAG	CTTAA	G
HaeIII	GGCC	GG	CC
	CCGG	CC	GG
HindIII	AAGCTT	A	AGCTT
	TTCGAA	TTCGA	A
MboI	GATC	xx	GATCxx
	CTAG	xxCTAG	xx
PstI	CTGCAG	CTGCA	G
	GACGTC	G	ACGTC
ThaI	CGCG	CG	CG
	GCGC	GC	GC

[a]Arrows indicate site of cleavage. A, Adenine; C, cytosine; G, guanine; T, thymine.
Source: Ref. 294.

once in every 4^6 (4096) pairs, assuming random distribution of DNA base pairs. The endonucleases cleave the DNA molecules either at the axis of symmetry, yielding blunt double-stranded ends, or at positions offset from the center, giving fragments of DNA with one protruding single-stranded end known as "sticky" ends. DNA from different sources acted on by the same restriction endonuclease will produce complementary termini. In some cases, different restriction endonucleases with different recognition sequences will produce complementary termini as well. These ends can join with complementary ends on a different fragment to form new molecules.

Deoxyribonuclease

Deoxyribonuclease (DNase) is an enzyme that cleaves double-stranded or single-stranded DNA randomly, yielding fragments with 5'-phosphate termini. Depending on the conditions of the reaction, either the double strand of DNA is cleaved at approximately the same site or each strand is cleaved independently. Under certain conditions this enzyme creates nicks in double-stranded DNA which does not destroy the unity of the molecule. The concentration of DNase in the solution will affect the extent of nicking. Fragments of DNA containing regions of interest can be inserted into other molecules. Nicked regions on DNA permit insertion of nucleotides. The order of insertion can be controlled to permit creation of a defined DNA strand, and the nucleotides can be radiolabeled to permit tracking of the constructed strain during other manipulations.

Polymerases

Polymerases are enzymes that add nucleotides to the 3'-hydroxyl terminus created when the double-standed DNA molecule is nicked by DNase. The enzyme can also remove nucleotides from the 5'-phosphate ends. Both of these processes acting at the same time result in movement of the nick along the intact strand of DNA (nick translation). If the nucleotides being added are radioactive, such as with ^{32}P, labeled DNA with a high specific activity can be prepared. Normally the replacement nucleotides are distributed uniformly along the DNA molecule, since the nicks occur randomly throughout the DNA.

Reverse Transcriptase

This enzyme, also known as RNA-dependent DNA polymerase, uses mRNA as the template for transcription to form double-stranded DNA. Single-stranded DNA or RNA can also be utilized by this enzyme to make probes for use in hybridization experiments. Initiation of the action of reverse transcriptase requires a short DNA primer sequence base-paired to the template. Primers can be generated by exhaust-

ively digesting DNA and retrieving the fragments. Since the fragments represent random portions of DNA, some fraction of them will
bind to the template and can be used as primers. The discovery that
eukaryotic mRNA contains multiple adenylate bases at the 3'-end allows construction of complementary thymine polymer sequences at
that site, which then will act as a primer.

The RNA template possessing a primer is then mixed with solutions of the four nucleotides. Usually only one nucleotide is radioactively labeled. Reverse transcriptase then catalyzes synthesis of
the complementary DNA (cDNA). Following completion of the reaction,
the RNA strand can be selectively degraded. Complementary DNA to
be used as hybridization probes are retained in single-stranded form.
However, the cDNA can form a hairpin loop at its 5'-end which acts
as a self primer for synthesis of the complementary strand, or a
second primer can be added to the cDNA to initiate synthesis, resulting in double-stranded cDNA.

Ligases

T4 DNA ligase links together complementary fragments of double-
stranded DNA by forming a phosphodiester bond between adjacent
3'-hydroxy and 5'-phosphate ends. The ends may be either "sticky"
(one strand of the double-stranded molecule extends beyond the
other) or "blunt" (both strands end at the same place). T4 RNA
ligase joins single-stranded RNA or DNA. These enzymes can also
catalyze circularization of DNA molecules if the concentration of DNA
in solution is low.

Transferases

The enzyme terminal deoxynucleotidyl transferase adds deoxynucleotides to the 3'-hydroxyl end of DNA. By adding homopolymer sequences of one type of nucleotide to one set of DNA fragments, and
a series of complementary homopolymer nucleotides to a second set
of DNA molecules, the two populations can be joined by their newly
formed complementary ends.

Methylases

These enzymes add a methyl group to particular nucleotides. Some
restriction endonucleases will fail to recognize a sequence which differs only by addition of a methyl group. This is an important component of cellular defense systems, in which host DNA is methylated
to protect against host restriction endonucleases which will attack
foreign (nonmethylated) DNA.

MECHANICAL SHEARING OF DNA

Double-stranded DNA can be broken by the shearing forces present in solutions. Very small fragments (approximately 300 base pairs in length) can be obtained by subjecting the solution to sonication with ultrasound. Larger fragments of about 8000 base pairs result from stirring the solution at high speed in a blender. The DNA molecule is sheared randomly along its length, producing fragments with short single-stranded ends.

CLONING VEHICLES

In order for a fragment of DNA to be replicated, it must contain a specific sequence called an origin of replication. Plasmids and prokaryotic chromosomes each usually contain one origin of replication. DNA that possesses an origin of replication is called a replicon. If a fragment of DNA in a cell cannot replicate, it will be diluted out of the population after several generations. Therefore, DNA fragments of interest must be attached to replicons, called vectors or cloning vehicles, before insertion into the cell. Replicons which are not native to the host cell may not be functional after insertion. The combination of the replicon and the foreign DNA fragment creates a hybrid molecule often called a chimera. The process of constructing a hybrid DNA molecule is known by several names, including genetic engineering or gene manipulation, to acknowledge the potential for creating new combinations of genes, and gene cloning or molecular cloning because this method allows amplification of the chimera via growth of the host population of organisms, each carrying the identical piece of genetic information. The DNA can be extracted from the new population and the chimeras recovered.

Plasmids

Plasmids are stable extrachromosomal circular double-stranded DNA replicons which are inheritable, but are also dispensable. Under constant selective pressure the plasmids will be replicated in the daughter cells, but when not essential for cell function many plasmids are lost from the cell. Plasmids contain from 1000 to 200,000 base pairs. Conjugative plasmids carry a set of genes that promotes bacterial conjugation; nonconjugative plasmids lack these genes. The term relaxed plasmids refer to plasmids which are present as multiple copies (10 to 200 copies) within a single cell, while stringent plasmids are limited to one to three copies per cell. Generally, plasmids of relatively high molecular weight are conjugative and stringent, while low-molecular-weight plasmids are nonconjugative and are present in multiple numbers. If cellular protein synthesis is blocked, chromosomal and stringent plasmid replication ceases, while relaxed plasmids

continue to replicate and can increase their copy number to several
thousand per cell. Therefore, relaxed plasmids are most useful for
molecular cloning processes. Some plasmids ("promiscuous plasmids")
can be transferred into a wide range of Gram-negative bacteria.
These plasmids are potentially useful in transferring genetic information to diverse bacterial hosts. Some plasmids are incompatible with
others and cannot coexist within the same cell. Plasmids have been
grouped into incompatibility classes on the basis of mutual incompatibility.

Plasmids useful as cloning vectors generally are small and under
relaxed control. They carry an easily selectable marker (such as antibiotic resistance) which allows identification of transformants which
have acquired the plasmid. These plasmids also contain single
recognition sequences for a given restriction endonuclease which permits insertion of DNA into a region of the plasmid not essential for
replication. A restriction site located within the marker genes will
inactivate the gene when foreign DNA is inserted, providing a tracer
for successful DNA insertion.

Some plasmids have been modified to include polylinkers, segments of DNA that contain closely spaced recognition sites for several restriction endonucleases. Generally, plasmids so modified are
small and lack natural restriction sites. Use of small plasmids is ad-
vantageous in that they are less likely to be damaged physically
during handling. Small plasmids also tend to generate higher copy
numbers.

Construction of plasmid vectors involves cleaving both foreign
DNA and plasmid DNA with the same restriction endonuclease to form
complementary ends. Both types of DNA are mixed and are ligated.
In some of the resulting molecules, the foreign DNA will be ligated to
the plasmid DNA, and a circular recombinant plasmid recovered. Use
of a restriction site within a marker gene simplifies the process of
detecting recombinant plasmids. The inserted DNA inactivates the
gene. Plasmids which recircularize without insertion of foreign DNA
will express the marker characteristic and can be rejected during
the screening process.

Recircularization of plasmid DNA can be minimized using a procedure called directional cloning. This method takes advantage of
the fact that most plasmid vectors carry single recognition sites for
more than one restriction enzyme. A plasmid is digested with two
such endonucleases. The larger fragment is separated and ligated
with foreign DNA containing ends complementary with the two dissimilar ends generated by the two restriction enzymes. The plasmid
fragment itself does not contain complementary ends and therefore
will not circularize.

Another method of preventing recircularization involves treating
the linear plasmid DNA with alkaline phosphatase. This enzyme removes the 5'-terminal phosphates. Ligation requires both a 3'-hy-

droxyl and a 5'-phosphate end. The foreign DNA combines with the treated plasmid DNA to create a circular molecule with a single nick on each strand where the phosphates have been removed. This open circular molecule can be inserted into cells much more efficiently than linear DNA and so most of the transformants will carry recombinant plasmids.

Bacteriophages

Bacteriophages (phages) are viruses which attack bacteria. The most extensively studied phage is bacteriophage λ which contains a double-stranded linear DNA molecule about 50,000 base pairs long with single-stranded complementary sticky ends. The intact phage consists of the genomic material surrounded by a protein coat with a protruding tail. The tail attaches to the bacterial cell and injects only the DNA into the cell. One of two processes of replication, either lysogeny or the lytic cycle, can be initiated within the cell.

During lysogeny, the viral DNA is integrated into the host chromosome and is replicated and transmitted to progeny cells along with the host chromosome. This can happen indefinitely. At some point the lysogenic cell is triggered to begin the second process of replication, the lytic cycle. Alternatively, upon infection the viral DNA can initiate the lytic cycle immediately.

The lytic cycle begins with viral adsorption and DNA penetration. These steps require specific conditions and the phages are host specific. After entry into the host cell, the linear viral DNA circularizes via its complementary sticky ends and replicates as an independent molecule. Copies of the DNA are continuously made. Transcription of the molecule is initiated soon after replication begins. One of the earliest proteins produced is a regulatory element which acts to prevent defensive activities of the host which might otherwise prevent further transcription. During the late phase of transcription, proteins involved in assembly of the head and tail and in cell lysis are produced. As many as 200 copies of a phage can be replicated within a single host cell.

During the assembly phase of the lytic cycle, a linear copy of the DNA becomes coiled into a phage prehead. When the head is filled, an additional protein attaches and locks the DNA into the head. The head finally attaches to the preassembled tail unit to form a complete phage particle. Progeny phage particles are released in a single burst when the host cell lyses. Each phage is then able to infect another cell.

Bacteriophage λ Vectors

About one-third of the phage λ genome is nonessential for virus replication and can be replaced by foreign DNA so that the total length

of the genome is conserved. Although phage λ contains several recognition sites for each of the restriction endonucleases of interest,
derivatives of phage λ have been developed which no longer carry
restriction sites in critical areas of the genome, but carry only one
or two such sites in nonessential regions. The phage λ thus has
been manipulated to become a useful cloning vector while still re-
taining its infective and lytic properties.

Cosmid Vectors

Cosmids are constructed vectors designed for cloning large fragments of DNA. They consist of a drug-resistance marker, a plasmid
origin or replication, one or more restriction sites, and the ligated
sticky end of phage λ (the *cos* site). They are very small, so
that large amounts of foreign DNA can be added to the molecule.
The complete cosmid DNA is packaged into a bacteriophage coat
which mediates its injection into the host cell. Inside the cell the
cos site allows circularization and the plasmid origin of replication
initiates replication.

Single-Stranded Bacteriophage Vectors

Bacteriophages containing single-stranded DNA replicate in a different manner from phage λ. After penetration, the single-stranded
form is converted to double-stranded DNA which can be isolated
and used as a cloning vector. The double-stranded form replicates
until 100 to 200 copies are made. Then DNA replication shifts to
produce large amounts of only one of the two DNA strands. Single
strands are incorporated into the phage coats and the mature phage
particles are continually extruded from the cell without lysing the
host cell. The single-stranded DNA thus produced can be recovered
and used as a template for DNA sequencing or for generating DNA
probes.

METHODS OF MANIPULATING DNA

Isolation and Purification of Plasmid DNA

Plasmids inserted into cells can be amplified by growing the bacteria
to high cell yields in the presence of an antibiotic providing selective pressure for the plasmid. The cells can be harvested by centrifugation and then lysed by several methods. Lysis procedures include boiling, treatment with alkali, and treatment with the surfac-
tant sodium dodecyl sulfate. Lysozyme is added to help break apart
the cell walls. The treated solution is centrifuged to remove DNA
from other cellular material.

The DNA preparation contains chromosomal DNA as well as plasmid DNA. These two types of DNA can be separated by taking advantage of several differences in the properties of chromosomal and
plasmid DNA. For some applications, the crude DNA preparation from
as little as 10 ml of culture may be used successfully. After treatment by one of the lysis procedures, plasmid DNA is recovered from
cells in intact circular form, while chromosomal DNA generally is extracted in short linear pieces. When plasmid DNA of high purity is
required, centrifugation at very high speeds in a solution of cesium
chloride and ethidium bromide will separate the DNA according to
density. The linear chromosomal DNA takes up more of the inter-
calating agent ethidium bromide than the plasmid DNA, and therefore will be a less dense molecule. When subjected to ultracentrifugation, the two types of DNA form narrow bands in separate regions of the centrifuge tube. Contaminating protein will form a third
band, and RNA will form a pellet. The ethidium bromide can be removed after this step or it can be retained during subsequent procedures.

Isolation and Purification of Bacteriophage λ DNA

Bacteriophages which are lysogenic can be recovered by inducing
the bacterial culture to begin the lytic cycle. One method useful
for phages containing a temperature-sensitive repressor is to raise
the temperature of the culture briefly.

Phages that are not lysogenic may be amplified by infecting the
host bacterial culture with a low number of phages. Much of the
bacterial culture will replicate for several generations, increasing
the number of host cells, before successive rounds of the lytic cycle
infect the entire culture.

Cell debris remaining after completion of the lytic cycle is removed by centrifugation. The remaining solution can be subjected
to density gradient ultracentrifugation, after which the intact bacteriophage particles appear as a thin band. Crude bacteriophage
preparations useful for many purposes can be obtained without density gradient ultracentrifugation.

DNA can be recovered from the phage particles by treatment
with a solution of a protein-digesting enzyme and sodium dodecyl
sulfate. The protein components can be extracted into phenol and
removed by centrifugation.

Separation and Purification of DNA Fragments

Gel electrophoresis is a sensitive method for resolving mixtures of
DNA. Samples containing DNA are loaded onto a slab of agarose or
polyacrylamide gel. The gel is submerged into a buffer solution of
nearly the same electrical resistance and a current is applied. Var-

ious types of DNA—linear, nicked circular, and closed circular—will
migrate through the gel at different rates depending on the molecular size of the DNA, the concentration of agarose or polyacrylamide,
the applied voltage, and the DNA base composition and temperature.
Under some conditions single strands of DNA can be separated. The
DNA is stained with the fluorescent dye ethidium bromide and can
be detected directly. Bands of DNA can be cut from the gel and the
DNA recovered.

The DNA is then packaged into a vector and transformed into a
bacterial culture for amplification. Colonies containing plasmids are
grown on nitrocellulose or nylon filters seated on agar Petri plates,
while bacteriophage plaques are formed in a lawn of indicator bacteria on agar media and then eluted into a liquid suspension which
is stable indefinitely.

Identification of Recombinant Clones

In situ hybridization of bacterial colonies or bacteriophage plaques
on agar media is rapid and can efficiently screen large numbers of
potential clones. Colony hybridization involves lysing the colonies
on the nitrocellulose or nylon filter and then fixing the DNA to the
filter in situ. Bacteriophage plaques are transferred to the nitrocellulose filters following plaque formation. Filters are placed in contact with plaques on the agar and then removed. Some portion of
the viruses in the plaques will be removed with the filter. The DNA
is then fixed, in some cases by heat treatment. The DNA probe of
choice is labeled with ^{32}P and the hybridization reaction between
the probe and the fixed DNA carried out.

Hybridization of Probes to Immobilized DNA

Hybridization reactions are governed by such factors as solvent used,
temperature, length of hybridization, concentration and specific activity of the ^{32}P-labeled probe or density of fluorescent probe, and
washing procedures after hybridization. Prior to hybridization, the
filters are treated with one of a number of compounds to saturate
sites on the filter with nonspecific affinity for single- or double-
stranded DNA, in order to ensure that the probe DNA will not bind
directly to the filter. The ^{32}P-labeled probe DNA is denatured
(double-stranded molecule separated into its single-strand components) and added to the filters. During incubation, the single strands
of the probe DNA will join to complementary DNA strands on the filter. Following hybridization the filters are washed thoroughly to remove unbound DNA and dried. An autoradiograph is made by placing
the filters in contact with x-ray film. Following development, radioactive signals representing positive homology of the probe with plas-

mid or phage DNA can be correlated with colonies or plaques on the agar plates. Those colonies can be retrieved and the recombinant DNA contained therein amplified.

Other procedures employ DNA probes with induced fluorescence and/or antigenic properties for use with fluorescent antibodies to detect positive hybrids. Both methods have the sensitivity to detect as little as 1 picogram of DNA or as few as 10,000 copies of a single gene.

MAPPING OF RESTRICTION ENDONUCLEASE RECOGNITION SITES

The order of bases on a strand of DNA can be determined using a number of techniques; usually more than one is required to obtain a detailed map. By cleaving DNA with restriction endonucleases having known recognition sites, the presence and number of such sequences can be resolved.

The relative positions of dissimilar endonuclease recognition sequences can be determined by first labeling one end of linear DNA with radioactive nucleotides to obtain a reference point. Digestion of separate aliquots of the DNA by different restriction enzymes is followed by gel electrophoresis and autoradiography. Resulting fragments of different lengths define the distance of each recognition site from the labeled end, and the relative distances between each site can be determined as well.

Small fragments of DNA can be mapped using an exonuclease which digests the DNA from each end by single nucleotides. Samples of the digestion reaction are withdrawn at time intervals and are treated with restriction endonucleases. As the DNA digestion progresses, the recognition sites disappear in specific order related to their position along the molecule.

IDENTIFICATION OF DNA SEQUENCES WITHIN FRAGMENTS

The Southern transfer technique is an effective method for identifying particular sequences of DNA. Fragments of DNA created by one of the previously described methods are separated by size on agarose gels using electrophoresis. The DNA within the gel is stained with ethidium bromide and denatured (strands separated) in situ. The DNA is then eluted from the gel directly onto nitrocellulose filters by placing the gel on absorbent paper whose edges are trailing in a salt solution. The filter is placed on the gel and more absorbent paper placed above the filter. Wicking action will cause the DNA to migrate from the gel to the filter with the relative positions of the fragments intact. The filter is treated with ^{32}P-labeled probe DNA of known base composition and then washed well. Fragments containing complementary bands will hybridize to the probe and can be visualized after autoradiography.

EXPRESSION OF PROKARYOTIC GENES IN FOREIGN HOSTS

Most genetic engineering studies have used *E. coli* as the host with
introduction of either *E. coli* genes or genes from other prokaryotes
or eukaryotes. Other studies have involved introduction of genes
from one strain to another strain of the same species. Little research
has been conducted on expression of *E. coli* genes into other hosts.
Some studies have noted that *E. coli* genes cloned into a *B. subtilis*
host were not expressed, although the genes themselves when recovered and inserted back into *E. coli* were functional (334). This
has been attributed to differences in the specificity of the RNA polymerases of the two hosts. There are differences as well as between
the translation mechanisms of *E. coli* and *B. subtilis*. Thus, *B. subtilis* genes function in *E. coli* but the reverse is not true.

GENE CLONING IN YEASTS

The yeast *Saccharomyces cerevisiae* has received the most attention
with respect to application of genetic engineering techniques. This
species contains a plasmid which replicates with high copy number
âlthough it has no known function (334). Fragments of yeast DNA
as well as an *E. coli* plasmid vector have been cloned into this plasmid and the recombinant molecule transforms yeast with high frequency and replicates in both *E. coli* and yeast. Some yeast genes
replicate autonomously and these can be used to construct vectors
which transform yeasts with high efficiency.

EXPRESSION OF EUKARYOTIC GENES IN A PROKARYOTIC HOST

Genes from many eukaryotes including humans have been cloned
into and expressed in *E. coli* (334). However, many other genes
from eukaryotic sources have been cloned into prokaryotic hosts
but have not been expressed. This has been explained in part by
the differences in the mechanism of expression of the genes (protein
synthesis). The steps involved in synthesis of a functional protein
include transcription of the DNA, translation of the mRNA, and posttranslational modification of the newly formed polypeptide. Transcription and translation require a promoter or a binding site recogniz-
able by the host RNA polymerase. Further, the protein produced is
often subject to rapid degradation unless it is protected by the modified amino acids or three-dimensional configuration of the native proteins. Even if all of these components are present, genes which are
expressed in a foreign host may not necessarily give rise to a stable
gene product.

3

Methods of Biodegradation Assessment

A decade ago biodegradation was measured at a worst-case level by substrate disappearance. Many reports in the area of environmental microbiology and wastewater engineering documenting biodegradation were accompanied by gas chromatographic analysis showing the net loss over time of a parent compound. In some cases, even visual disappearance of insoluble crystalline organics, such as naphthalene, was used as a gross measure of biodegradation. Most often, abiotic or sterile control samples were used in such assays of biodegradation. However, little insight was developed into nonmetabolic interactions among organisms and pollutant substrates and the measured biodegradation. In the area of wastewater treatment, BOD or COD removal measured by comparing influent and effluent concentrations was used as a measure of pollutant biodegradation, with the assumption that recalcitrant organics had essentially the same fate as labile organics in wastewater treatment.

Such approaches to measuring biodegradation have been replaced by more stringent parameters to give accurate estimates of microbial catabolic potential under laboratory conditions and to determine more accurately the environmental fate of chlorinated aromatic pollutants. The most stringent criteria for accurately estimating biodegradation include mass or material balance approaches and mineralization approaches. In actual practice, both approaches are frequently integrated to give the best estimate and predictive capability of determining biodegradative fate. In either instance, the use of radiolabeled (primarily ^{14}C) substrates complements the approach, especially at environmentally realistic, low concentrations. These trace concentrations may make conventional analytical approaches more difficult and/or expensive. The approaches are summarized in Table 2.

Table 2   Material Balance and Mineralization Approaches to Biodegradation Assessment

Biodegradation approach	Process examined
Material balance	Recovery of parent substrate. Recovery of radiolabeled parent substrate and metabolic products
Mineralization	Production of carbon dioxide, methane, or their carbon-radiolabeled congeners from the parent substrate
	Release of substituent groups, e.g., chloride or bromide ion

Laboratory assays for biodegradation, using labeled or unlabeled compounds, require a determination of physical processes which contribute to the overall loss of the substrate. Accurate material balances are determined by an accounting of substrate loaded onto biomass and suspended particulates, aqueous phase substrate, residual substrate sorbed to glassware and reaction vessels, and volatilized substrate. Assuming efficient recovery for each component phase, the difference between input and accounted-for residual and comparison to abiotic control samples should give a reasonable approximation of true biodegradation. However, even under these circumstances, biodegradation may be poorly understood if aqueous phase and cellular-associated substrate is transformed to polar oxidized products. In such cases ^{14}C analysis without conventional analysis (HPLC) may underestimate biodegradation, or conventional analysis may fail to detect transformation products that in themselves are resistant to further microbial degradation. Joint conventional and ^{14}C analysis (where available) provide excellent material balance analyses for biodegradation of parent substrate and accumulated metabolic products.

Mineralization assays as a measure of total biodegradation (oxidation or reduction to terminal decomposition products) have enjoyed utility as unambiguous measures of biodegradation. In the event that non-^{14}C-labeled substrates are employed, mineralization assays must include an absolute material balance for the system. Such approaches have been used to study anaerobic biodegradation resulting in methane (CH_4) production as the terminal mineralization product (417).

In more general practice, CO_2 production is the most common measure of pollutant mineralization. With the commercial availability or custom synthesis of ^{14}C-labeled organic pollutants, measurements

of $^{14}CO_2$ production indicating both the extent and rates of substrate degradation have become common practice.

In cases involved with biodegradation of aromatic and chlorinated
aromatic pollutants, the parent substrate is generally chosen with
aromatic ring-labeled ^{14}C atoms. Production of $^{14}CO_2$ during biodegradation is therefore the result of aromatic ring oxidation and cleavage, representing virtually total destruction of the parent aromatic
molecule and associated bioactive properties that are of initial concern from environmental health and ecological perspectives. In alternative mineralization approaches where ^{14}C-parent substrate may
not be available, the release of halogen ions, generally Cl^- or Br^-,
from the aromatic ring during ring oxidation and cleavage is a good
measure of biodegradation. However, if the goal is to determine terminal
decomposition, care must be taken to assure that halogen ion release follows ring cleavage rather than preliminary reductive or oxidative dehalogenation of the aromatic ring.

An integrated flow diagram (Figure 14) describes biodegradation
assessment. Carbon-radiolabeled parent substrate is added to a reaction system containing the microbiological population of interest.
In a time course fashion, samples are withdrawn or replicate samples
are sacrificed. At each time point, material balances for the parent
substrate and ^{14}C are prepared using a combination of conventional
procedures (most conveniently analyzed by HPLC) and liquid scintillation analysis of radioactive decay of ^{14}C. Mineralized products
such as $^{14}CO_2$ or $^{14}CH_4$ can be collected and analyzed by liquid
scintillation analysis and confirmed by conventional analytical methods.
Where specific identification and confirmation are required, mass
spectrometry (MS) or GC/MS can be employed for isolated products.

The resulting data, when compared to appropriate abiotic con-
trols to accommodate nonspecific sorption, volatilization or stripping,
and photolytic or other abiotic degradation mechanisms, provide the
information to determine the potential and extent of degradation. A
relative rank, in terms of rates of degradation, may be assigned if
multiple comparisons are being made.

Biodegradation assessment may be divided into two broad categories, one comprising studies of the environment or laboratory
studies which simulate the environment, and a second category which
includes pure culture studies under defined environmental conditions.
Environmental studies yield information on disappearance or movement of the substrate. Pure culture systems permit studies at the
molecular level, including information about specific enzyme systems
and gene involvement in the manifestation of degradative capability.
Studies of degradation must include both mechanisms of induction
and the mode of action of the enzyme. However, results from laboratory studies are not necessarily an indication of results to be expected in the environment. These studies must be correlated with
environmental conditions for an assessment of in situ biodegradation.

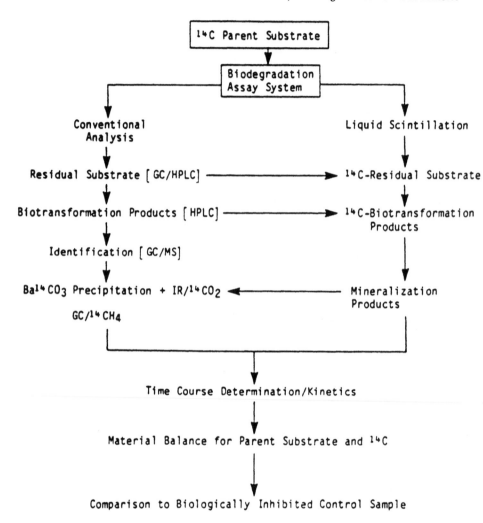

Figure 14 Integration of material balance and mineralization approaches in biodegradation assessment.

Primary degradation or biotransformation is considered to be disappearance of the substrate, without consideration of metabolite formation or mineralization. While primary degradation is important evidence to assess the potential biodegradability of the molecule, only knowledge of the metabolites formed or of complete mineralization will enable confirmation of the biodegradability of the substrate under the specific environmental conditions used.

The biodegradation of a chlorinated compound is considered complete when the chloride ion is returned to its mineral state (HCl or Cl⁻) and the carbon skeleton converted to cellular products (83). Appearance of the chloride ion is most conveniently measured by using an ion-selective electrode.

Disappearance of the substrate may be due to a number of factors in addition to biodegradation. These include photolytic decomposition, volatilization, chemical degradation, and sorption and irreversible binding to soils, clays, or organic matter including cells. In addition to their separate effects, these factors may work in concert with the biota to degrade the substrate. In studies of biological degradation, these factors must be controlled, eliminated, or accounted for.

CHEMICAL ANALYTICAL TECHNIQUES

The ability to quantify the amount of substrate in a given experimental system is of prime importance. To this end, sophisticated analytical techniques have been developed which allow the unambiguous identification of the substrate or its metabolites.

Gas-liquid partition chromatography (GC) is a separation technique which combines high sensitivity, accuracy, and repeatability. Low concentrations of the sample are required. The sample may be solid, liquid, or gaseous as long as the sample can be volatilized at the operating temperature of the instrument. Samples which are insufficiently volatile can sometimes be derivatized and converted to a more volatile ether or ester compound.

The sample to be analyzed is injected along with a gas which carries the sample along a column packed with inert particles coated with a liquid. The solutes in the sample are distributed between the liquid and gas phases according to the relative solubility of the solute in the liquid. Solutes of lower solubility or high volatility move through the column at faster rates. As the bands exit from the column they are recorded as roughly symmetrical peaks with retention times related to their relative partition coefficients. These can be compared with the retention times of standard materials. This does not constitute absolute proof of identity, however, as two or more solutes can elute from the column with the same retention time.

The solutes separated by GC can be analysed directly by a mass spectrometer (MS) for determination of the molecular structures of the compounds (GC/MS). In gas chromatography-mass spectrometry, the separated sample effluent from the GC is highly ionized and accelerated in a vacuum, causing it to fragment into smaller ions. These ions are separated according to their mass-to-charge ratio (m/e) and then collected according to their relative abundance in the sample. The fragmentation pattern results from the molecular structure and is characteristic of the type of compound. The mass spec-

trum of the compound can be used to recreate its molecular structure and often it can be compared with mass spectra in computer-
ized libraries for quick identification of unknown compounds.

The use of GC/MS has become widespread because of the potential unequivocal identification and quantification of metabolites and
residual substrate. Materials present in minute quantities can be
identified within a mixture of other materials.

Additional information regarding the types of bonds between
functional groups can be obtained through infrared spectroscopy.
The different types of bonds absorb at specific frequencies when infrared radiation is passed through the molecule, and the resulting
spectrum is unique to the particular material. Infrared spectra are
most useful when used in conjunction with other methods of compound
identification and when looking for specific functional groups.

Liquid chromatography can be an effective technique for the
separation of materials of either lower volatility or ionic structure.
The sample, in a solvent, is passed through a column containing an
absorbing material dispersed on an inert support. The solutes partition between the liquid and the absorbing material according to their
relative affinities and solubilities. Each solute elutes from the column
at characteristic time intervals. The great variety of both solvent
and stationary phases makes this a very versatile and sensitive technique for both separation and identification of compounds by comparisons with known standards. Increased resolution can be obtained by
decreasing the size of the particles in the stationary phase. This
formerly required high pressures to achieve the desired flow rates,
leading to the term high pressure liquid chromatography (HPLC).
Newer developments have reduced the required pressure and this technique is now widely referred to as high performance liquid chromatography. Solutes eluted from a liquid chromatography column can be
used for other applications including mass spectrometry.

A convenient way to monitor a substrate subjected to biodegradation testing is to use a radiolabeled compound. When the exposure
of a compound to biodegrading agents is halted, the residual quantity
of the compound can be measured by counting the radioactive emission of the solution in a liquid scintillation counter. If the compound
is uniformly labeled (all atoms of the given element are radioactive),
then the metabolites separated by liquid chromatography can be examined for radioactivity and a determination of the fate of the original substrate can be made. This identifies the metabolites arising from
the presence of metabolites which may have arisen from sources other
than the substrate of interest.

Radioactive-carbon (^{14}C) labeled compounds are particularly useful in mineralization experiments in which carbon dioxide is evolved.
Mineralization of the substrate will result in radioactive carbon dioxide which can be quantified. The advantage of this method is that
it is extremely specific and gives unequivocal evidence of the ex-

tent of mineralization for the radiolabeled compound. However, few compounds are routinely available in labeled form and custom synthesis is expensive. Stringent regulations govern the use and disposal of radioactive compounds.

ANALYSES OF METABOLIC ACTIVITY

Determination of the fate of a substrate in a biological community gives specific evidence of metabolic activity. However, several nonspecific methods of analysis are available which measure the metabolic activity of the microorganisms. If the substrate of interest is the only available source of a required nutrient or energy, metabolic activity is considered to be directly related to the presence of the substrate. Lack of metabolic activity is considered to be evidence of inability of the microbial population to utilize the substrate as long as all other essential nutrients and growth factors are present. These methods do not give information about the metabolic products arising from substrate utilization.

The biochemical oxygen demand (BOD) is a measure of the oxygen required to biochemically oxidize all the carbonaceous and nitrogenous matter in a sample. This test is subject to many variations in the methods of sample collection and measurement of data, and results are comparable only between tests performed by the same protocol. The BOD test is most useful for sewage and other samples with high organic content. The test is not very sensitive as a measure of the bio-oxidation of recalcitrant substrates.

A more sensitive measurement of oxygen uptake as a substrate is utilized can be obtained with such manometric devices as a Warburg respirometer. The procedure is based on the ideal gas law which states that, at constant temperature and constant gas volume, a change in the amount of a gas can be measured by the change in its pressure. The utilization of a substrate by microorganisms usually involves utilization of oxygen and production of carbon dioxide. If the carbon dioxide is absorbed in alkali, the only change in gas volume or pressure will be uptake of oxygen. Both the rate and amount of oxygen uptake can be determined by this method. Measurements may be taken at specified time intervals and a graph of oxygen uptake vs. time can be constructed. When the graph indicates a straight line function with time, the enzyme systems are usually considered to be saturated with respect to substrate, although exceptions exist and in some cases higher levels of substrate may result in an increased rate. Determination of the amount of oxygen in moles taken up per mole of substrate yields information on the completeness of substrate oxidation.

Alternatively, carbon dioxide evolution as measured by trapping in alkali gives a measure of the complete mineralization of a compound supplied as the sole source of carbon. If the substrate contains radioactively labeled carbon, the liquid scintillation counter can be used to measure carbon dioxide evolution as captured in the alkali trap.

Warburg respirometry has been used to develop the technique of simultaneous adaptation (407) for determination of the involvement of specific compounds in the pathway of substrate metabolism. Simultaneous adaptation is based on the theory that cells adapted to metabolize the primary substrate will also be adapted to metabolize all the intermediates of the pathway, but will not attack other substrates. In respirometry tests, this is seen as immediate uptake of oxygen after addition of the substrate or its metabolites. When other substrates are introduced, no uptake or uptake only after a lag period is observed. The only prerequisite for this system is that the enzymes be largely adaptive, such that they are not induced until the specific substrate is present. Therefore, a limitation to this test is involvement of nonspecific enzymes.

Assays have been developed for determining the presence of a specific enzyme in a solution or culture. The assays are usually designed so that a substrate which is acted on directly and specifically by the enzyme of interest is made available. Enzyme activity is monitored by measuring loss of the substrate or appearance of a product by spectrophotometric, chemical or chromatographic methods. These assays are most useful for enzymes which are specific for a single substrate. Some enzyme assays require that a cell-free culture filtrate be prepared, while others work with intact cells.

Enrichment cultures are widely used to search for organisms with degradative ability. The substrate of interest is supplied as the growth-limiting nutrient in a culture medium to which a mixed culture of microorganisms is added. The substrate most commonly is the sole source of carbon, and the inoculum may be sewage sludge, sediment samples, or river or ocean water. An inoculum is often selected from environments thought to be contaminated with the substrate of interest. Only organisms with ability to degrade the substrate will be able to grow in the culture medium and eventually will become the dominant population. These cells then can be recovered and isolated from the other cells added in the original inoculum. Evidence of substrate utilization is obtained indirectly by measuring growth of the population of the degradative organism, indicating incorporation of the substrate into cellular material. Further tests, usually with the isolated culture, are required to determine if the substrate is completely mineralized or whether intermediate metabolites remain. As some bacteria can grow in the absence of specifically added carbon, control experiments must be performed to ensure that the substrate is necessary for growth.

PARAMETERS FOR PURE CULTURE STUDIES

Bacteria used for pure culture studies may be selected from enrichment cultures. Identification of these isolates permits the results of such studies to be analyzed in the context of other such studies. Investigations of specific enzymatic or genetic features of degradative bacteria are more easily integrated with other studies when reference bacteria are used. These reference bacteria represent isolates which have been identified and then deposited in culture libraries such as the American Type Culture Collection (ATCC) or the National Collection of Industrial Bacteria (NCIB). Bacteria registered with such libraries are always indicated by a reference number which permits other researchers to obtain the same strain for subsequent studies. However, bacteria frequently mutate during repeated subculturing in a laboratory, and a strain studied for a length of time may no longer resemble the original culture. For this reason, the conditions under which a strain is maintained should be reported. Of particular importance is the frequency with which bacteria lose their degradative capability when removed from the substrate of interest. Such cultures must be maintained on media containing the substrate.

Some general parameters for biodegradation in solution are:

The concentration of the substrate is an important consideration in all studies of biodegradation. The capability of bacteria to degrade substrates supplied at trace levels may be very different from the response to high concentrations.

Chemicals used in formulating culture media should be of the highest purity possible, particularly when the contaminating chemical may be implicated in the degradative strategy of the organism.

Parameters such as pH, temperature, and dissolved oxygen should be monitored as fluctuations may affect not only the metabolic activities of the bacteria but also the chemical nature of the substrate.

As newer techniques of analysis become available, broad studies comprising many of the techniques discussed here become feasible. Knowledge of both the environmental degradative behavior of the bacterial culture and its genetic structure can lead to manipulation of the culture toward increased degradative capability.

4

Metabolism of Nonchlorinated Aromatic Compounds

CHEMISTRY OF BENZENE AND SUBSTITUTED BENZENES

The basic chemical structure of most aromatic compounds is the benzene ring. Benzene (molecular formula C_6H_6) is the simplest six-membered aromatic carbocycle. The entire molecule is structurally flat, i.e., the six carbon atoms and attached hydrogens lie in the same plane. The six aromatic electrons are delocalized and thereby confer the stability which is inherent to aromatic structures. The stability of benzene refers to the availability of the aromatic electrons for bonding. When the reactivities of aromatic and aliphatic carbocycles are compared under identical conditions, the aromatic systems are found to be less reactive, hence more stable.

The naming of substituted benzenes can be quite confusing due to several completely different sets of nomenclature rules. For instance, there is a set of trivial names such as toluene, phenol, and aniline. These trivial names are enigmatic to the casual observer since there are no rules, only a historical selection. Positional isomers have at least two nomenclature systems. The terms *ortho, meta,* and *para* have been used to identify the position of the substituents attached to the benzene ring. Finally, the International Union for Pure and Applied Chemistry offers a numerical description for positional isomers (Figure 15). To show the interrelationships of these systems, catechol, for example, is the trivial name for 1,2-dihydroxy benzene and a more cumbersome *ortho*-hydroxy phenol.

Positional location of substituents (*ortho, meta, para* or 1,2; 1,3; 1,4) is important to the overall reactivity of an aromatic molecule. A substituent on the benzene ring substantially influences the mode of reaction for a given chemical or biochemical system, i.e., the where-

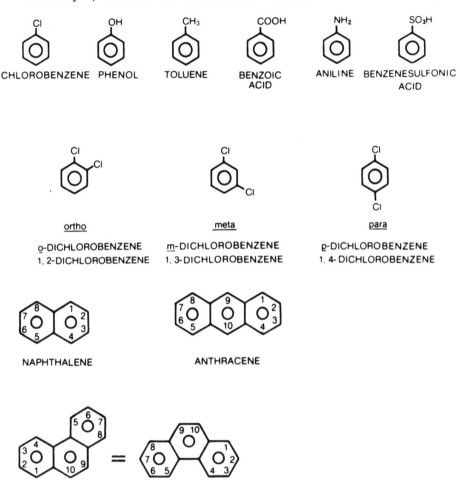

NAPHTHALENE

ANTHRACENE

PHENANTHRENE

Figure 15 Common names and conventional nomenclature for substituted benzenes.

and-how of the attack by another reactive molecule. Common substituents attached to chlorinated aromatic molecules, available common articles of commerce, are: the hydroxyl (-OH), amino (-NH$_2$), methyl (-CH$_3$), and phenyl (-C$_6$H$_5$) groups which can render the molecule more reactive to electrophilic ("electron-loving") reaction conditions. Additional halogen substituents serve to deactivate the aromatic ring for electrophilic attack. Each of the cited substituent groups has a directing influence on subsequent electrophilic substitution which occurs mainly at the *ortho* and *para* positions (316).

The chemical nature and structural position of a substituent on the benzene ring affects the mode and ease of microbial attack on the compounds. The biological "recalcitrance" of a molecule, i.e., the resistance of a molecule to microbial degradation, is a direct consequence of the chemical nature and structural position of a substituent on the aromatic ring (2). Mention throughout this work of the recalcitrance of a compound will refer to biological activity rather than chemical or photooxidative processes.

MICROBIAL ATTACK ON BENZENE STRUCTURES

The first step in oxidative microbial attack on benzene involves hydroxylation of the ring. This is accomplished by different mechanisms in prokaryotes (bacteria) and eukaryotes (fungi, mammals, etc.). Bacteria employ a dioxygenase which incorporates two atoms of oxygen from an oxygen molecule simultaneously into the ring (Figure 16) (15,175,205). Molecular oxygen is a required substrate for this enzyme (84). The resultant intermediate compound is a *cis*-1,2-dihydroxy-1,2-dihydrobenzene which loses two hydrogens to become catechol (1,2-dihydroxybenzene), a process mediated by the enzyme *cis*-benzene glycol dehydrogenase (14). The stereospecific nature ("*cis*") of the intermediate compound was identified in 1968 (181) and confirmed in 1970 (178) for *Pseudomonas putida,* and since then has been shown to be true for all bacterial species studied (14,176,177,205).

Substituted benzenes similarly are oxidized to *cis*-dihydrodiols (substituted catechols) by bacteria. Some species oxidize the substituent before hydroxylation of the aromatic ring while others attack the ring yielding a substituted catechol. For example, toluene is attacked by *P. aeruginosa* with oxidation of the methyl group to benzyl alcohol, benzaldehyde, and benzoic acid, followed by ring hydroxylation to form catechol (254,331), while *Achromobacter* spp. and *Pseudomonas* spp., including *P. putida,* hydroxylate toluene directly to form 3-methylcatechol (Figure 17) (82,175,176,179,182,332). Other alkylbenzenes are also subject to these two types of oxidative degradation, either oxidation of the aromatic ring to form an alkylcatechol or oxidation of the alkyl substituent to form an aromatic carboxylic acid which is dihydroxylated to catechol (84,177,177a,181).

Benzoic acid is metabolized by a number of different pathways, depending on the bacterial strain. *Alcaligenes eutrophus* oxidizes benzoic acid to catechol. This mechanism in *A. eutrophus* is by way of the 3,5-cyclohexadiene-1,2-diol-1-carboxylic acid intermediate (74, 368) catalyzed by benzoic acid 1,2-dioxygenase, a two-protein enzyme composed of NADH—cytochrome *c* reductase and another protein (475). The intermediate is converted to catechol by a single protein

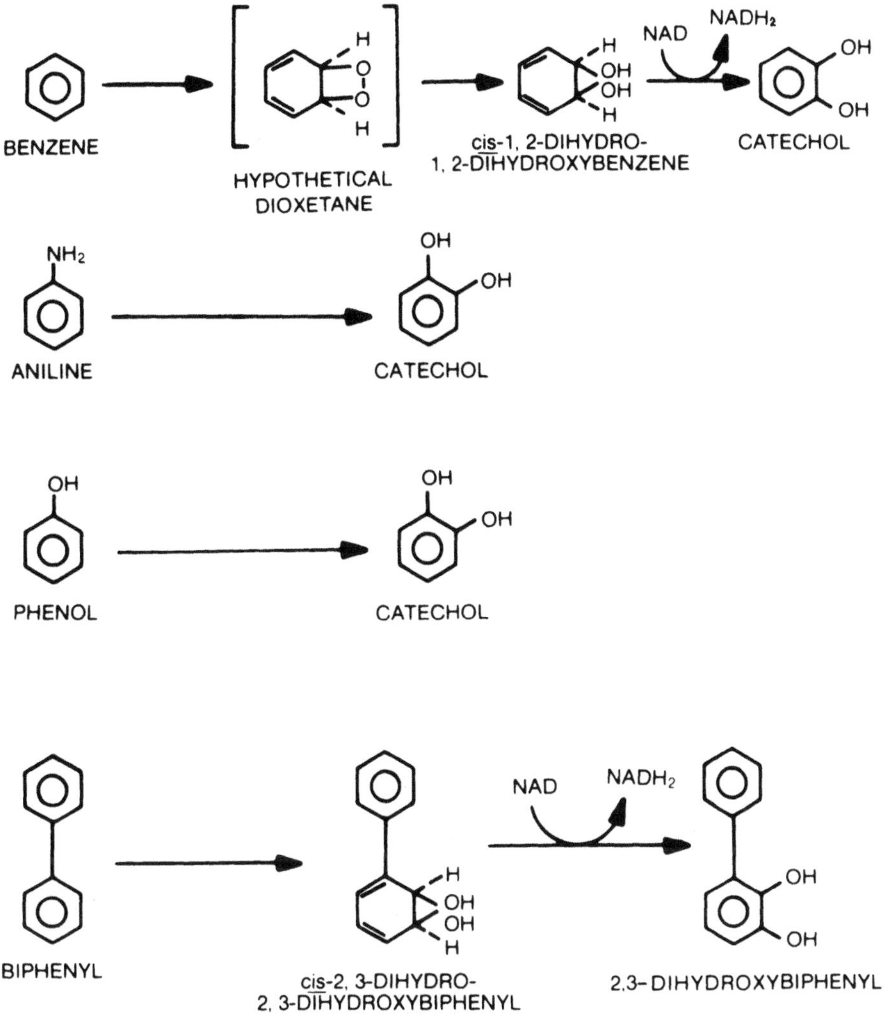

Figure 16 Oxidation of aromatic molecules by bacteria. (Adapted from Ref. 14,16,63,71a,142,176,178,183,196,287,326.)

(Figure 18). The fluorescent pseudomonad group also oxidizes benzoic acid to catechol (410,467). In contrast, the acidovorans pseudomonad group monohydroxylates benzoic acid to *m*-hydroxybenzoic acid and subsequently to either gentisic acid (*P. acidovorans*) by *m*-hydroxybenzoic acid 6-hydroxylase or protocatechuic acid (*P. testosteroni*) by *m*-hydroxybenzoic acid 4-hydroxylase (467). Other

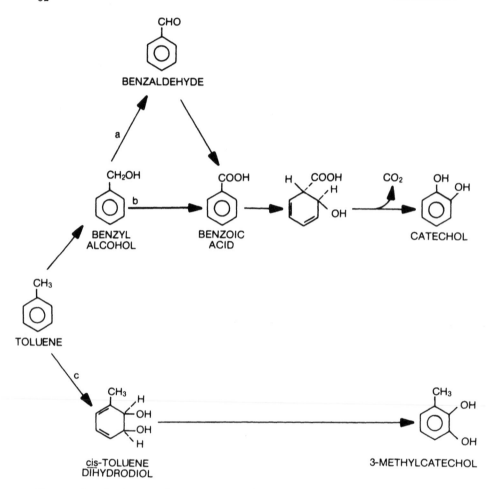

Figure 17 Pathways for the bacterial metabolism of toluene. Pathway (a) *P. putida* mt-2; (b) *P. aeruginosa*; (c) *P. putida, Pseudomonas* sp., *Achromobacter* sp. (Adapted from Ref. (a) 471a; (b) 254; (c) 92,179.)

Pseudomonas spp. monohydroxylate benzoic acid to *p*-hydroxybenzoic acid by utilizing benzoate 4-hydroxylase, and metabolize this intermediate further to protocatechuic acid by the enzyme *p*-hydroxybenzoate 3-hydroxylase (Figure 18) (82).

   Pseudomonas PN-1 metabolizes benzoic acid, *p*-hydroxybenzoic acid, and *m*-hydroxybenzoic acid to protocatechuic acid aerobically. However, under anaerobic conditions of nitrate respiration both pro-

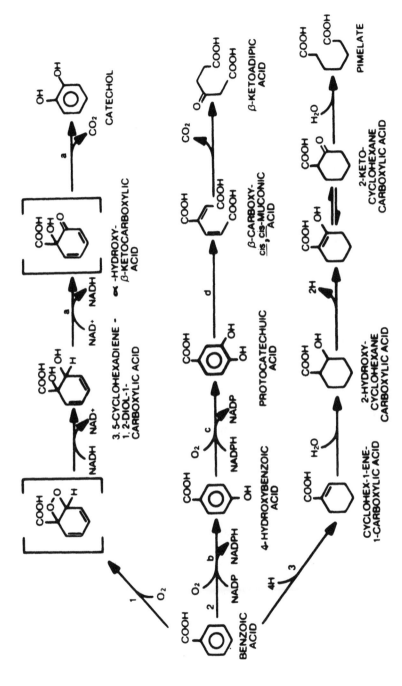

Figure 18 Pathways for the metabolism of benzoic acid. (1) *A. eutrophus*; (2) soil pseudomonad; (3) *R. palustris* (anaerobic photometabolism). Enzymes: (a) 3,5-cyclohexadiene-1,2-diol-1-carbox-ylate dehydrogenase; (b) benzoate 4-hydroxylase; (c) *p*-hydroxybenzoate 3-hydroxylase; (d) protocatechuate oxygenase. (Adapted from Refs. (1) 368; (2) 82; (3) 124,125,191.)

tocatechuic acid and *m*-hydroxybenzoic acid are metabolized through
p-hydroxybenzoic acid to benzoic acid.   The mode of attack resulting in ring cleavage has not been elucidated (430,431).

Members of *Streptomyces* spp. metabolize benzoic acid via catechol, *m*-hydroxybenzoic acid via gentisic acid, and *p*-hydroxyben-
zoic acid via protocatechuic acid (421). Two pathways have been
shown in facultatively thermophilic *Bacillus* spp. In one species, benzoic acid, *m*-hydroxybenzoic acid and *p*-hydroxybenzoic acid are all
metabolized via gentisic acid (64). The conversion of *p*-hydroxyben-
zoic acid to gentisic acid by this organism requires an "NIH shift"
of the carboxyl group. In another species, benzoic acid is metabolized through salicylic acid to 2,3-dihydroxybenzoic acid (406). The
latter pathway is analogous to that shown for *Azotobacter* sp. (450).

The phototrophic bacterium *Rhodopseudomonas palustris* is unable to use benzoic acid as a substrate for aerobic growth (191), although early work postulated an aerobic pathway via protocatechuic
acid and catechol (357). The organism can metabolize *p*-hydroxyben-
zoic acid aerobically via the protocatechuic acid path (125). Under
anaerobic photosynthetic conditions, however, benzoic acid is metabolized reductively to pimelic acid via cyclohexanol (Figure 18) (124,
125,191). Early decarboxylation does not occur. The enzymes involved are thought to be reductases such as ferredoxin coupled to
the light-induced electron transport system (125).

Rhodococcus sp. strain AN-117 utilizes aniline as a sole source
of carbon and energy and metabolizes it exclusively by conversion
to and *ortho* cleavage of catechol by inducible enzymes (235). In
contrast, strain SB3, thought to be a pseudomonad, utilizes a constitutive *meta* cleavage pyrocatechase and hydroxymuconic semialdehyde dehydrogenase. However, aniline degradation occurs only when
the cells are grown on aniline, indicating the presence of another
inducible enzyme. Aniline-grown resting cells of *Frateuria* sp. ANA-
18 oxidize aniline without a lag and oxidize catechol at a faster rate
than aniline (9). Metabolites resulting from incubation of a cell-free
extract with aniline include *cis,cis*-muconic acid, beta-ketoadipic
acid, and ammonia.

A mutant strain of a *Nocardia* sp. has been shown to convert
aniline to catechol via simultaneous dioxygenation (16). This pathway is partially corroborated by data indicating that a *Pseudomonas*
sp. grown on aniline oxidizes catechol rapidly, 4-aminophenol moder-
ately quickly, 2-aminophenol slowly, and phenol not at all (449).
However, only half the ammonia theoretically expected from the direct formation of catechol from aniline was recovered. This discrepancy has not been resolved. *P. multivorans* strain AN1 growing on
aniline was simultaneously adapted to oxidize catechol but not phenol
or 2-aminophenol (199). Transient formation of a catechol was noted
indicating replacement of the amine group with a hydroxyl.

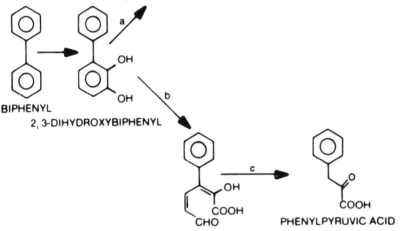

Figure 19 Metabolism of biphenyl by (a) *P. putida* and (b) *Beijerinckia* sp. (Adapted from Ref. (a) 70,71; (b) 183,287; (c) 287.)

Biphenyl (a benzene-substituted benzene) is dihydroxylated by *Beijerinckia* sp., *P. putida*, and an unidentified Gram-negative bacterium to *cis*-2,3-dihydro-2,3-dihydroxybiphenyl and subsequently to 2,3-dihydroxybiphenyl (Figure 19) (183,287).

Phenol is oxidized directly to catechol via phenol hydroxylase (63,141,196,326).

Polynuclear aromatic hydrocarbons are made up of fused aromatic rings. The three simplest compounds, naphthalene, anthracene, and phenanthrene, are metabolized by bacteria to form *cis*-dihydrodiols by the same mechanism as that shown for benzene (Figure 20) (72,227). A mutant strain of *Beijerinckia* sp. (strain B836), as well

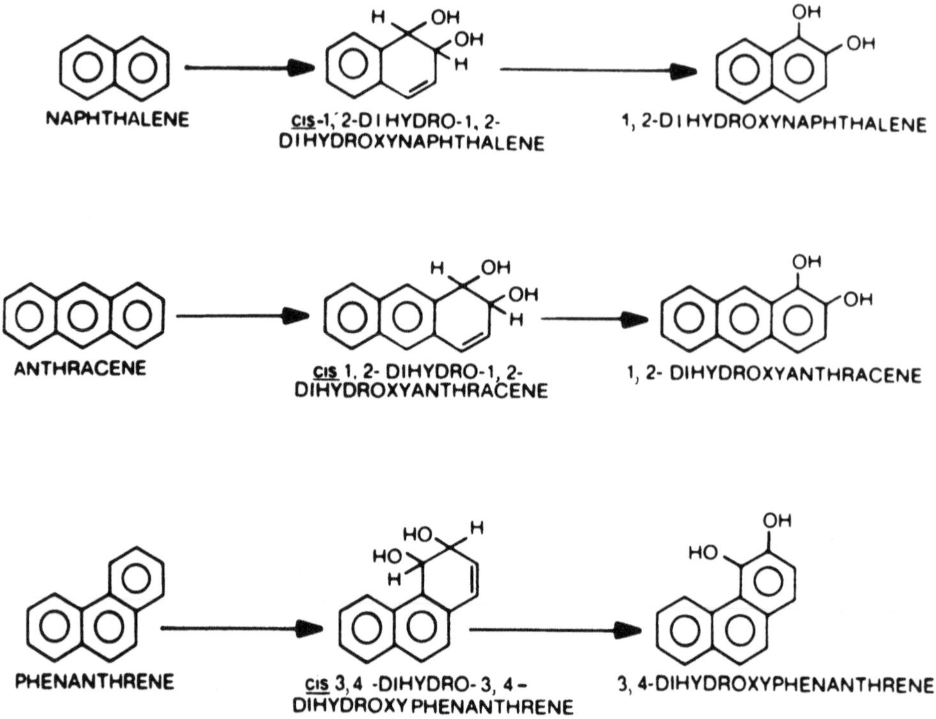

Figure 20   Mechanism of bacterial attack on naphthalene, anthracene and phenanthrene. (Adapted from Refs. 145,224.)

as *P. putida* strain 199, forms (+)-*cis*-1,2-dihydroxy-1,2-dihydro-anthracene from anthracene (226). These two organisms also convert phenanthrene to (+)-*cis*-3,4-dihydroxy-3,4-dihydrophenanthrene. A minor product formed is *cis*-1,2-dihydroxy-1,2-dihydrophenanthrene and there were no other diols recovered during these experiments. Pseudomonads also oxidize phenanthrene to 3,4-dihydroxyphenan-threne (Figure 21) and then to 1,2-dihydroxynaphthalene which is metabolized via the naphthalene pathway (Figure 22) (145,267,282, 373).

Naphthalene is metabolized by pseudomonads through *cis*-1,2-di-hydro-1,2-dihydroxynaphthalene, 1,2-dihydroxynaphthalene, sali-cylaldehyde, salicylic acid, and catechol (106). The enzyme which catalyzes the conversion of the 1,2-dihydro-1,2-dihydroxynaphthalene to 1,2-dihydroxynaphthalene is *cis*-naphthalenedihydrodiol dehydro-genase (224). Its proposed mechanism is stepwise, the first step bac-terial-enzyme catalyzed and the second step a nonenzymatic enoliza-tion (224).

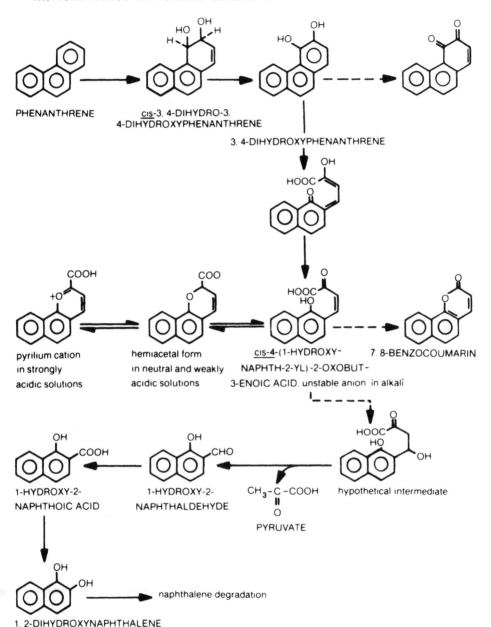

Figure 21 Pathway of phenanthrene metabolism by *Pseudomonas* sp. Dashed lines indicate hypothetical pathways. (Adapted from Ref. 145.)

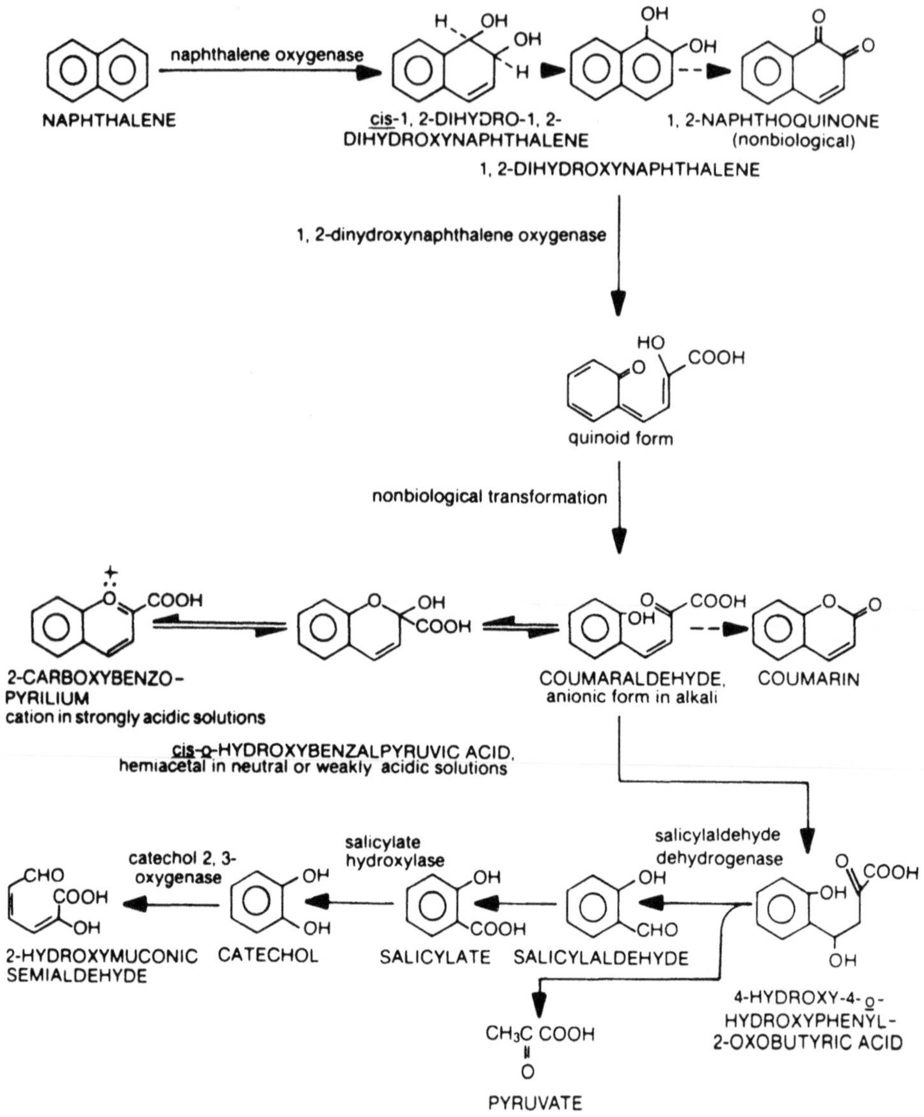

Figure 22 Pathway of naphthalene metabolism by *Pseudomonas* spp.
Dashed lines indicate hypothetical pathways. (Adapted from Ref. 106.)

Anthracene is oxidized to 1,2-dihydroxyanthracene by a naphthalene-grown *Pseudomonas* sp. (151,374). This intermediate is metabolized further to 2,3-dihydroxynaphthalene which follows an unknown
degradative pathway through salicylic acid (Figure 23) (145).

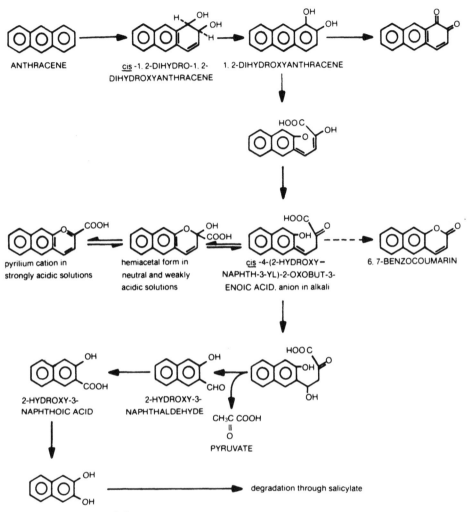

Figure 23 Pathway of anthracene metabolism by *Pseudomonas* spp. Dashed lines indicate hypothetical pathways. (Adapted from Ref. 145.)

In all the above examples with the exception of phenol, bacterial attack on the benzene ring proceeds via a dioxygenase with formation of a *cis*-dihydroxybenzene. This mechanism is consistent with almost all bacterial oxidations of all substituted benzenes studied to date. The phenol hydroxylase is a monooxygenase but results in an *ortho*-dihydroxylated molecule.

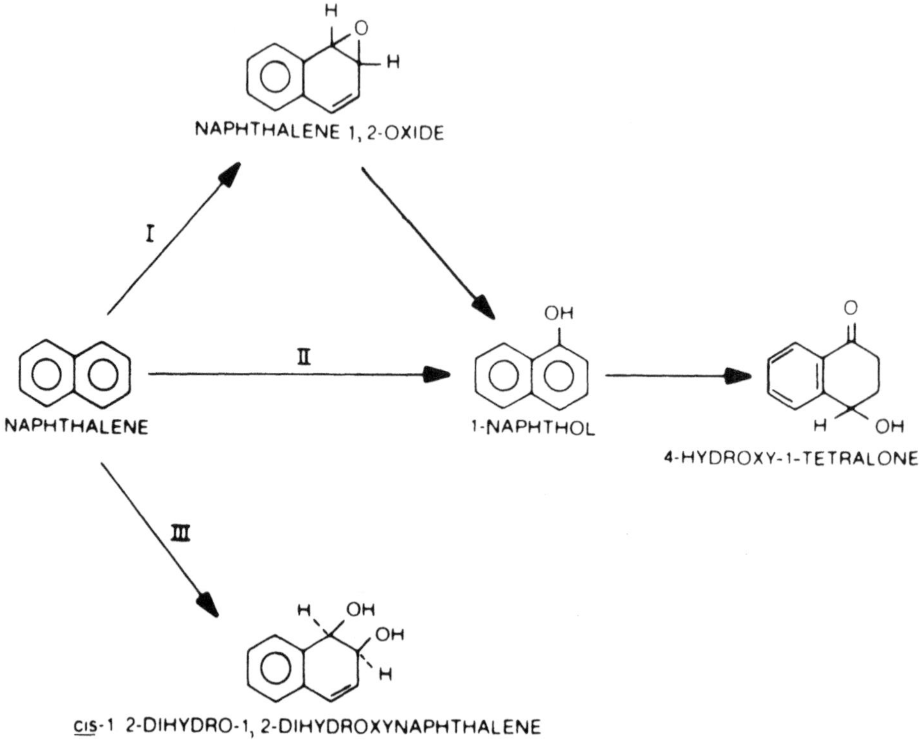

NAPHTHALENE 1, 2-OXIDE

I

NAPHTHALENE

II

1-NAPHTHOL

4-HYDROXY-1-TETRALONE

III

cis-1 2-DIHYDRO-1, 2-DIHYDROXYNAPHTHALENE

Figure 24   Pathways for naphthalene metabolism by *Oscillatoria* sp.,
strain JCM. (I) Metabolism via 1,2-oxide; (II) light-dependent direct hydroxylation of naphthalene; (III) metabolism via dihydrodiol.
(Adapted from Ref. 78a.)

ATTACK ON AROMATIC STRUCTURES BY CYANOBACTERIA

The cyanobacteria have not been studied as extensively with regard
to aromatic degradation. An investigation of the degradation of naphthalene by the cyanobacterium *Oscillatoria* sp. revealed the presence
of 1-naphthol as the major metabolite (Figure 24); 57% of the 1-naph-
thol formed involved the incorporation of molecular oxygen (76a,78a).
Small amounts of cis-1,2-dihydroxy-1,2-dihydronaphthalene were recovered which readily dehydrated to form 1-naphthol. *Trans*-1,2-di-
hydroxy-1,2-dihydronaphthalene was not recovered. There are three
possible pathways which suggest a mechanism for attack of cyanobacteria on the aromatic ring: (a) similar to that for mammalian systems
(path I, Figure 24), (b) unique to photosynthetic organisms (path
II), or (c) similar to that for heterotrophic bacteria (path III) (78a).

RING FISSION OF DIHYDROXY AROMATIC COMPOUNDS BY BAC-
TERIA

Bacteria can open an aromatic ring containing two hydroxyl groups if the groups are located *ortho* (adjacent) or *para* (opposite) to each other. If the groups are *ortho*, ring cleavage can occur either between the two (*ortho* cleavage) or next to one group (*meta* cleavage). The choice of which pathway is induced depends partly on the substrate and partly on the genetic constitution of the particular bacterial species. Both pathways can be induced independently of each other. An organism may utilize one pathway preferentially although it contains the enzymes for both pathways. For example, *P. putida (arvilla)* mt-2 metabolizes catechol via the *meta* pathway, although it contains the enzymes for both the *meta* and *ortho* pathways (323).

The *ortho* cleavage pathway of catechol leads to formation of 3-ketoadipic acid (Figure 25). Substituted catechols can be metabolized by an analogous path until either the substituent is expelled or the compound formed along the pathway cannot be metabolized further. Thus, the *ortho* pathway for protocatechuic acid dissimilation converges with that of catechol at 3-ketoadipate enol-lactone after the carboxyl group is expelled (Figure 26) (66,84,104). This latter compound is converted to 3-ketoadipic acid, which picks up coenzyme-A from succinyl-CoA to form the intermediate 3-ketoadipyl-CoA. Cleavage of this compound results in acetyl-CoA and succinyl-CoA, which in turn exchanges its coenzyme-A with 3-ketoadipyl-CoA leaving one molecule of succinic acid (408). These aliphatic molecules enter the cell's tricarboxylic acid cycle.

The *meta* cleavage pathway (Figure 25) results in the formation of pyruvic acid and acetaldehyde from catechol (102,103,180,330). Substituted catechols usually form pyruvic acid and another aldehyde or an acid (Figure 26). The substrate, 3-methyl catechol, yields acetate while 4-methyl catechol yields formate (27). The nonfluorescent pseudomonad group metabolizes protocatechuate by the *meta* pathway, utilizing protocatechuate 4,5-oxygenase with subsequent production of oxaloacetic acid and pyruvic acid. *Meta* cleavage of protocatechuic acid in *Bacillus* spp. is catalyzed by protocatechuate 2,3-oxygenase and yields pyruvic acid and acetaldehyde via the 2-hydroxymuconic semialdehyde intermediate (95). *Meta* cleavage degradation of catechol results in two possible routes of metabolism as demonstrated in *Azotobacter* sp. and *P. putida* NCIB 10015 (382, 383), the major route involving an NAD^+-dependent dehydrogenase and resulting in formation of 4-oxalocrotonic acid and the minor route not requiring NAD^+ but rather employing a hydrolase to form 2-oxopent-4-enoic acid directly. The paths converge at this step. Degradation of 3-methylcatechol is constrained to follow the hydro-

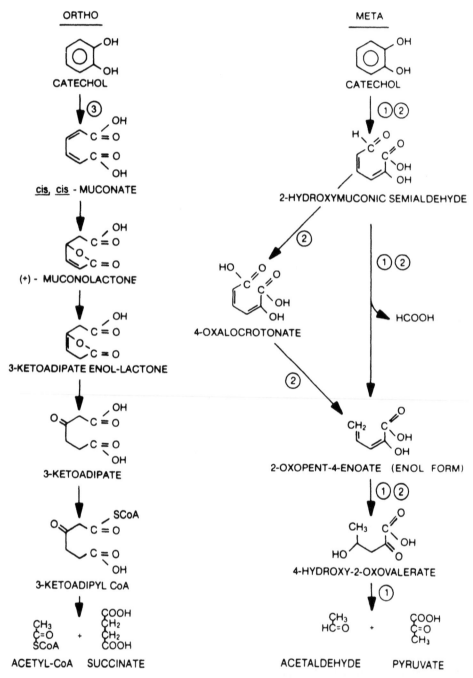

Figure 25 *Ortho-* and *meta*-cleavage pathways of catechol metabolism by bacteria. (Adapted from Refs. (1) 27,103; (2) 382,382a,383,468a; (3) 340,408.)

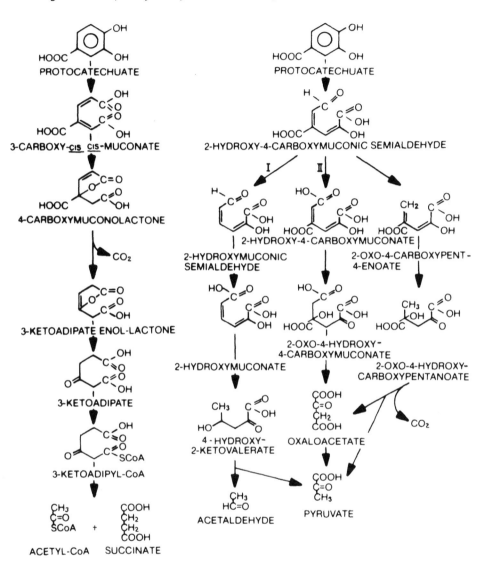

Figure 26 *Ortho-* and *meta*-cleavage pathways of protocatechuate metabolism by bacteria. (I) *Bacillus* sp.; (II) *P. testosteroni.* (Adapted from Refs. 66a,95,340,408.)

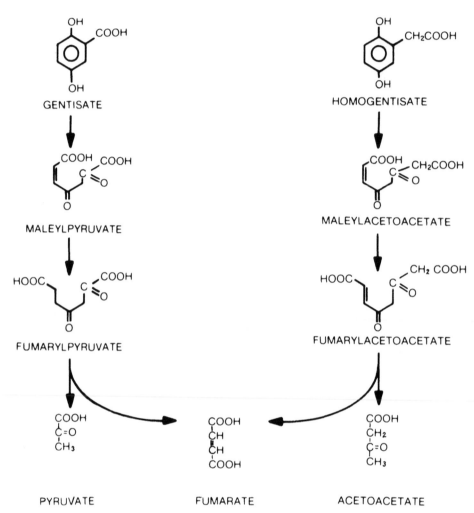

Figure 27 Pathways of gentisic acid and homogentisic acid metabolism by bacteria. (Adapted from Refs. 84,84a,266a.)

lase pathway only, as the NAD^+-dependent path requires the presence of an aldehyde group on the molecule. Other compounds which undergo *meta* cleavage include naphthalene, 2,3-dihydroxyphenylpropionic acid, 2,3-dihydroxybenzoic acid, and homoprotocatechuic acid (84,106).

Compounds that contain two hydroxyl groups located opposite to each other (*para* substitution), such as homogentisic acid and gentisic acid, are usually cleaved at the bond between one hydroxyl

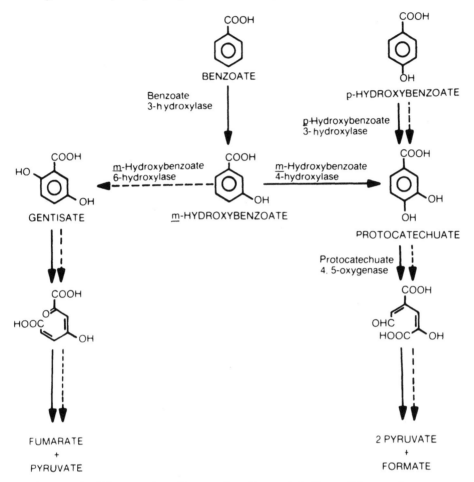

Figure 28 Divergent pathways for the metabolism of benzoate, *p*-hydroxybenzoate, and *m*-hydroxybenzoate by *P. testosteroni* (solid arrows) and *P. acidovorans* (broken arrows). (Adapted from Ref. 467.)

and the adjacent side chain, leading to fumaric acid and acetoacetic acid or fumaric acid and pyruvic acid, respectively (Figure 27) (84).

Metabolism of benzoic acid by *P. testosteroni* and *P. acidovorans* follows divergent pathways with a different hydroxylase being induced in each one (Figure 28), resulting in *meta* and *para* cleavage pathways, respectively. *P. testosteroni* produces two moles of pyruvic acid and one of formic acid, while *P. acidovorans* induces the gentisic acid pathway and produces one mole each of fumaric acid and pyruvic acid (467).

Biphenyl hydroxylated at the carbon-2 and carbon-3 positions may be cleaved in the *meta* position in two ways. An unidentified Gram-negative organism and a mutant strain of *Beijerinckia* sp. metabolize this compound by cleavage between the carbon-3 and carbon-4 positions to yield 2-hydroxy-3-phenylmuconic semialdehyde and subsequently phenylpyruvic acid (183,287). However, *P. putida* cleaves 2,3-dihydroxybiphenyl between the carbon-1 and carbon-2 positions to form 2-hydroxy-6-oxo-6-phenylhexa-2,4-dienoic acid and then benzoic acid (70,71).

The dihydroxy fused ring compounds are cleaved initially by *meta* cleavage between the angular carbon and the adjacent hydroxyl. Hydroxynaphthalene is cleaved to form *cis-o*-hydroxybenzalpyruvic acid and subsequently salicylaldehyde, salicylic acid, and catechol (Figure 22). Catechol may be degraded by *ortho* or *meta* cleavage (106,224,255,267,353). In some bacteria, phenanthrene is dihydroxylated to *cis*-4-(1-hydroxy-naphth-2-yl)-2-oxobut-3-enoic acid, 1-hydroxy-2-naphthaldehyde, 1-hydroxy-2-naphthoic acid, and 1,2-dihydroxynaphthalene (Figure 21) (145,151,267,373). Subsequent steps follow the pathway through salicylic acid. Other bacteria, including fluorescent and nonfluorescent pseudomonad groups, vibrios, and *Aeromonas* spp., metabolize phenanthrene to 1-hydroxy-2-naphthoic acid, 2-carboxybenzaldehyde, *o*-phthalic acid, and protocatechuic acid, which may undergo *ortho* or *meta* cleavage (255,256). The first ring cleavage product of 1,2-dihydroxyanthracene is *cis*-4-(2-hydroxynaphth-3-yl)-2-oxobut-3-enoic acid, which is degraded to 2-hydroxy-3-naphthaldehyde, 2-hydroxy-3-naphthoic acid, and 2,3-dihydroxy-naphthalene (Figure 23) (145,151,267,374). Degradation of 2,3-dihydroxynaphthalene continues through salicyclic acid by an unknown pathway (145).

ATTACK OF AROMATIC BENZENE STRUCTURES BY EUKARYOTES

The fate of aromatic organic substances in mammalian systems has been studied extensively, both in vivo (injecting animals directly

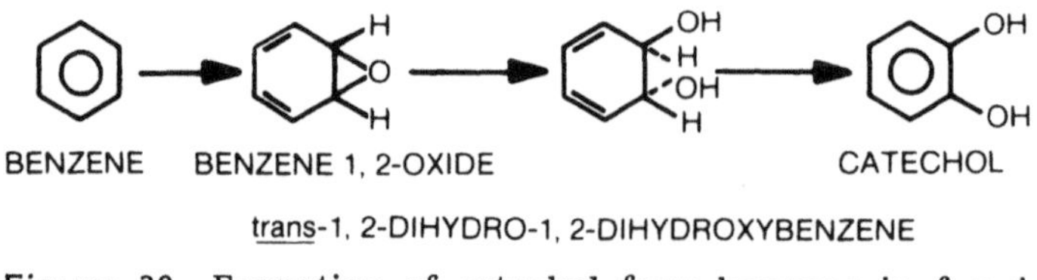

BENZENE   BENZENE 1, 2-OXIDE   CATECHOL

trans-1, 2-DIHYDRO-1, 2-DIHYDROXYBENZENE

Figure 29   Formation of catechol from benzene in fungi, yeasts, and mammals.

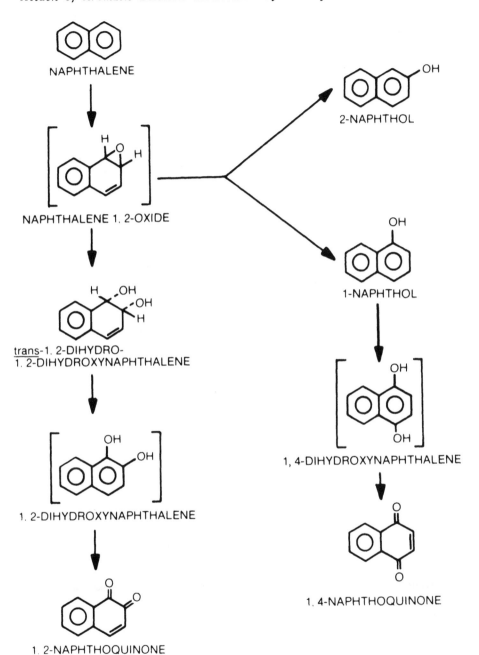

Figure 30 Pathway of naphthalene metabolism by *C. elegans*. (Adapted from Ref. 76.)

and recovering metabolites in body fluids or tissues) and using extracts of liver (or other) cells which contain the enzymes active in
degradation of compounds (the microsomal enzymes). The mechanisms
by which fungi and yeasts degrade aromatic compounds have been
shown to be analogous to that of mammalian systems (404).

In contrast to bacteria, fungi utilize a monooxygenase which incorporates one atom of molecular oxygen into the benzene ring while
converting the other to water. The resulting intermediate is an epoxide, which undergoes hydration with water to form a *trans*-1,2-dihy-
droxy-1,2-dihydro intermediate and subsequently a *trans*-dihydroxy
compound (Figure 29) (175). Both of these mechanisms operate in
the attack on naphthalene by *Cunninghamella elegans* in which the
primary metabolite is 1-naphthol (Figure 30) (76). Anthracene is
oxidized by *C. elegans* predominantly to trans-1,2-dihydroxy-1,2-di-
hydroanthracene with formation of 1-anthryl sulfate (sulfate conjugation of 1-anthrol) as well (73). Other unidentified metabolites are
also produced. Other compounds which have a hydroxyl group added
during fungal metabolism include acetanilide, aniline, anisole, benzene, benzoic acid, biphenyl, and toluene (404). In some cases the
position of the hydroxyl is variable or more than one hydroxyl group
is added to the ring. For example, *C. elegans* hydroxylates biphenyl
to 2-hydroxybiphenyl, 3-hydroxybiphenyl, 4-hydroxybiphenyl, 2,4'-
dihydroxybiphenyl, and 4,4'-dihydroxybiphenyl (116).

DEGRADATION OF DIHYDROXYLATED AROMATIC COMPOUNDS BY YEASTS AND FUNGI

In general, fungi and yeasts lack many of the ring fission dioxygenases characteristic of bacteria (5,274). In most fungi and yeasts,
catechol and hydroxyquinol are cleaved only by the *ortho* mechanism,
utilizing 1,2-dioxygenases only (5,327,395). Phenol is metabolized
through catechol by the *ortho* pathway (395). *Aspergillus niger* converts benzoic acid to benzaldehyde (359). A single strain of *Penicillium* sp. only has been found to utilize the *meta*-fission pathway
(67). However, certain catabolic enzymes of the yeasts have broader
substrate specificities than the equivalent bacterial enzymes. In addition, a third hydroxyl group can be introduced into the aromatic
ring (5). Thus, although the yeast *Trichosporon cutaneum* lacks dioxygenases for protocatechuic acid, gentisic acid, and homoprotocatechuic acid, it can metabolize these substrates by means of NADH-dependent hydroxylases (Figure 31).

Methoxylated aromatic compounds are demethylated and converted
to the corresponding hydroxybenzoic acids by microfungi (42) but
are reduced to their corresponding aldehydes or alcohols by the wood-

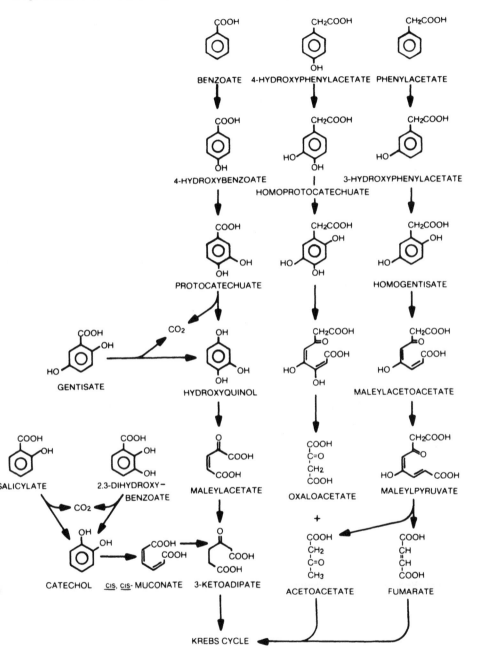

Figure 31 Metabolism of aromatic compounds by *T. cutaneum*. (Adapted from Ref. 5.)

rotting basidiomycete *Polystictus versicolor* (147). The metabolism
of protocatechuic acid though 3-carboxymuconic acid and 3-carboxy-
muconolactone to 3-ketoadipate by *Neurospora crassa* is typical of
many fungi (67,188). Protocatechuic acid is formed from *p*-hydroxy-
benzoic acid or *p*-methoxybenzoic acid. The protocatechuic acid 3,4-
oxidase of *Rhodotorula mucilaginosa* was used to identify the first
metabolite of protocatechuic acid as 3-carboxy-*cis,cis*-muconic acid
(67). Fungi contain a lactonizing enzyme which converts this compound to 3-carboxymuconolactone (93). The product of 3-carboxymu-
conolactone degradation is 3-ketoadipic acid.

Some groups of fungi form catechol from protocatechuic acid (67)
via catechol 1,2-oxygenase. Further degradation leads to *cis,cis*-
muconate and (+)-muconolactone via a *cis,cis*-muconate-lactonizing
enzyme and a muconolactone isomerase. Eventually 3-ketoadipic acid
is formed and subsequently metabolized to succinic acid and acetyl-
CoA. The catechol pathway is similar to that of bacteria (67). Fungal
metabolism of 1,2-dihydroxynaphthalene or 1-naphthol leads to 1,2-
naphthoquinone or 1,4-naphthoquinone, respectively (75,76,78).
These pathways are similar to those of mammalian microsomal extracts
and are due to the cytochrome P-450 present in some fungi as well
as in mammals (75). Monohydroxylated biphenyl compounds are hydroxylated further to various dihydroxy compounds (184a). These
transformations are similar to those of mammals (184a,404).

SUMMARY

Bacteria and eukaryotes differ fundamentally in the mechanism of
primary oxidation of aromatic compounds. Bacteria usually add two
atoms of molecular oxygen from the same oxygen molecule using a
dioxygenase enzyme. The mechanism of the oxidative addition results
in a *cis*-1,2-dihydrodiol intermediate. In a few cases, such as when
the aromatic ring is already monooxygenated (as for phenol), a
hydroxylase (a monooxygenase enzyme) is utilized. The *ortho*-dihy-
drodiol molecules are subject to ring cleavage by either of two mechanisms. *Ortho* cleavage enzymes such as catechol 1,2-oxygenase open
the ring between the adjacent hydroxyl groups. The molecule subsequently is metabolized via *cis,cis*-muconic acid and 3-ketoadipic acid
to acetyl-CoA and succinic acid. In contrast, the *meta* cleavage pathway opens the ring adjacent to one hydroxyl group using enzymes
such as catechol 2,3-oxygenase. The intermediate compounds of this
pathway include 3-hydroxymuconic acid, and the end products include pyruvic acid and an aldehyde. Compounds such as benzoic acid
initially may be monohydroxylated in the *meta* position. A second hydroxylase may then form a *para*-dihydroxylated molecule which is
metabolized via the gentisic acid pathway to fumaric acid and pyru-

Table 3 Comparison of Metabolites Formed by Eukaryotes and Mammalian Systems

		Mammalian metabolites	
Substrate	Fungal metabolites	In vitro	In vivo
Aniline	Acetanilide, 2-OH-acetanilide, and 4-OH-aniline	4-OH-aniline	Acetanilide, 2-,3-, and 4-OH-aniline
Anisole	2- and 4-OH-anisole, phenol	2- and 4-OH-anisole, phenol	2- and 4-OH anisole
Benzene	Phenol	Phenol	Phenol
Benzoic acid	2- and 4-OH-benzoic acid, 3,4-$(OH)_2$-benzoic acid	3-OH-benzoic acid	2-, 3-, and 4-OH-benzoic acid
Biphenyl	2- and 4-OH-biphenyl, 4,4'-$(OH)_2$-diphenyl	2- and 4-OH biphenyl	4-OH-, 3,4-$(OH)_2$-, and 4,4'-$(OH)_2$-biphenyl
Chlorobenzene	2- and 4-OH-chlorobenzene	2-, 3-, and 4-OH-chlorobenzene	2-, 3-, and 4-OH chlorobenzene
Naphthalene	1- and 2-OH-naphthalene	1- and 2-OH-naphthalene	1- and 2-OH-naphthalene
Toluene	2- and 4-OH-toluene	2- and 4-OH-toluene, benzyl alcohol	Benzoic acid and conjugates

Source: Adapted from Ref. 404.

vic acid. An alternative pathway of benzoic acid dissimilation attributable to bacteria is demonstrated in anaerobic reductive metabolism by *R. palustris*, which forms pimelic acid via the cyclohexanol intermediate. Fused-ring compounds are *ortho*-dihydroxylated at the positions adjacent to the angular carbon, and are cleaved by the *meta* pathway.

Cyanobacteria appear to attack aromatic structures by a mechanism similar to that of heterotrophic bacteria, which results in a *cis*-dihydrodiol intermediate. However, the exact mechanism has not been elucidated. Fungi and other eukaryotes attack benzene structures by monooxygenases which incorporate one atom of molecular oxygen, forming an epoxide. Subsequently the epoxide may undergo hydration with water to form a *trans*-1,2-dihydrodiol intermediate. Alternatively, a monohydroxylated compound may be formed.

The *ortho* mechanism is most commonly utilized by fungi for ring fission. The protocatechuic acid pathway in fungi differs from that of bacteria in that 3-carboxy-*cis,cis*-muconic acid is lactonized to the 3-lactone in fungi and the 4-lactone in bacteria. Fungi lack many of the substrate-specific dioxygenases characteristic of bacteria, but some of their catabolic enzymes have broader substrate specificities than the equivalent bacterial enzymes. A third hydroxyl group may be added to the ring by fungi to facilitate metabolism of a molecule. The similarity in metabolism by fungi and mammals of many aromatic compounds has been demonstrated by comparison of the resultant metabolites (Table 3) (404). Thus, fungi may serve as models for many mammalian metabolic mechanisms.

5

Chlorobenzoic Acids

Chlorobenzoic acids are introduced into the environment as degradative products of polychlorinated biphenyls (1,334a) and herbicides as well as in direct application as herbicides (41). For example, the multi-substituted herbicide 2,3,6-trichlorobenzoic acid is a growth regulator similar in function to 2,4-dichlorophenoxyacetic acid (208).

BACTERIAL METABOLISM OF CHLOROBENZOIC ACIDS

The degradation of 2,3,6-trichlorobenzoic acid (Figure 32) has been investigated using resting cell suspensions of *Brevibacterium* sp. grown on benzoic acid (208,210). The major resulting product is 3,5-dichlorocatechol which appears with stoichiometric release of one mole of chloride and one mole of CO_2 per mole of herbicide metabolized (208). The initial oxidation of 2,3,6-trichlorobenzoic acid takes place at the unsubstituted carbon-4 position. This is followed by a one-step oxidation-dechlorination at the adjacent chlorinated carbon. The pathway thus proceeds through 2,3,6-trichloro-4-hydroxybenzoic acid to 2,3,5-trichlorophenol and subsequently to 3,5-dichlorocatechol (Figure 23). The dichlorocatechol accumulates in the medium and is toxic to the *Brevibacterium* sp. cells. However, resting cell suspensions of *Achromobacter* sp. grown on benzoic acid will cleave 3,5-dichlorocatechol by the *meta* pathway to form 2-hydroxy-3,5-dichloromuconic semialdehyde (210). This metabolite accumulates and is toxic to the *Achromobacter* sp. cells.

Several species of bacteria have been shown to metabolize 3-chloro- and 4-chlorobenzoic acid. Resting cell suspensions of *Arthrobacter*

Figure 32 Formation of 3,5-dichlorocatechol from 2,3,6-trichlorobenzoate by *Brevibacterium* sp. (Adapted from Ref. 208.)

sp. grown on benzoic acid oxidize 3-chlorobenzoic acid to 4-chloro-
catechol, which is not inhibitory to growth or oxygen uptake (210,211).
Pseudomonas aeruginosa strain B23 accumulates 3-chlorocatechol from
the metabolism of 3-chlorobenzoic acid (216). *Acinetobacter calcoaceticus* strain Bs5 grown on succinic acid or pyruvic acid will cometabolize
3-chlorobenzoic acid to both 3-chloro- and 4-chlorocatechol, which accumulate and are toxic (362). When mixtures of chlorocatechols can be
formed, 3-chlorocatechol is the major metabolite (261,362). *Meta* cleavage of 3-chlorocatechol results in an acylhalide which acts as an acy-
lating agent and inactivates the *meta* pyrocatechase (catechol-cleaving
enzyme) irreversibly, resulting in the lethal accumulation of catechols
(261). Inefficient *ortho* cleavage will also result in the accumulation
of chlorocatechols. *Meta* cleavage of 4-chlorocatechol yields 5-chloro-
2-hydroxymuconic semialdehyde. Corresponding chlorocatechols are
also formed from 3-chlorobenzoic acid and 4-chlorobenzoic acid by *Azotobacter* sp. grown on benzoic acid (414) and by *Pseudomonas* sp.
WR912 (195).

Cells grown on chlorinated compounds including 3-chlorobenzoic
acid are induced to produce high levels of pyrocatechase II, which has
high activity against chlorocatechols as compared to catechols (118).
Cells grown on nonchlorinated substrates express only pyrocatechase
I, which does not function in chlorocatechol oxygenation. Pyrocatechases I and II are separate catechol 1,2-dioxygenases (118). Pyrocatechase I is involved in the degradation of catechol via the 3-keto-
adipic acid pathway (117). Pyrocatechase II is similar to the *Brevibacterium* spp. pyrocatechase (321) and is unusual in its broad substrate specificity (118). The ability to cleave chlorocatechols, which
are toxic, appears to be a crucial factor in the ability of microorganisms to degrade chloroaromatic compounds (118).

In cells of *Pseudomonas* sp. WR912, pyrocatechases I and II are
both induced when the growth substrate is unsubstituted benzoic acid
(195). This organism can use benzoic acid, 3-chloro-, 4-chloro-, and
3,5-dichlorobenzoic acids as sole sources of carbon and energy. Each
is metabolized to the corresponding chlorocatechol, which undergoes
ortho cleavage to form the chlorinated muconic acid. The muconic acid
in each case is cycloisomerized with coincident or subsequent stoichiometric elimination of the chloride ion (Figure 33). Because of the wide
range of substrates, the benzoic acid 1,2-dioxygenase of *Pseudomonas*
sp. WR912 is characterized as being of low substrate specificity and
also not stereospecific, similar to the corresponding enzyme of *P. putida* mt-2 (195).

In contrast, the benzoic acid 1,2-dioxygenase of *Pseudomonas* sp.
B13 (Figure 34) shows narrow substrate specificity, as this enzyme
metabolizes only 3-chlorobenzoic acid (195). The isomer 2-chloroben-
zoic acid does not induce oxygen uptake, and 4-chlorobenzoic acid is
oxidized only at very low reaction rates (363). Cells grown on 3-chloro-

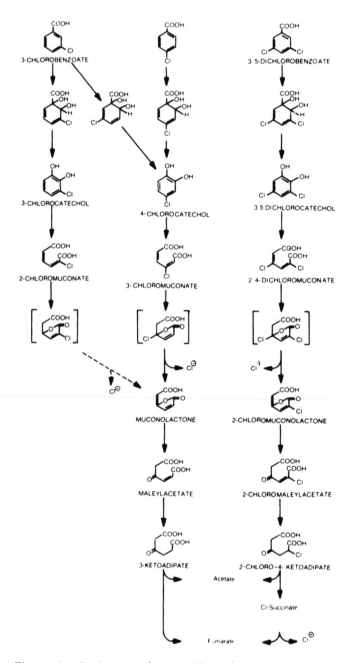

Figure 33 Pathways of metabolism of chlorobenzoates by *Pseudomonas* sp WR 912. Compounds in brackets are hypothetical intermediates (Adapted from Ref. 195.)

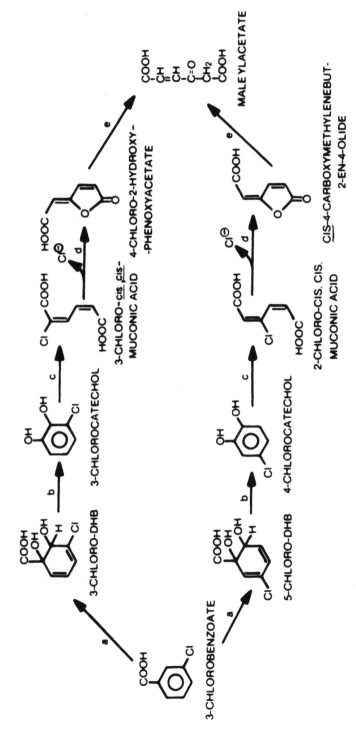

Figure 34 Metabolism of 3-chlorobenzoate by *Pseudomonas* sp. B13. DHB = 2,5-cyclohexadiene-1,2-diol-1-carboxylic acid. Enzymes: (a) benzoate dioxygenase; (b) DHB dehydrogenase; (c) pyrocatechase II; (d) muconate isomerase II; (e) 4-carboxymethylenebut-2-en-4-olide hydrolase. (Adapted from Refs. 119,386,386a.)

benzoic acid are adapted simultaneously to metabolize benzoic acid,
but the reverse is not true (119). The enzymes of the 3-ketoadipic
acid pathway are induced in cells grown on benzoic acid, and the
chlorinated catechols accumulate. In contrast, resting cell suspensions
of cultures grown on 3-chlorobenzoic acid produce both 3-chloro- and
5-chlorodihydroxybenzoic acid in almost equal quantities. A branched
pathway thus exists for metabolism of 3-chlorobenzoic acid by *Pseudomonas* sp. B13 (Figure 34). Along one branch 3-chlorodihydroxyben-
zoic acid is metabolized to 3-chlorocatechol and along a parallel branch
5-chlorodihydroxybenzoic acid is converted to 4-chlorocatechol. The
common enzyme involved, 3,5-cyclohexadiene-1,2-diol-1-carboxylic
acid dehydrogenase, has the same relative activity in both benzoic
acid-grown and 3-chlorobenzoic acid-grown cells (365). The chlorocatechols are metabolized to muconic acids by pyrocatechases which
are induced only in 3-chlorobenzoic acid-grown cells. The muconate
cycloisomerase II enzyme which acts on the muconic acids to perform
cycloisomerization has much higher activity in 3-chlorobenzoic acid
grown cells (119). Combined dechlorination and regeneration of the
diene system is a spontaneous secondary reaction (386). *P. putida*
strain 87 isolated from soils treated with pesticides also contains two
pyrocatechases, one specific for the nonchlorinated catechol and the
other specific for chlorinated catechols (187). The former is controlled
by chromosomal genes and the latter is plasmid mediated. Chloromuconic acid was detected upon incubation of this strain with 3-chloro-
benzoic acid.

The mutant strain *Alcaligenes eutrophus* B9 also produces 3-chloro-
and 5-chlorodihydroxybenzoic acid from cooxidation of 3-chlorobenzoic
acid (367). A strain of *P. putida*, harboring a plasmid termed pAC25,
degrades 3-chlorobenzoic acid via 3-chlorocatechol and 3-chloromuconic
acid (86). Chloride is released and maleylacetic acid rather than 3-ke-
toadipic acid is produced. This pathway is analogous to one path of
3-chlorobenzoic acid metabolism by *Pseudomonas* sp. B13.

Four strains of *Pseudomonas* spp. which utilize 3-chlorobenzoic
acid as the sole source of carbon and energy for growth were isolated
from sewage which had been enriched with the substrate (194). One
isolate studied in detail, *Pseudomonas* sp. strain H1, resembles *Pseudomonas* sp. B13 in metabolism of 3-chlorobenzoic acid and benzoic
acid, and therefore seems to possess both pyrocatechases. A black
color, resulting from oxidation and polymerization of unmetabolized
catechols, forms when *Pseudomonas* sp. strain H1 is incubated with
both 3-chlorobenzoic acid and benzoic acid without prior adaptation to
3-chlorobenzoic acid. In contrast, another species isolated from the
sewage enrichment culture metabolizes catechols and chlorocatechols
rapidly enough to prevent occurrence of the black color.

A hybrid strain has been developed which combines the *ortho*
cleavage pathway of *Pseudomonas* sp. B13 with the relatively nonspe-

cific toluate 1,2-dioxygenase of *P. putida* mt-2 (364). This derivative of *Pseudomonas* sp. B13 acquired the TOL plasmid from *P. putida* mt-2. In the resulting cells, both 4-chlorobenzoic acid and 3,5-dichlorobenzoic acid as well as 3-chlorobenzoic acid are metabolized (Figure 35). The enzyme, toluate 1,2-dioxygenase from the genes on the plasmid, is induced and slightly modified to result in increased turnover of the chlorinated compound used as the selective substrate. Dihydrodihydroxybenzoic acid dehydrogenases from both plasmid and chromosomal sources are induced. In addition, *ortho* pyrocatechases I and II are induced, but not the unproductive *meta* pyrocatechase.

A *Bacillus* sp. grown on benzoic acid uses a unique pathway to cometabolize 3-chlorobenzoic acid to 5-chloro-2,3-dihydroxybenzoic acid via 5-chlorosalicylic acid (406). Another unique pathway involving enzymatic rather than spontaneous elimination of chloride ion was demonstrated in *Bacillus brevis* isolated from polluted river water (96). This organism utilizes 5-chloro-2-hydroxybenzoate (5-chlorosalicylic acid) as a sole carbon and energy source. The first step in metabolism is cleavage between the carbon-1 and carbon-2 by a specific 5-chlorosalicylate 1,2-dioxygenase. This enzyme requires a halogen (except iodine) or a methoxyl group (but not a hydroxyl) at the carbon-5 position. Only one hydroxyl group is present on the molecule. After loss of the chloride ion and formation of maleylpyruvic acid, metabolism continues along the steps of the gentisic acid pathway (Figure 36).

A novel pathway for 3-chlorobenzoic acid metabolism in which the chloride is replaced by a hydroxyl group in the first step has been demonstrated in a *Pseudomonas* sp. isolated from soil (231). This organism utilizes 3-chlorobenzoic acid as a sole source of carbon for growth and metabolizes it to 3-hydroxybenzoic acid and subsequently to 2,5-dihydroxybenzoic acid.

Similarly, an *Arthrobacter* sp. growing on 4-chlrobenzoic acid as the sole source of carbon produces 4-hydroxybenzoic acid and 3,4-dihydroxybenzoic acid (protocatechuic acid) (380). A strain of *Arthrobacter globiformis* also utilizes 4-chlorobenzoic acid as the sole carbon source and metabolizes it via 4-hydroxybenzoic acid and protocatechuic acid, with release of chloride (486). Direct replacement of the chloride by the hydroxyl group precludes formation of the potentially toxic chlorocatechol. *Pseudomonas* sp. strain CBS 3 also utilizes 4-chlorobenzoic acid as the sole source of carbon for growth (257) by the same pathway. The enzymes induced by growth with this substrate have been identified as 4-chlorobenzoate 4-hydroxylase, 4-hydroxybenzoate 3-hydroxylase, and protocatechuate 3,4-dioxygenase. The first enzyme was not induced by growth with 4-hydroxybenzoic acid or any of several other chlorinated and nonchlorinated substrates. The mechanism of action of this enzyme has not been elucidated. Protocatechuic acid was metabolized by the 3-ketoadipic acid pathway following *ortho* cleavage.

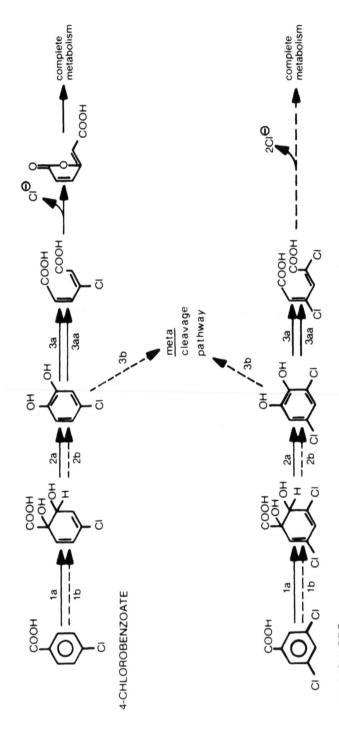

Figure 35 Metabolism of 4-chlorobenzoate and 3,5-dichlorobenzoate by *Pseudomonas* sp. B13 transconjugants. Enzymes marked (a) from *Pseudomonas* sp. B13; those marked (b) from donor *P. putida* mt-2. Enzymes: (1a) benzoate 1,2-dioxygenase; (1b) toluate 1,2-dioxygenase; (2a, 2b) dihydrodihydroxybenzoate dehydrogenase; (3a) pyrocatechase I; (3aa) pyrocatechase II; (3b) *meta*-pyrocatechase. (Adapted from Ref. 464.)

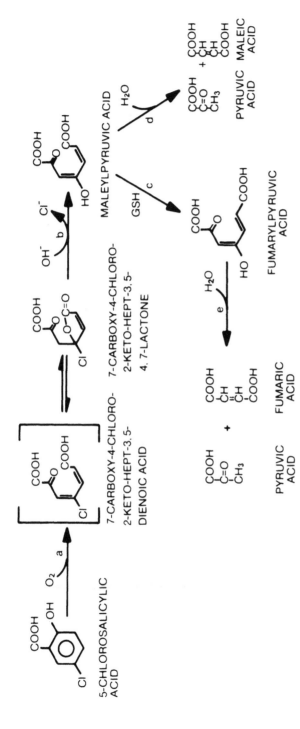

Figure 36 Metabolism of 5-chlorosalicylic acid by *B. brevis*. Enzymes: (a) 5-chlorosalicylic acid 1,2-dioxygenase; (b) "lactone hydrolase"; (c) maleylpyruvate isomerase; (d) maleylpyruvate hydrolase; (e) fumarylpyruvate hydrolase. GSH = reduced glutathione. (Adapted from Ref. 96.)

ALGAL METABOLISM OF CHLOROBENZOIC ACIDS

The only reference to algal metabolism of chlorobenzoic acids obtained involves a monoalgal culture of *Chlamydomonas* sp. strain A2 isolated from sewage (221). The nonaxenic culture (bacteria present) transforms 4-chloro-3,5-dinitrobenzoic acid to 3-hydroxymuconic semialdehyde, indicating a *meta*-cleavage pathway. Approximately 20% chloride release was reported. Since the culture contained bacteria along with the algae, dechlorination might have been due to the action of the algae or it might have been a bacterial process with the algae providing fixed carbon or growth factors.

FUNGAL METABOLISM OF CHLOROBENZOIC ACIDS

Aspergillus niger cultures utilized both 2-chloro- and 3-chlorobenzoic acid as sole sources of carbon and energy (390). Protocatechuic acid and 4-hydroxybenzoic acid were isolated from both samples. Cells grown on these substrates oxidize all four compounds as well as benzoic acid more rapidly than do cells grown on glucose. Adapted cultures dechlorinate either substrate, while glucose-grown cells do not have this capability. Dehalogenating activity was also noted in the cell-free extracts of cultures grown on 2-chlorobenzoic acid.

METABOLISM OF CHLOROBENZOIC ACIDS IN SOILS AND BY CONSORTIA

Under cometabolic conditions with glucose as the additional carbon source, a sewage plant effluent inoculum metabolized 3-chlorobenzoic acid with production of 3-chlorocatechol (212). Upon continued incubation this metabolite disappeared with concomitant appearance of 2-hydroxy-3-chloromuconic semialdehyde. After a 29-day period of no discernible metabolism, the semialdehyde was metabolized with appearance of stoichiometric amounts of inorganic chloride. There was no additional increase in cell numbers due to the presence of chlorobenzoic acid until degradation of the semialdehyde occurred.

Diluted wastewater sludge supernatant fluid mediated disappearance of 16 mg/L 3-chlorobenzoic acid within 14 days, although no degradation of 2- or 4-chlorobenzoic acid was seen after 25 days (193). Readaptation of the sludge inocula to 3-chlorobenzoic acid greatly reduced the time required for metabolism of both 3- and 4-chlorobenzoic acid. Soil suspension also did not degrade 4-chlorobenzoic acid in 25 days, although the other two substrates were metabolized within 14 days. Application of 2,3,6-trichlorobenzoic acid to soil resulted in 30% chloride release within one month (113). No intermediate metabolites were detected.

A sewage microcosm enriched with chlorinated benzoic acids resulted in development of a consortium of Gram-negative motile rods and Gram-positive pleomorphic rods which could utilize as sole carbon and energy sources benzoic acid, 2-chloro-, 3-chloro-, 4-chloro-, and 3,4-dichlorobenzoic acids but not 2,4-dichloro- or 2,3,6-trichlorobenzoic acid (114). Addition of biodegradable benzoic acids did not lead to decomposition of any of the substrates. Degradation of 3-chlorobenzoic acid led to formation of both 4-chlorocatechol and 5-chlorosalicylic acid, the latter compounds disappearing from solution. Formation of both metabolites indicates two separate pathways of metabolism within the consortium.

Pronamide [3,5-dichloro-N(1,1-dimethyl-2-propynyl)benzamide] is an herbicide used for weed control of crops of lettuce and alfalfa and other legumes (153). Pronamide was metabolized in soils to $^{14}CO_2$ from both ^{14}C(carbonyl)- and ^{14}C(ring)-labeled substrate. In addition, seven other metabolites were found, none of which was dechlorinated. The potential metabolite 3,5-dichlorobenzoic acid was also metabolized with 80% of ^{14}C(carbonyl)- and 50% of ^{14}C(ring)-labeled substrate recovered as $^{14}CO_2$ within 28 days.

Dicamba (3,6-dichloro-2-methoxybenzoic acid) is used on wheat, oats, barley and flax crops for weed control. Studies with a temperature gradient table showed that an increase in temperature from 5 to 35°C resulted in a linear increase in the disappearance rate for the herbicide in silty clay and heavy clay, while the range for a linear relationship of temperature to disappearance was 9 to 20°C for sandy loam, a soil with a lower moisture holding capacity (401). Studies with Regina heavy clay have shown that only 10% of the herbicide remains five weeks after application (399). The only radioactive metabolite recovered from application of carboxyl-group labeled dicamba was 3,6-dichlorosalicylic acid. About 20% of the substrate was recovered as $^{14}CO_2$, and metabolites which were not radioactive were not identified. Dicamba applied at a rate of 1.1 kg/ha to a silty clay with high organic content almost entirely disappeared within two weeks (399a). At the same application rate onto sandy loam and heavy clay, there was less than 10% substrate remaining after 4 weeks when the moisture content was high, although residues were still detected after 6 weeks in low-moisture soils. After 4 weeks, over 90% of the substrate applied to sterile soils was recovered. Studies with ^{14}C(carboxyl)- and ^{14}C (ring)-labeled dicamba revealed that 18% was released as $^{14}CO_2$ in 6 weeks and 45% in 17 weeks from ^{14}C(carboxyl)-dicamba, and 9% in 6 weeks from ^{14}C(ring)-dicamba (398). The only metabolite that could be recovered was 3,6-dichlorosalicylic acid.

REDUCTIVE DECHLORINATION

The chlorobenzoic acids have served as model structures for the elucidation of anaerobic reductive dechlorination by consortia of bacteria

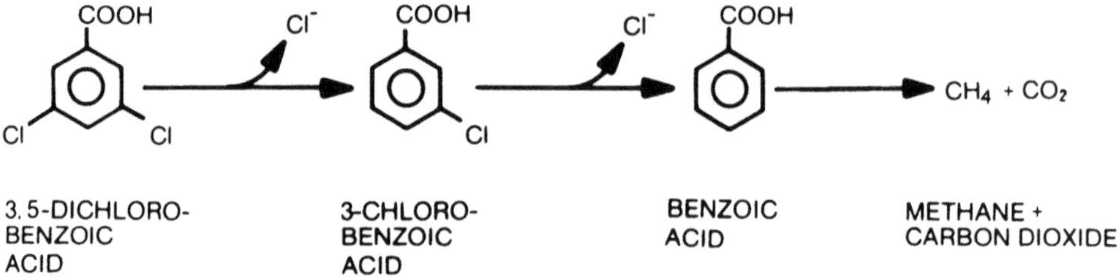

3,5-DICHLORO-   3-CHLORO-   BENZOIC   METHANE +
BENZOIC   BENZOIC   ACID   CARBON DIOXIDE
ACID   ACID

Figure 37   Representative pathway for the reductive dechlorination of
chlorobenzoates by anaerobic microbial consortia. (Adapted from Refs.
206,417.)

from anaerobic sediment or sludge environments. This pathway for dechlorination of aromatic compounds involves removal of the aryl halide
from the aromatic ring (Figure 37) (417). A consortium resulting from
enrichment with 3-chlorobenzoic acid mineralizes this substrate through
benzoic acid to methane and CO_2. The substrate 4-amino-3,5-dichloro-
benzoic acid is converted to 4-amino-3-chlorobenzoic acid by replacement of one chlorine atom with a hydrogen atom. No chloride shift
takes place. Neither 2-chloro- nor 4-chlorobenzoic acid was metabolized
in experiments lasting for one year of incubation (206). In multisub-
stituted compounds, the *meta* substituent is utilized preferentially.
Thus, 2,5-dichlorobenzoic acid is reduced to 2-chlorobenzoic acid,
3,4-dichlorobenzoic acid to 4-chlorobenzoic acid, and 2,3,6-trichloro-
benzoic acid to 2,6-dichlorobenzoic acid (417). In these experiments,
persistence was not correlated with the number of halogens present
on the molecule.

    Although benzoic acid was always an intermediate in anaerobic
mineralization of the chlorobenzoic acids, acclimation to the chlorinated
substrate did not result in acclimation to benzoic acid (206). This
phenomenon was explored further with the consortium acclimated to 3-
chlorobenzoic acid. Metabolism of 3,5-dichlorobenzoic acid to 3-chloro-
benzoic acid proceeded with accumulation of 3-chlorobenzoic acid until
the parent substrate concentration fell to a low level. Only then was
3-chlorobenzoic acid metabolized to benzoic acid followed by production
of methane and CO_2. Similarly, 4-amino-3,5-dichlorobenzoic acid was
metabolized to 4-amino-3-chlorobenzoic acid which accumulated until
the concentration of parent substrate decreased to a low level, after
which the intermediate metabolite was converted to 4-aminobenzoic
acid which was not degraded further. These events were postulated
to be due to competitive substrate inhibition, in which one enzyme involved in multiple steps of a degradative pathway acts only on the
parent compound until its concentration falls below a threshold level
(418). Under these conditions, bacteria from environments receiving

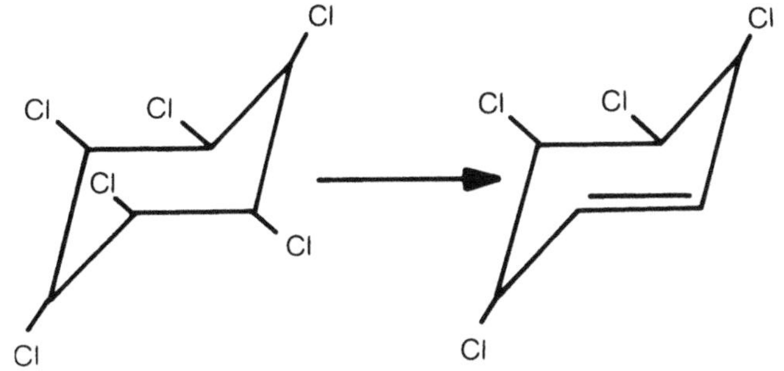

δ -HEXACHLOROCYCLOHEXANE   δ -TETRACHLOROCYCLOHEXENE

Figure 38 Primary metabolic reductive dechlorination of γ-hexachloro-cyclohexane by anaerobic microorganisms. (Adapted from Refs. 192, 267.)

several structurally related chemicals may metabolize substrates selectively due to competitive substrate inhibition.

The consortium could be acclimated to degrade 4-amino-3,5-dichlorobenzoic acid to its metabolites, even though this substrate was not used as a sole carbon and energy source and was not mineralized (206). Aliquots of the substrate added after acclimation were metabolized without a lag. Partial dechlorination of several compounds by sewage microflora acclimated to nonchlorinated products was reported (221). Substrates attacked include 2-chlorotoluene, 3-chlorobenzoic acid, 4-chloro-2,5-dinitrobenzoic acid, 4-chloroaniline, 4-chlorobiphenyl, 4-chlororesorcinol, and 4-chlorobenzonitrile. From 2 to 47% of the chlorine from these substrates was removed within 2 days. The substrate 4-chloro-3,5-dinitrobenzoic acid was 13 to 45% dechlorinated in sewage but was only 13 to 20% dechlorinated within 20 days by single isolates of bacteria from the sewage.

Reductive dechlorination of pentachlorophenol has been demonstrated in anaerobic soils (217,266,320). Resultant products include isomers of tetrachlorophenols, trichlorophenols, dichlorophenols and 3-chlorophenol. The methylated chloroanisole analogues of these isomers have been detected as well. These investigations led to the conclusion that the chloride ions *ortho* and *para* to the hydroxyl group are utilized preferentially (217). Pentachlorophenol metabolism is discussed further in a later chapter. In contrast, the reductive dechlorination of DDT to DDD described in another chapter involves the alkyl chlorides but not the chlorides attached to the ring (107,189a,202, 234). This has been demonstrated in yeasts as well as bacteria and in pure cultures as well as consortia. Lindane (γ-hexachlorocyclohexane) is a nonaromatic molecule which is also reductively dechlorinated

(Figure 38). The major metabolite is γ-tetrachlorocyclohexene, followed by benzene, monochlorobenzene, and small amounts of tri- and tetrachlorobenzenes (192).

SUMMARY

The chlorobenzoic acids can be metabolized to several different intermediate products in bacteria. The most common mechanism results in the conversion of the chlorobenzoic acids to chlorocatechols. If the *meta* cleavage pathway is the only route induced in the bacteria, then chlorocatechols are metabolized usually to chloromuconic semialdehydes, which are not metabolized further. In addition, the *meta* cleavage product of 3-chlorocatechol inactivates the *meta* pyrocatechase, causing an accumulation of the toxic chlorocatechols. This pathway does not lead to mineralization of the chlorinated substrate, and ultimately results in cell death and release of chlorinated intermediates into the environment. In contrast, *ortho* cleavage of chlorocatechols is a successful pathway for chlorobenzoic acid mineralization. The *ortho* pyrocatechase results in the formation of chlorinated muconic acids, which are cycloisomerized with coincident or subsequent elimination of the chloride ion. After release of the chloride ion, the compound is fully metabolized by established cellular mechanisms.

One block to utilization of chlorinated aromatic compounds in organisms expressing the *ortho* pyrocatechase is specificity of the enzyme required for the first oxygenation step. The hybrid strain constructed from *Pseudomonas* sp. B13 and *P. putida* mt-2 incorporates the relatively nonspecific oxygenase of *P. putida* mt-2, carried on the TOL plasmid, into *Pseudomonas* sp. B13 which metabolizes chlorinated compounds via the *ortho* pathway. The hybrid strain is capable of mineralizing a wider range of chlorobenzoic acids than either parent.

Another pathway which has been discovered replaces the chloride ion by a hydroxyl group directly, resulting in a nonchlorinated hydroxybenzoic acid which can be metabolized by established cellular mechanisms. A third series of metabolic pathways results in hydroxylation of the chlorobenzoic acid without loss of prior substituents. A specific and unique enzyme opens the ring and an enzyme has been identified which specifically dechlorinates the resulting aliphatic molecule. Data on algal and fungal pathways for metabolism of chlorobenzoic acids are lacking. The relaxed substrate specificities of fungi suggest that these organisms may be of prime importance in metabolism of the chlorobenzoic acids.

6

Chlorobenzenes

MICROBIAL METABOLISM OF CHLOROBENZENES

Chlorobenzenes are used as industrial solvents and diluents for poly-chlorinated biphenyl compounds (PCBs) and thus have a complemen-tary distribution as pollutants in the environment, including capaci-tor and transformer storage and disposal (295). They are solvents for paints and appear as byproducts in the textile dyeing industry and in other industries. Pentachloronitrobenzene is used as a fungi-cide for seeds and soils.

A chemostat seeded with a mixture of soil and sewage samples was used to enrich for an organism capable of utilizing chloroben-zene as a sole growth substrate (366). After nine months an uniden-tified bacterium, strain WR1306, was isolated which degrades chloro-benzene with stoichiometric release of chloride ion. Detection of the enzymes cis-1,2-dihydroxycyclohexa-3,5-diene (NAD^+)-oxidoreduc-tase, catechol 1,2-dioxygenase, muconate cycloisomerase, 4-carboxy-methylenebut-2-en-4-olide hydrolase, and NADH-dependent maleyl-acetate reductase, and isolation of the metabolite 3-chlorocatechol, enabled construction of a pathway for chlorobenzene dissimilation (Figure 39). The proposed pathway is similar to that demonstrated for the metabolism of other nonphenolic benzene compounds such as 3-chlorobenzoic acid. Chlorobenzene is converted to 3-chlorocate-chol, which is cleaved by the $ortho$ pathway. The substituted mu-conic acid thus formed is cycloisomerized with coincident or subse-quent elimination of chloride. The nonchlorinated intermediate is me-tabolized to 3-oxoadipate, which enters the cell's tricarboxylic acid cycle.

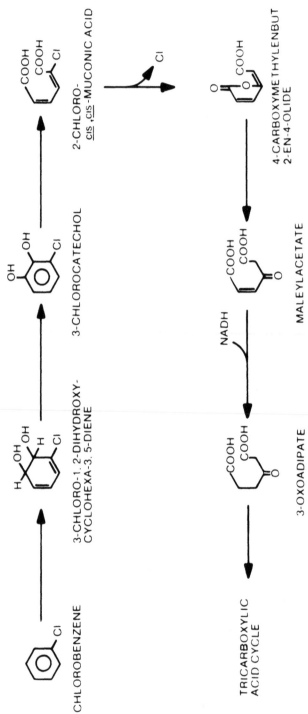

Figure 39 Pathway of chlorobenzene mineralization by bacterial strain WR1306. (Adapted from Ref. 366.)

Attempts to find other strains of bacteria capable of using chlorobenzene as a sole source of carbon and energy for growth have been
hampered by the lipophilicity of the compound (366). Strain WR1306,
although capable of growth on chlorobenzene, was inhibited by high
concentrations of the compound. Accumulation of the toxic metabolite
3-chlorocatechol is lethal for cells which have not evolved a mechanism for efficient metabolism of this compound. The mechanism of
catechol cleavage must be by the *ortho* rather than the *meta* pathway for production of metabolites useful in cell biosynthesis (discussed further in chapter on chlorobenzoates).

Resting cells of *P. putida* grown on toluene oxidize chlorobenzene and to a lesser extent all three isomers of chlorotoluene (182).
Cometabolic growth of *P. putida* on toluene and chlorobenzene results in formation of 3-chlorocatechol. Cells grown on toluene and
4-chlorotoluene metabolize the latter through *cis*-4-chloro-2,3-dihy-
droxy-1-methylcyclohexa-4,6-diene to 4-chloro-2,3-dihydroxy-1-me-
thylbenzene.

Bacteria were isolated which utilize 1-chloronaphthalene for
growth (451a). Chloronaphthalene diol (8-chloro-1,2-dihydro-1,2-di-
hydroxynaphthalene) and 3-chlorosalicylic acid were recovered as
metabolites.

An alternative pathway has been developed which proposes chlorophenols as intermediates in the degradation of chlorobenzenes (19).
However, the formation of phenolic products from the dihydrodiol
metabolite may occur spontaneously under mild acid conditions as
well as enzymatically (366). Confirmation of this pathway requires
isolation of the enzymes involved.

Pure cultures of bacteria isolated from pond water and pond sediment samples metabolize dichlobenil (2,6-dichlorobenzonitrile) to
2,6-dichlorobenzamide, 2,6-dichlorobenzoic acid and several other
metabolites which appeared in trace quantities (131,132). An *Arthrobacter* sp. which was grown on benzonitrile metabolized more than
70% of ^{14}C-dichlobenil to 2,6-dichlorobenzamide and an additional
20% to other metabolites within 6 days (314).

The degradation of chloronitrobenzenes by fungi has been studied and the mechanism is believed to result in detoxification of the
fungicides (94). The yeast *Rhodosporidium* sp., when grown in a
complex nutrient medium containing 4-chloronitrobenzene, produces
several metabolites. This information enabled a branched pathway
to be proposed (Figure 40) (94). The common early steps of 4-chloro-
nitrobenzene metabolism involve sequential reduction of the substrate to form 4-chloronitrosobenzene and subsequently 4-chloro-
phenylhydroxylamine. This product may be metabolized by two mechanisms. The main pathway is further reduction of the hydroxylamine to 4-chloroaniline, followed by acetylation to produce 4-chloro-
acetanilide, the major metabolite. This product accumulates in the
culture medium. An alternative mechanism involves a shift of the

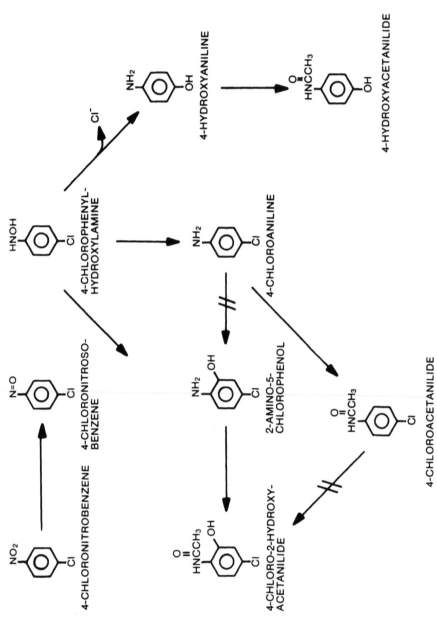

Figure 40 Pathways of metabolism of 4-chloronitrobenzene by *Rhodosporidium* sp. (Adapted from Ref. 94.)

hydroxyl group from nitrogen to carbon (called a Bamberger rearrangement) resulting in conversion of the hydroxylamine to 4-hydroxyaniline and 2-amino-5-chlorophenol. Formation of 4-hydroxyaniline involves loss of the chloride ion and subsequent acetylation results in formation of 4-hydroxyacetanilide.

Pentachloronitrobenzene is metabolized to pentachloroaniline by *Streptomyces aureofaciens*, *Rhizoctonia solani*, *Fusarium oxysporum*, and many other genera of fungi and actinomycetes (80,322). In addition, *F. oxysporum* also metabolizes the substrate to pentachlorothioanisole (Figure 41) (322). The introduction of a methylthio group was also noted in the metabolism of 2,4-dichloro-1-nitrobenzene by *Mucor javanicus* AHU6010 (Figure 41) (428). The source of the sulfur atom has not been established. The metabolite may be formed by secondary degradation of a glutathione degradation product, as proposed for rhesus monkey metabolism, or the methylthio group may be transferred from S-adenosylmethionine as demonstrated by cells of *Mycobacterium* sp. (428).

Metabolism of 1,4-dichloro-2,5-dimethoxybenzene (chloroneb) to 2,5-dichloro-4-methoxyphenol by *R. solani* is a detoxification mechanism that results in tolerance by the organism to at least twice the concentration of metabolite as product (204). This conversion occurred only at a high ratio of mycelia to growth medium. *Sclerotium rolfsii* and *Saccharomyces pastorianus* did not metabolize chloroneb. *Neurospora crassa* converts chloroneb to an unidentified product.

In a broad study of 23 species, conversion of chloroneb to 2,5-dichloro-4-methoxyphenol occurred only in cultures actively growing in nutrient medium (468). The most active species was a *Fusarium* sp. which demethylated 60 to 80% of a 5 ppm solution within 10 days. Thirteen other species demethylated the fungicide as well, although *R. solani* neither grew in the presence of nor demethylated chloroneb. *Fusarium* sp. also converted the metabolite back to chloroneb at the rate of 4% in 7 days. Eleven other species performed the same transformation. Demethylation of 2,5-dichloro-4-methoxyphenol to form 2,5-dichlorohydroquinone occurred in four species, but at concentrations of 10% or less of the applied 2,5-dichloro-4-methoxyphenol.

METABOLISM OF CHLOROBENZENES BY MICROBIAL COMMUNITIES

The chlorobenzenes are largely volatile and attempts to study their biodegradation have been hampered by disappearance of the substrates. A mesocosm experiment using tanks of seawater amended with mixtures of volatile organic compounds showed that at 3 to 7°C both 1,4-dichlorobenzene and 1,2,4-trichlorobenzene disappeared at rates explainable by volatilization (446). At warmer temperatures

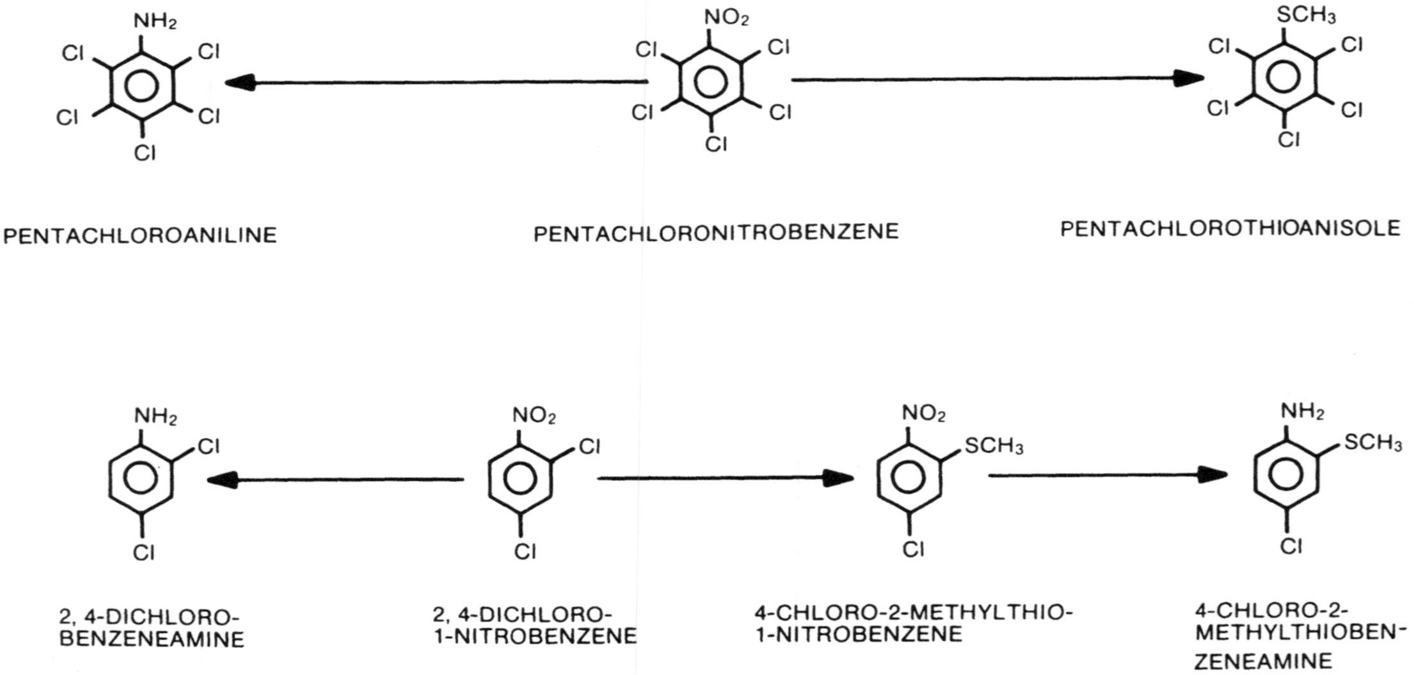

Figure 41 Metabolism of pentachloronitrobenzene by *F. oxysporum* and 2,4-dichloro-1-nitrobenzene by *M. javanicus* AHU6010. (Adapted from Refs. 322,428.)

(20 to 22°C), the rate of disappearance was much more rapid, indicating biodegradation by the planktonic and microbial communities.

Application of 1,2,3- and 1,2,4-trichlorobenzene at a rate of 50 mg/g soil resulted in $^{14}CO_2$ evolution of more than 10% after several weeks (295). Addition of high levels of organic matter increased only 1,2,3-trichlorobenzene mineralization. Extracts of soil dosed with 1,2,3-trichlorobenzene yielded 2,3- and 2,6-dichlorophenol and 3,4,5-trichlorophenol, while 1,2,4-trichlorobenzene samples yielded 2,4-, 2,5-, and 3,4-dichlorophenol.

A mixed population of soil bacteria precultured on benzene metabolized benzene and chlorobenzenes (20 to 200 mg/L) to chlorophenols (19). Mono- through tetrachlorobenzenes were monohydroxylated at a position *ortho* to the chloride. No phenol was detected in media containing pentachlorobenzene. Diphenyls eventually were detected in the media.

A granular activated carbon column was seeded with a mixed culture of bacteria (primary sewage) and supplied with acetate as a carbon source (52). A biofilm was formed which after acclimation metabolized more than 90% of a 10 to 30 mg/L solution each of chlorobenzene, 1,2-di-, 1,4-di, and 1,2,4-trichlorobenzene. Partial disappearance of 1,3-dichlorobenzene was noted. Studies with ^{14}C-1,4-dichlorobenzene confirmed that these substrates were mineralized to $^{14}CO_2$.

Hexachlorobenzene was applied to zoysia plots at a rate resulting in 6 ppm in the upper 2 cm of the soil (28). The bulk of the material was volatilized with 24% remaining after 29 days and 3.4% after 19 months. The remaining material was unchanged substrate. The original application resulted in 0.11 ppm in the 2 to 4 cm layer which did not change during the course of the experiment.

Several experiments concerning the metabolism of dichlobenil in soils have shown that the major metabolite is 2,6-dichlorobenzamide (88,314,314a,442a). Other metabolites appeared in trace quantities and could not be identified. After 61 days' incubation of the substrate ^{14}C-labeled in the nitrile group, only trace amounts of $^{14}CO_2$ or volatile ^{14}C-compounds could be recovered. More than 85% of the substrate added at 1 ppm remained unaltered.

Formation of 2,6-dichlorobenzamide in a pond water and sediment system was followed by a decrease in its concentration, indicating further transformation (314). Carbonyl-^{14}C-labeled 2,6-dichlorobenzamide was metabolized with 5.6% converted to $^{14}CO_2$ after 40 days and 28% recovered as metabolites.

Application of 1 mg/L dichlobenil to a farm pond resulted in initial sorption of the herbicide to the soil with subsequent disappearance from both water and soil (371). Less than 10% remained 90 days after treatment. Soils that had been pretreated with 2,6-dichloro-4-nitroaniline showed evolution of $^{14}CO_2$ when treated with this radiolabeled fungicide (189). No $^{14}CO_2$ evolution was noted in soils that had not been pretreated. Some unidentified metabolites were also seen. A pure culture of rod-shaped bacteria was isolated which also converted the fungicide to $^{14}CO_2$.

Greenhouse soils treated with pentachloronitrobenzene were analyzed for the presence of metabolites (108). Products recovered included pentachloroaniline, pentachlorothioanisole, and tetrachloronitrobenzene. Hexachlorobenzene and pentachlorobenzene were detected but are known to be present as impurities in technical grade pentachloronitrobenzene.

Extracts of soils amended with 1000 ppm pentachloronitrobenzene revealed the presence of pentachloroaniline but no polychloroazobenzenes (62). Soils treated perodically for 11 years still showed residual pentachloronitrobenzene and the technical grade impurities tetrachloronitrobenzene, pentachlorobenzene, hexachlorobenzene, pentachloroaniline, and methythiopentachlorobenzene when analyzed one to five years later (29).

Anaerobic flooded or moist Hagerstown silty clay loam was treated with 10 ppm pentachloronitrobenzene (319). After 40 days' incubation there was no evolution of $^{14}CO_2$ and only slight volatilization of the unchanged substrate. The main metabolite formed was pentachloroanisole, and lesser amounts of pentachlorothioanisole and pentachlorophenol were also detected.

Chloroneb (^{14}C-ring labeled) was applied to soil plots at the rate of 2 lb/acre (370). Another layer of soil was applied to the plots. After 12 months 40% of the original activity was recovered, of which 90% was unchanged substrate and 10% an unidentified metabolite.

SUMMARY

Chlorobenzenes containing less than five chlorines can be mineralized by acclimated populations under permissive conditions. High concentrations of these compounds are toxic to the bacteria. Most of the information regarding metabolism of the chlorobenzenes has come from studies with soil or mixed culture consortia. There is little information available on pathways of metabolism by pure cultures.

A single study indicates that' hexachlorobenzene is not metabolized in soils. Pentachlorobenzene was not oxidized in a sole study, although under anaerobic conditions pentachloronitrobenzene is converted to several metabolites.

The available evidence indicates that chlorocatechols or chlorophenols are the primary degradation products of chlorobenzenes. These metabolites can be metabolized by mechanisms dicussed in the appropriate chapters.

7

Chlorophenols

The chlorophenols are used extensively as antifungal agents and are often applied as a preservative to freshly sawn lumber. They have found some use as herbicides and in food processing plants to control mold (99). Chlorophenols are also common degradation products of chlorophenoxy herbicides. The wood shavings from lumber processes have been used for litter in chicken houses and contain high levels of these chlorophenols, especially 2,3,4,6-tetrachlorophenol and pentachlorophenol (101). The chlorophenols degrade to volatile chloroanisoles via methylation of the oxygen atom and the resulting compounds have been implicated in the "musty taint" of chicken eggs and meat (101).

BACTERIAL METABOLISM OF CHLOROPHENOLS

Studies with *Arthrobacter* spp. have confirmed methylation as a mechanism for chlorophenol metabolism. Conversion of 2,4,6-trichlorophenol to 2,4,6-trichloroanisole has been demonstrated. Methylation is also the predominant reaction in the conversion of guaiacols (*o*-methoxyphenol) to veratroles (1,2-dimethoxybenzene) by *Arthrobacter* spp. (325). For example, 4-chloroguaiacol, 4,5-dichloroguaiacol, and 3,4,5-trichloroguaiacol are converted to the corresponding veratroles by dense cell suspensions of cultures grown on hydroxybenzoic acid (Figure 42). Low concentrations of catechols have also been found in the culture medium. An exception to this mechanism has been demonstrated in the metabolism by *Arthrobacter* sp. strain 1395 of 3,4,5-trichloroguaiacol to 3,4,5-trichlorosyringol, which re-

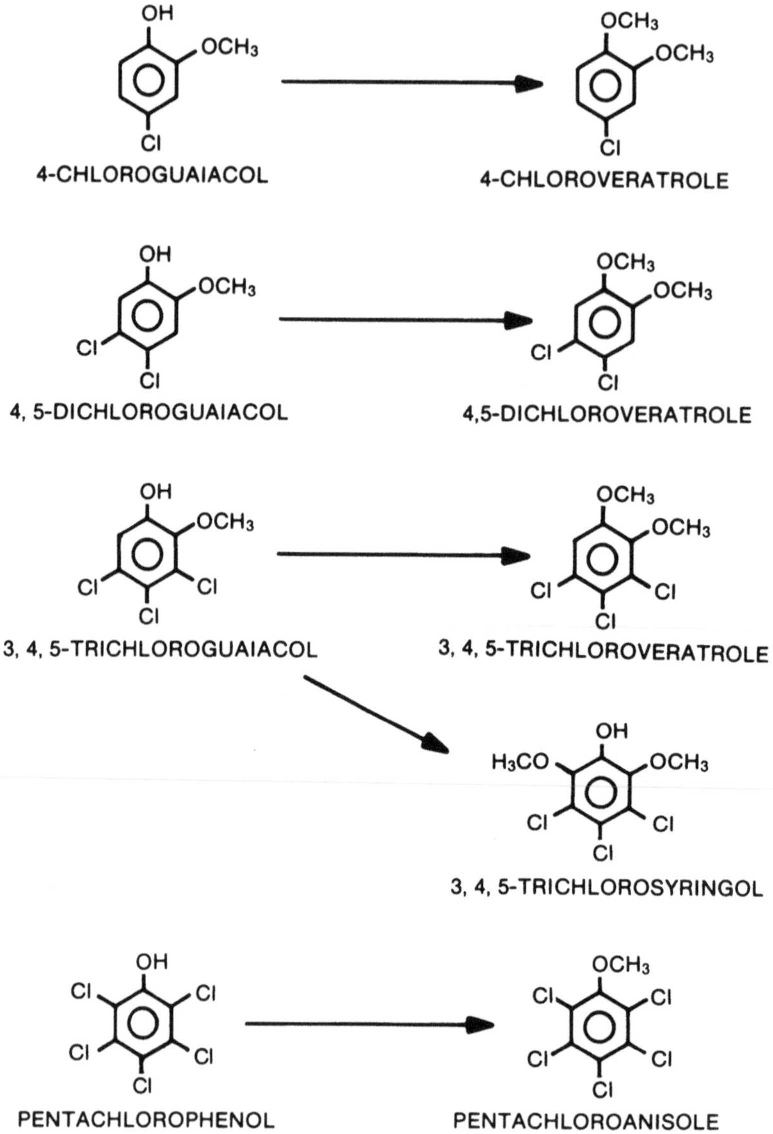

Figure 42 Methylation of chlorophenols by *Arthrobacter* spp. (Adapted from Ref. 325.)

quires hydroxylation and subsequent methylation of the previously unsubstituted carbon-6 (Figure 42). This latter compound is resistant to further degradation by this species.

An alternative mechanism results in the formation of chlorocatechols from chlorophenols. Cells of a *Nocardia* sp. grown on phenol metabolize 2-chlorophenol to 3-chlorocatechol, 3-chlorophenol to 4-chlorocatechol, and 4-chlorophenol to 4-chlorocatechol (406). Similarly, phenol-grown *Pseudomonas* sp. B13 or *Alcaligenes eutrophus* cells metabolize 2-chlorophenol to 3-chlorocatechol and 4-chlorophenol to 4-chlorocatechol (262). *Pseudomonas* sp. B13 can utilize 4-chlorophenol as the sole source of carbon and energy, and with this substrate can cometabolize 2-chlorophenol and 3-chlorophenol completely without accumulation of metabolites (262). A phenylcarbamate-degrading *Arthrobacter* sp. also metabolizes 4-chlorophenol to 4-chlorocatechol (466a). While the same initial enzyme is used for the first step in both phenol and 4-chlorophenol degradation, phenol-grown cells contain a muconate-lactonizing enzyme which has little activity for 3-chloromuconic acid, the metabolite of 4-chlorophenol (118).

Resting cell suspensions of *Achromobacter* sp. metabolize 4-chlorocatechol to 4-chloro-2-hydroxymuconic semialdehyde and 3,5-dichlorocatechol to 3,5-dichloro-2-hydroxymuconic semialdehyde via a unique catechol 1,6-oxygenase which differs from the more common catechol 2,3-oxygenase (Figure 43) (209). Neither product is metabolized further. *Pseudomonas putida* metabolizes 4-chlorophenol to 4-chlorocatechol, and then employs the *meta* cleavage enzyme catechol 2,3-dioxygenase to produce 2-hydroxy-5-chloromuconic semialdehyde, which accumulates to 10% of the starting substrate. Free chloride amounting to 85% of the substrate is recovered, although a pathway for liberation of the chloride has not been elucidated (333).

Two species of bacteria were utilized to produce a genetically constructed strain with altered ability to metabolize aromatic compounds (388). *Pseudomonas* sp. B13 with ability to metabolize chlorophenols, and *Alcaligenes* sp. A7, which degrades phenol by the *meta* path and has no activity against chlorophenols, were combined to produce a mutant (designated A7-2) which utilizes phenol by the *ortho* path and also metabolizes 2-, 3-, and 4-chlorophenol as well as 3-chlorobenzoic acid. Three enzymes were isolated, pyrocatechase II and cycloisomerase II, which have high activity for chlorinated substrates, and a third enzyme which functions exclusively in the chloroaromatic pathway to perform a dehalogenating cycloisomerization of chloromuconic acids (Figure 44).

The 2,4,5-trichlorophenoxyacetic acid-degrading strain of *P. cepacia* AC1100 can dechlorinate a wide variety of chlorophenols (237). Resting cell suspensions can dechlorinate within 3 hours at 0.1 mM substrate concentration, the following chlorophenols: 2,3-,

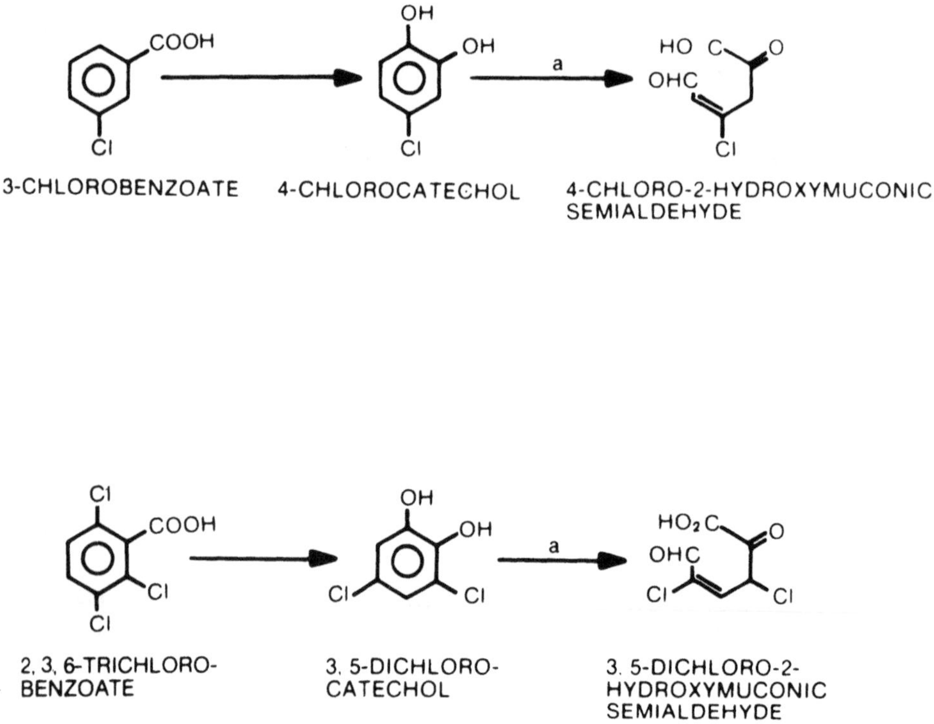

3-CHLOROBENZOATE   4-CHLOROCATECHOL   4-CHLORO-2-HYDROXYMUCONIC
                                                SEMIALDEHYDE

2, 3, 6-TRICHLORO-   3, 5-DICHLORO-   3, 5-DICHLORO-2-
BENZOATE   CATECHOL   HYDROXYMUCONIC
                                                  SEMIALDEHYDE

a = catechol 1, 6-oxygenase

Figure 43   Cometabolism of chlorocatechols via catechol 1,6-oxygenase by resting cell suspensions of *Achromobacter* sp. (Adapted from Ref. 209.)

2,4-, and 2,5-dichlorophenol, 2,3,4- and 2,4,5-trichlorophenol, 2,3,4,6- and 2,3,5,6-tetrachlorophenol and pentachlorophenol. The strain has less activity against 2,4,6-trichlorophenol and 2,3,4,5-tetrachlorophenol and metabolizes 2,6- and 3,5-dichlorophenol and 2,3,5-, 2,3,6-, and 3,4,5-trichlorophenol poorly.

METABOLISM OF CHLOROPHENOLS BY FUNGI

Fungal metabolism of chlorophenols often involves methylation in a manner analogous to that of bacteria (172). A study of 116 fungal isolates from chicken house litter revealed that 59% produce 2,3,4,6-tetrachloroanisole from 2,3,4,6-tetrachlorophenol at conversion ef-

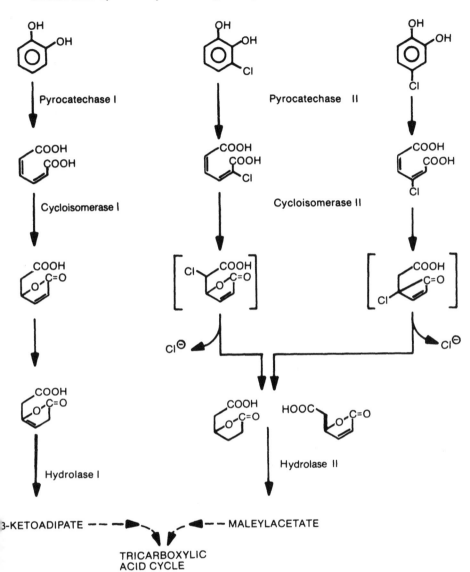

Figure 44 Action of aromatic and chloroaromatic enzymes from *P. putida* B13 and *P. putida* derivative strains. (Adapted from Ref. 261.)

ficiencies of from 1 to 83%. Flask cultures in this study were sealed and incubated for five days, so the transformation may have occurred under aerobic or anaerobic conditions. The fungi demonstrating this ability include *Aspergillus* spp., *Paecilomyces* spp., *Penicillium* spp., and *Scopulariopsis* spp. (100,101,172). Some strains metabolize 2,3,4,6-tetrachlorophenol without formation of the anisole, suggesting an alternate mechanism for chlorophenol metabolism (172). The yeast *Rhodotorula glutinis* grown on phenol converts 3-chlorophenol to 4-chlorocatechol (448).

There is little additional information available on fungal metabolism of chlorophenols, although there is evidence to indicate that a *Penicillium* sp. produces 2,4-dichlorophenol as a natural metabolite (8).

METABOLISM OF CHLOROPHENOLS BY MIXED MICROBIAL CULTURES

Soil populations exhibited enhanced rates of metabolism of 2-chlorophenol after prior acclimation by soil percolation (447). Following an initial decrease in 4-chlorophenol concentration during soil percolation, however, additional applications of that substrate were not metabolized. In other experiments, soil inocula mediated the complete disappearance of 4-chlorophenol within 25 days, although neither 2- nor 3-chlorophenol was degraded during that time (193).

Wastewater sludge supernatant liquid required 14 to 25 days for complete disappearance of 16 mg/L 2-chlorophenol and 3-chlorophenol, although 4-chlorophenol was faster in polluted acclimated or non-acclimated river water than in diluted sewage inocula experiments (139). The substrates were not degraded by the sewage microorganisms after 25 days, while less than 15 days was required for disappearance of the substrates from river water.

An acclimated sludge culture exposed to 100 mg/L substrate was able to metabolize the following compounds within 5 days with chloride release as noted (218): 2-, 3-, or 4-chlorophenol, 100%; 2,4-dichlorophenol, 100%; 2,5-dichlorophenol, 16%; 2,4,6-trichlorophenol, 75%, and sodium pentachlorophenate, 0%.

The fate of the monochlorophenols when incubated anaerobically with a 10% municipal sewage sludge inoculum was determined (54). At a concentration of 50 mg/L, 2-chlorophenol required 3 weeks, 3-chlorophenol 7 weeks, and 4-chlorophenol 16 weeks for complete disappearance. Mineralization was monitored by measuring net methane production, and results indicated nearly complete mineralization of 2-chlorophenol. Methane was produced from 3-chlorophenol after a lag period of 5 weeks, and 4-chlorophenol was not mineralized. During the degradation of 2-chlorophenol, phenol was recovered as the initial metabolite, followed by methane production. This is consistent with other studies in which dechlorination was shown to be the initial step in the reductive metabolism of 3-chlorobenzoic acids (417).

Fresh undiluted sludge also reductively dechlorinated several dichlorophenols with removal of the *ortho* chloride, such that 2,6-dichlorophenol was converted to 2-chlorophenol, 2,3- and 2,5-dichlorophenol to 3-chlorophenol, and 2,4-dichlorophenol to 4- chlorophenol (53). Two isomers that lack an *ortho* substituent, 3,4- and 3,5-dichlorophenol, were not metabolized during the 6 weeks of the experiment.

Undiluted sludge samples were acclimated to the monochlorophenols by repeated inoculations with 20 mg/L substrate, until the cultures could metabolize 25 mg/L within one week (53). Each acclimated sludge culture was then challenged with a 20 mg/L solution of other chlorophenols. Cultures acclimated to 2-chlorophenol metabolized both 2- and 4-chlorophenol at equal rates and 2,4-dichlorophenol somewhat more slowly. However, 3-, 2,3-di-, 2,5-di-, and 2,6-dichlorophenol were not metabolized. Sludge inocula acclimated to 3-chlorophenol also metabolized 4-chlorophenol, and 3,4- and 3,5-dichlorophenol also permitted metabolism of 3-chlorophenol and at much slower rates 2-chlorophenol and 2,4- and 3,4-dichlorophenol as well. The population acclimated to 4-chlorophenol seemed to have a broader substrate range than the other acclimated populations. Incubation of 2- and 4-chlorophenol-acclimated sludge inocula with uniformly ring-^{14}C labeled 2- and 4-chlorophenol and 2,4-dichlorophenol showed that in all cases nearly complete mineralization of the substrates to ^{14}C-methane and ^{14}CO$_2$ occurred.

Experiments in which tainted litter was incubated with sawdust, pentachlorophenol, and 2,3,4,6-tetrachlorophenol, showed nearly quantitiative conversion of the latter substrate to 2,3,4,6-tetrachloroanisole (101). There was virtually no conversion in the absence of the litter inoculum. Pentachlorophenol was 50% converted to pentachloroanisole after 29 days.

Aspergillus sydowi, *Scopulariopsis brevicaulis*, and a *Penicillium* sp. were isolated from the litter and each species was also found to be capable of the above substrate conversions.

SUMMARY

Many of the chlorophenol compounds have been shown to be metabolizable by pure cultures or mixed natural populations of microorganisms both aerobically and anaerobically. In most cases complete mineralization occurs. However, the rates of disappearance of the isomers vary widely, depending upon degree of acclimation of the population and other environmental factors. Mixed cultures seem to be required for complete mineralization of the chlorophenols.

8

Pentachlorophenol

Pentachlorophenol (PCP) is widely used in a variety of agricultural and industrial applications as a fungicide, bactericide, insecticide, herbicide, and molluscicide (36,97,466). It is most widely used in the United States and elsewhere for wood preservation, both for newly cut timber and for slime control in pulp and paper production. PCP is usually used as a 5% solution in petroleum solvents or as the water-soluble sodium or potassium salt.

BACTERIAL METABOLISM OF PCP

In spite of its use as a fungicide and bactericide, PCP is metabolizable by a variety of microorganisms. Reports of decomposition of PCP in rice paddy soil and other soils or aquatic environments (90, 217,253,266,282,320,354) were followed by experiments with consortia and pure cultures of bacteria which demonstrated chloride release and $^{14}CO_2$ formation from labeled PCP (89,354,369,411,423, 459). However, few studies have identified metabolites arising from pure culture metabolism of PCP.

Cultures of *Pseudomonas* spp. produce both tetrachlorocatechol and tetrachlorohydroquinone from PCP (Figure 45) (423,424). These are metabolized rapidly soon after they are produced. There is no evidence of methylation of PCP to form pentachloroanisole. Amino acid analyses with hydrolysates of bacterial cells indicate incorporation of ^{14}C derived from PCP into the cell constituents (423). Pentachlorophenol is metabolized by *Arthrobacter* spp. to pentachloroanisole at levels of less than 0.5% conversion at approximately 44 mM substrate concentration (325).

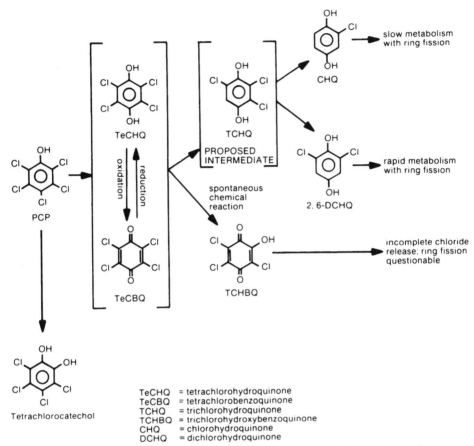

Figure 45 Proposed pathway for pentachlorophenol (PCP) metabolism by the bacterial culture KC-3 and by *Pseudomonas* sp. (Adapted from Refs. 369,423.)

A bacterium identified as *Mycobacterium* sp. which cannot use PCP as a growth substrate methylates PCP to pentachloranisole (424,425). Further methylations by washed cell suspensions of this culture result in the formation of tetrachloro-1,2-dimethoxybenzene and tetrachloro-1,4-dimethoxybenzene. Additional metabolites include tetrachlorocatechol, tetrachlorohydroquinone, tetrachloro-2-methoxy-phenol and tetrachloro-4-methoxyphenol. The formation of these products indicates that the main metabolite is the methylated derivative of PCP, but in addition, PCP is hydroxylated in the *ortho* or *para* positions and subsequently methylated at these positions. As pentachloroanisole is less toxic to the bacteria than PCP, methylation is suggested as a detoxification mechanism (424). The methylation of PCP was also demonstrated in cell-free systems of *Mycobacterium* sp. (422). The mechanism appears to involve the enzymes that transfer the methyl group from S-adenosylmethionine to the hydroxyl groups of these substrates.

A saprophytic soil corynebacterium was isolated which utilizes PCP as a sole source of carbon and energy for growth (89,369). By measuring $^{14}CO_2$ evolution, the conversion rate was calculated to be 10 mg PCP per mg of dry cell weight per hour (90). Cells of this isolate, referred to as KC3, when grown on PCP also show immediate uptake, as measured by ultraviolet spectrophotometry, of a wide variety of chlorophenol isomers, including 2,3,5-, 2,3,6-, and 2,4,6-trichlorophenol, 2,3,4,6- and 2,3,5,6-tetrachlorophenol, and pentachlorophenol (89). Uptake of 3,4,5-trichlorophenol and 2,3,4,5-tetrachlorophenol is delayed. While the *para* and *meta* monochlorophenols are oxidized, as measured by manometric techniques, the *ortho* isomer is oxidized poorly and phenol itself not at all (89). Chloride is not released to an appreciable degree from any of the monochlorophenols. In general, release of chloride is greatest from the 2,6-substituted di-, tri-, and tetrachlorophenols. Isomers with chloride substitutions in other positions are less well attacked by culture KC3.

Substantial investigations into the metabolism of PCP by the KC3 isolate failed to show accumulation of metabolites in the medium. However, mutants were developed which failed to grow in a PCP-minimal salts medium (369). One of these mutants, designated ER-47, converts PCP primarily to 2,6-dichlorohydroquinone rapidly and without a lag. A trace of monochlorohydroquinone also appears but its role in the pathway of biological degradation is uncertain, as it is only slowly attacked by parent KC3 cells. A second mutant, ER-7, accumulates several metabolites, including tetrachlorohydroquinone, tetrachlorobenzoquinone, and trichlorohydroxybenzoquinone (Figure 45). These three products are converted rapidly and spontaneously from the hydroquinone through the benzoquinone to the more stable hydroxybenzoquinone. The latter product is metabolized by KC3 but dechlorination is not complete and the ring is not ruptured. Tetrachlorohydroquinone is rapidly metabolized to 2,6-dichlorohydroquinone, but this metabolic transformation must compete with the rapid spontaneous transformation to the trichlorohydroxybenzoquinone.

Another series of experiments explored the metabolism of sodium pentachlorophenate by a wide variety of bacteria metabolically active for phenols, chlorophenols, or chlorobenzenes (379). Metabolites were identified by detecting acetyl derivatives using combined gas chromatography and mass spectrometry. With few exceptions metabolites occurred in concentrations of less than one percent of the starting material. Reported metabolites included PCP-acetate, pentachloroanisole, tetrachloroanisoles, tetrachlorophenols, and tetrachlorodihydroxybenzenes.

FUNGAL METABOLISM OF PCP

The role of fungi in detoxifying PCP has been studied to some degree. Fungi associated with PCP-treated wood reduce PCP to a less toxic metabolite (122,288). Three *Trichoderma* spp. metabolized sodium pentachlorophenate (Na-PCP) within 2 weeks in a malt extract medium as well as on wood treated with Na-PCP (98). Pentachloroanisole was detected in the culture medium of *T. virgatum* after 5 days' incubation at levels corresponding to 10—20% of the starting Na-PCP. It is unclear whether this is an integral step of the pathway or whether methylation is a side reaction (98).

During a comparison of the growth of several species of fungi on PCP, *Trichoderma* spp. were the only ones that reduced PCP levels after 12 days' incubation at 5 to 10 mg/L concentration (98). Fungi inactive against PCP were *Cephaloascus fragrans*, *C. pilifera*, *Graphium* spp., and *Penicillium* sp. Another experiment showed that *Trichoderma viride* and *Coniophora puteana* reduced the concentration of PCP in treated wood blocks, although *C. puteana* was much more sensitive to PCP in liquid culture (441). It was postulated that the presence of an alternative substrate of wood or the binding capacity of PCP to wood reduced exposure of the fungus to below the toxic level, thus permitting metabolism of PCP.

DISAPPEARANCE OF PCP IN ENVIRONMENTAL SAMPLES

The standard procedure of applying PCP in a carrier solvent to wood products has complicated subsequent analyses of disappearance and biodegradability. A carrier that is too volatile will carry PCP with it as it evaporates (277). A carrier which retains liquidity at ambient temperatures will bleed from the wood until an equilibrium is established. PCP in solution will be carried along, reducing the final concentration in the wood and increasing the amount reaching the surrounding environment.

Extraction and analysis procedures are also subject to error, including incomplete extraction due to poor choice of extracting solvent, and use of procedures which extract pure PCP but not polymerized molecules, which may be present in technical grade PCP at levels as high as 18 percent (277).

Degradation of PCP with release of chloride and CO_2 has been demonstrated in a number of environments. In a waste stream continuously contaminated with PCP, acclimation occurred after 3 weeks (354). The microflora, particularly the attached bacteria, metabolized up to 0.43 ppm influent concentration. Pure cultures were isolated from the waste stream which were capable of mineralizing 100 mg/L PCP in 90 hours with almost complete chloride release.

A soil perfusion apparatus using rice paddy soil effected disappearance of PCP with more than 90% liberation of chloride (459). A *Pseudomonas* sp. isolated from the enrichment culture degraded 40 mg/L PCP in 10 days with complete chloride release.

PCP added to moist garden soil at 150 to 200 mg/L soil-water concentration was 25% metabolized after 12 days when the experiment was conducted using outdoor shaded test plots (128). When a culture of *Arthrobacter* sp. ATCC 33790 was mixed into the soil, about 85% of the PCP disappeared during the same time. Under laboratory conditions addition of the bacterial culture reduced the half-life for PCP disappearance from 12 to 14 days to 1 day. This *Arthrobacter* sp. utilizes PCP as the sole source of carbon and energy with complete release of chloride (130).

Comparisons of aerobic and anaerobic metabolism of PCP show that aerobic metabolism is much more efficient (282). Enrichment cultures established in fermentors fed with 2 mg/L PCP revealed a half-life of 0.36 days under aerobic conditions and 192 days under anaerobic conditions. Addition of glucose or 4-chlorophenol as cometabolic substrates depressed the rate of PCP metabolism.

Soils treated with 10 mg/L PCP and incubated for 24 days under aerobic conditions revealed considerable loss of ^{14}C-labeled material from the system (320). Of 59% total recovered material, 51% was identified as pentachloroanisole. Volatile products and CO_2 were not measured. In the same system maintained under anaerobic conditions, 7% of the material was converted to metabolites and no $^{14}CO_2$ was detected. Metabolites included about 5% pentachloroanisole and lesser amounts of 2,3,6-trichlorophenol, 2,3,4,5-tetrachlorophenol, and 2,3,5,6-tetrachlorophenol.

PCP applied to flooded paddy soil, simulating anaerobic conditions, was metabolized after 3 weeks to the following products: 3-chlorophenol, 3,4-, and 3,5-dichlorophenol, 2,3,5- and 2,4,5-trichlorophenol, and 2,3,4,5-, 2,3,4,6-, and 2,3,5,6-tetrachloroanisole (217).

The rate of PCP metabolism in 11 soils was found to be related to the organic matter content of the soils (266). Degradation products included a mixture of tri- and tetrachlorophenols.

A major PCP spill on the Mississippi River Gulf Outlet left PCP levels as high as 1.60 mg/g in the sediment (109). At 18 months there was no detectable PCP in the sediment. Studies arising from the spill indicated that the degradation rates increased with increasing sediment redox potential. Maximum degradation occurred at pH 8 at +500 mV. Less degradation occurred at pH 9 and at pH values less than 8.

PCP-degrading bacteria have been isolated both from polluted sites and from sites not known to be contaminated with PCP (411). An enrichment consortium established under continuous culture conditions became adapted to metabolize 500 mg/L PCP. *Arthrobacter* sp. strain NC was isolated from the culture and metabolized 100 mg/L PCP until the pH decreased to 6.15. Upon adjustment to pH 7.1 the residual PCP was metabolized. The strain was capable of growth at pH 6.0 upon other substrates. Other experiments showed a correlation of toxicity with the acid form of PCP.

SUMMARY

In summary, there is evidence that PCP is attacked by bacteria and fungi cometabolically or as sole source for growth with release of chloride and CO_2. The pathway involves dechlorination and hydroxylation either *ortho* or *para* to the phenolic hydroxyl group, forming a catechol or a quinone, respectively. However, the mechanism of this process is not understood and the enzymes involved have not been isolated. Further, the steps of the pathway leading to carbon incorporation into cell contents and CO_2 formation have not been elucidated. In fungi, methylation has been detected as a prominent metabolic process, but its role in PCP degradation has not been established.

9

Chlorophenoxy and Chlorophenyl Herbicides

Compounds with an arylcarboxylic acid parent structure have plant growth-regulating properties. They produce physiological effects such as morphogenic abnormalities, promote the rooting of cuttings, and aid in setting fruit in the absence of pollination (445).

There are three principal chlorine-substituted phenoxyacetic acids used widely as herbicides, 2,4-dichlorophenoxyacetic acid (2,4-D), 2,4,5-trichlorophenoxyacetic acid (2,4,5-T), and 4-chloro-2-methyl-phenoxyacetic acid (MCPA). They are selective against broadleaved weeds and woody broadleaved plants and are commonly used in lawns, grass pastures, and cereal crops (283). Other phenoxyalkanoic acids are useful in controlling weed species which are resistant to the phenoxyacetic acids. The phenoxybutyric acid herbicides have very low toxicity to plant species such as legumes which are damaged by exposure to phenoxyacetic acids. After application, they are activated by target plants (weeds) which β-oxidize them to their corresponding toxic phenoxyacetic acids. Other herbicides in this class include the phenoxyethyl esters, which are applied when deep soil penetration is required or in noncrop areas.

The phenoxyalkyl acid herbicides are detoxified in soils and aquatic environments due to microbial action (13,60,111,283,343,376, 377). Bacteria and fungi which metabolize the herbicides have been isolated from soils (Table 4). The products of microbial metabolism may be phytotoxic or they may result in inactivation of the herbicide (290,292). These products may be similar to those formed as a consequence of plant metabolism, such as 2,4-dichlorophenol from 2,4-D and 2,4,5-trichlorophenol from 2,4,5-T. There is evidence that 2,4,

Table 4 Microorganisms That Metabolize Phenoxy Acids

Phenoxy acid	Organism	References
2,4-D	*Achromobacter* sp.	31, 32, 412, 414
	Arthrobacter sp.	44, 45, 46, 126
		284, 285, 286, 434
	Arthrobacter globiformis	283
	Corynebacterium sp.	375
	Flavobacterium peregrinum	405a, 411a, 412, 414
	Mycoplana sp.	451
	Nocardia sp.	283
	Pseudomonas sp.	142, 146, 152, 169
		170, 171, 171a
	Sporocytophaga congregata	225
	(*F. aquatile*)	
	Streptomyces	51a
	viridochromogenes	
2,6-D	*Achromobacter* sp.	31, 32
2-Chloro-phenoxyacetate	*Achromobacter* sp.	31, 32, 412, 414
	Arthrobacter sp.	44, 45, 46, 126
		284, 285, 286, 434
	F. peregrinum	412, 414
	Pseudomonas sp.	142, 146, 152
4-Chloro-phenoxyacetate	*Achromobacter* sp.	31, 32, 412, 414
	Arthrobacter sp.	45, 284, 285, 286
		44, 46, 126, 434
	F. peregrinum	412, 414
	Mycoplana sp.	451
	Nocardia sp.	283
	Pseudomonas sp.	126, 142, 143, 146,
		152, 169, 170, 171
		171a
MCPA	*Achromobacter* sp.	31, 32, 412, 414
	Arthrobacter sp.	44, 45, 46, 126,
		284, 285, 286, 434
	F. peregrinum	45
	Mycoplana sp.	142, 146, 152
	Pseudomonas sp.	169, 170, 171, 171a
2,4,5-T	*Achromobacter* sp.	31, 32
	Brevibacterium sp.	207
	Mycoplana sp.	451
	S. viridochromogenes	51a

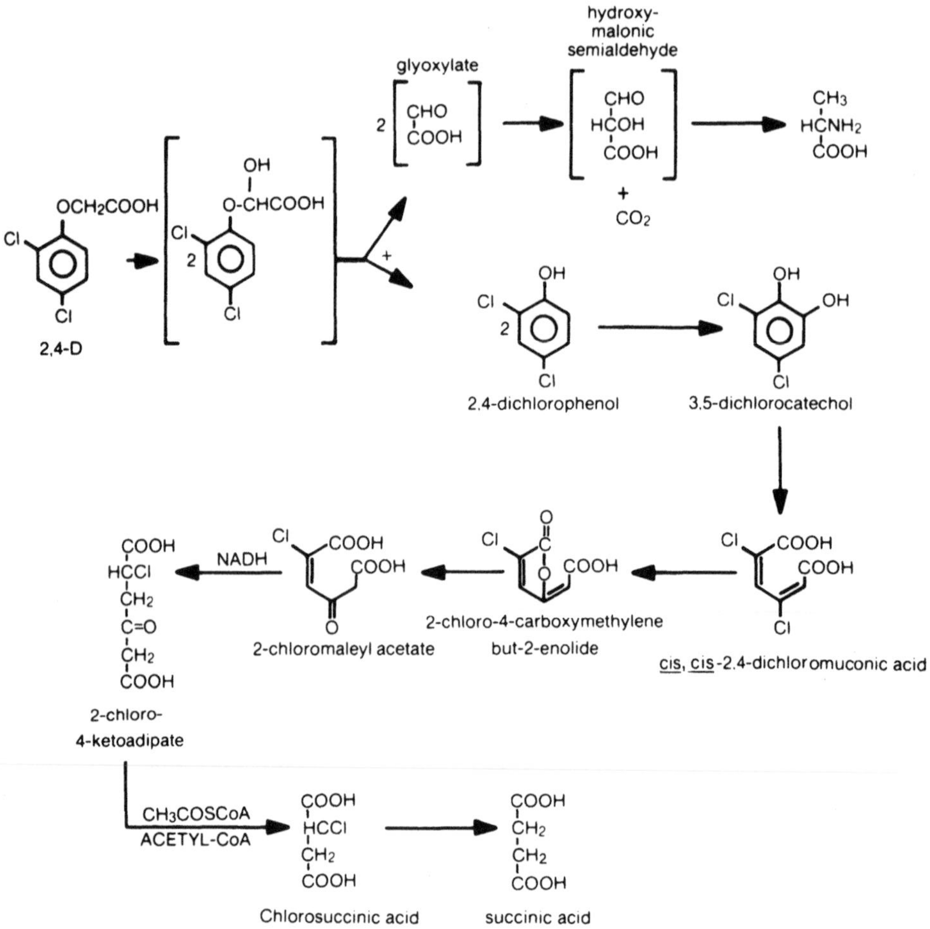

Figure 46 Bacterial metabolism of 2,4-D. (Adapted from Refs. 46, 142,433.)

5-trichlorophenol, rather than 2,4,5-T, is the active agent which damages the plants (283). Some processes of microorganisms prevent activation of the herbicides rather than actually detoxifying them (3). For this reason the pathways by which microorganisms metabolize the phenoxy herbicides are of importance in determining the choice of herbicide for a given application.

2,4-D

The herbicide most extensively studied has been 2,4-D. Bacteria including *Pseudomonas* sp., *Arthrobacter* sp., *Achromobacter* sp.,

Mycoplana sp., and *Flavobacterium peregrinum*, cleave the molecule at the ether linkage between the oxygen and the aliphatic side chain to form glyoxylic acid and 2,4-dichlorophenol (Figure 46) (31,32,40, 152,285,286,412,433,434,451). The latter compound is metabolized to 3,5-dichlorocatechol, *cis,cis*-2,4-dichloromuconic acid, 2-chloro-4-carboxymethylene but-2-enolide and 2-chloromaleylacetic acid (46, 393). Chloromaleylacetic acid is degraded further to succinic acid via 2-chloro-4-ketoadipic acid and chlorosuccinic acid. The entire pathway has also been demonstrated using cell-free extracts of *Arthrobacter* sp. (40,44,126,142,393,434). Cleavage of the ether-oxygen bond in phenoxyacetic acid to form the phenol has been proven to occur between the aliphatic side chain and the ether-oxygen in experiments with *Arthrobacter* sp. using $^{18}O_2$ (198).

The enzymes mediating the degradation of 2,4-D are relatively nonspecific. Oxygen and a reduced pyridine nucleotide (NADH or NADPH) are required, indicating that the enzyme(s) may be a mixed function oxidase (44). The broad specificity of the enzymes is reflected in the findings that *Pseudomonas* sp. cells grown on 2,4-D also oxidize MCPA, 2,4-dichlorophenol, 4-chloro-2-methylphenol, and 3,5-dichlorophenol. Neither 6-hydroxy-2,4-dichlorophenol nor 2,4-dichloroanisole is oxidized, indicating that neither is an intermediate in 2,4-D metabolism (284). The enzyme extracts also convert 2-chlorophenoxyacetic acid to 2-chlorophenol and 4-chlorophenoxyacetic acid to 4-chlorophenol, which subsequently forms 4-chlorocatechol. Catechol is converted to *cis,cis*-muconic acid and 4-chlorocatechol to *cis,cis*-3-chloromuconic acid. Chloride is released from 3-chloromuconic acid to form 4-carboxymethylene but-2-enolide, maleylacetic acid and, subsequently succinic acid (Figure 47) (40,44,434). This pathway is analogous to that demonstrated for 3,5-dichlorocatechol during 2,4-D degradation.

A *Corynebacterium* sp. isolated from soil by enrichment culture metabolized 2,4-D with nearly complete chloride release after 48 hours (375). An application rate to soil of 3000 ppm was metabolized but neither growth nor metabolism was noted upon application of 3500 ppm. No metabolites were seen during the incubation period.

A bacterial strain tentatively identified as *F. peregrinum* was isolated from enrichment culture with 2,4-D in soil (414). This strain metabolized 100 ppm 2,4-D in 25 days and upon addition of 0.1% yeast extract metabolized 0.1% 2,4-D in 12 to 16 days. Chloride release was estimated at 70% of that in 2,4-D within 39 days.

Flavobacterium aquatile metabolized 0.01% 2,4-D in sterile soil, nonsterile soil, and on solid agar plates, but not in a soil extract medium or on semisolid agar (225). MCPA was not metabolized in sterile soil. In similar experiments, *Corynebacterium* sp. metabolized both herbicides in sterile soil and on solid agar.

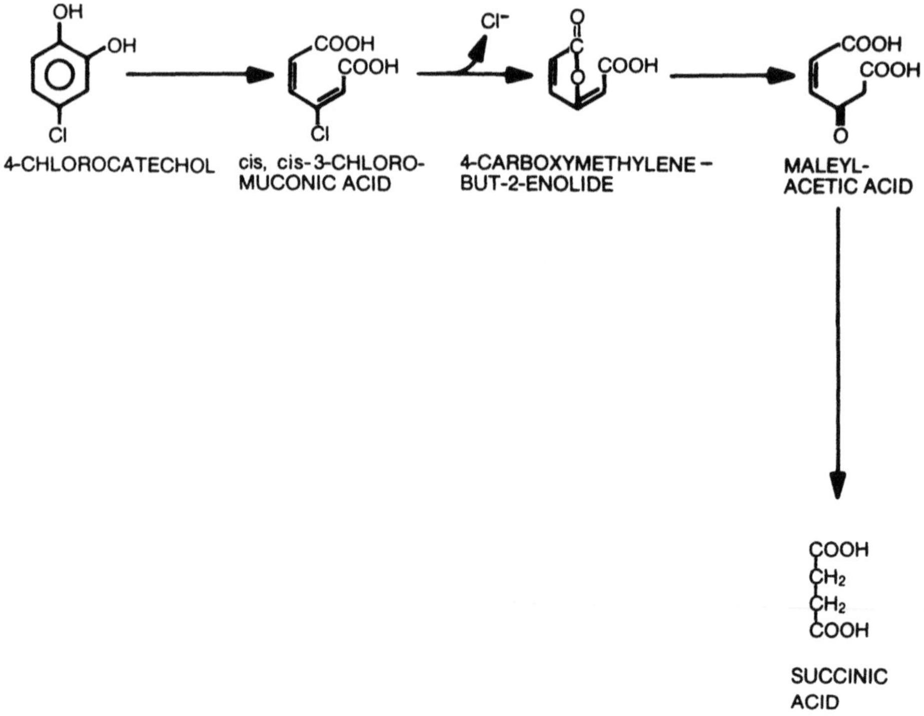
OH
OH
Cl
4-CHLOROCATECHOL
COOH
COOH
Cl
cis, cis-3-CHLORO-MUCONIC ACID
Cl⁻
O
C
O
COOH
4-CARBOXYMETHYLENE-BUT-2-ENOLIDE
COOH
COOH
O
MALEYL-ACETIC ACID
COOH
CH2
CH2
COOH
SUCCINIC ACID

Figure 47 Metabolism of 4-chlorocatechol by *Arthrobacter* sp. (Adapted from Refs. 46,434.)

A number of bacteria were isolated by enrichment culture from sewage or soil amended with 2,4-D or 2,4,5-T (376). None utilized either substrate as the sole source of carbon. Forty-one of fifty-two strains cometabolized 2,4-D only while 19 strains utilized both 2,4-D and 2,4,5-T. Experiments with these 19 isolates incubated with 2,4,5-T in nutrient medium showed that 12 isolates produced chloride ion and 8 produced a phenolic compound with or without concomitant production of free chloride.

MCPA

The metabolism of MCPA has been studied extensively in cultures of *Pseudomonas* sp. NCIB 9340 and *Arthrobacter* sp. (45,46,169,170, 412) as well as cell-free extracts of *Arthrobacter* sp. (40). Initial attack on the molecule results in oxidative cleavage of the ether linkage to form a phenol and glyoxylic acid (Figure 48) (169). The phenol thus formed is 4-chloro-2-methylphenol (5-chloro-*o*-cresol). In *Pseudomonas* sp. NCIB 9340 this product is metabolized to

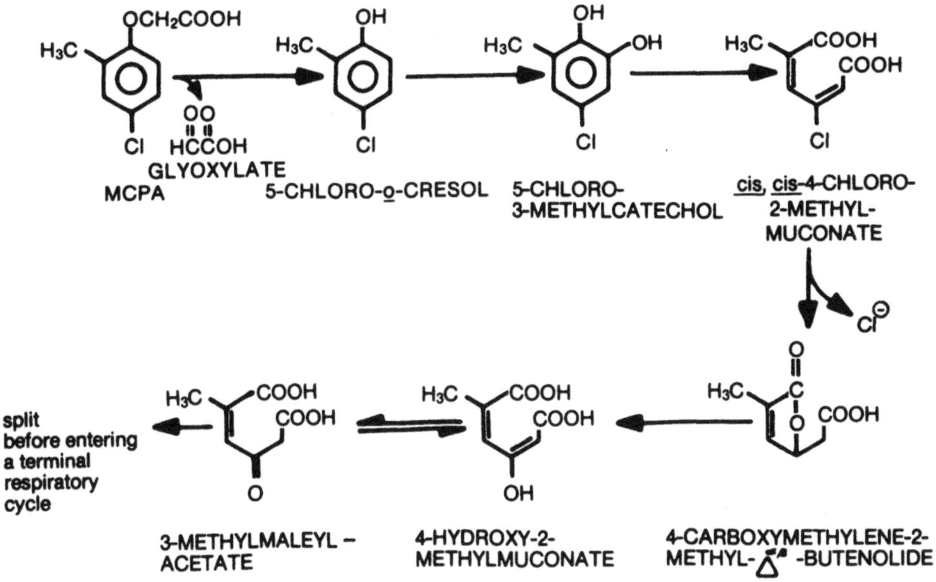

Figure 48 Pathway for MCPA metabolism by *Pseudomonas* sp. NCIB 9340. (Adapted from Refs. 169,170.)

5-chloro-3-methylcatechol and then to *cis,cis*-4-chloro-2-methylmuconic acid. The chloride ion is lost upon lactonization by dehydrochlorination to form 4-carboxymethylene-2-methyl-2,3-butenolide and subsequently 4-hydroxy-2-methylmuconic acid (170). Formation of the two double bonds of the lactone is an unusual feature in the metabolism of aromatic compounds by bacteria.

The three enzymes mediating the conversion of 5-chloro-3-methylcatechol to 4-hydroxy-2-methylmuconic acid have been isolated (171). These enzymes, responsible for ring cleavage, lactonization, and delactonization, confirm that lactonization and dehalogenation is a one-step process.

The enzymes that attack MCPA are also relatively nonspecific. Oxygen as well as NADH or NADPH are required for enzymatic activity. *F. peregrinum* cells grown in the presence of MCPA are induced to oxidize 2,4-D, as detected by manometric techniques, although MCPA is not metabolized (413). A strain thought to be an *Achromobacter* sp., isolated from enrichment culture with MCPA, metabolized 50 mg/L MCPA or 2,4-D with addition of 0.05% yeast extract in 4 days (414). Experiments showed a faster rate of oxygen uptake with 2,4-D than MCPA although the organism was cultured on MCPA. Nonacclimated cultures showed no oxygen uptake with either MCPA or 2,4-D.

2,4,5-T

The trichlorinated phenoxy herbicide, 2,4,5-T, has proven much
more difficult to degrade. The utilization of 2,4,5-T by a *Pseudomonas* sp. appears not to be plasmid encoded, unlike 2,4-D and
MCPA metabolism (154). The herbicide is metabolized to 2,4,5-tri-
chlorophenol by both *P. fluorescens* and *P. cepacia* AC1100 (249,
377). *Brevibacterium* sp. when grown on benzoic acid converts
2,4,5-T to 3,5-dichlorocatechol without a lag, indicating removal of
one chloride ion (207). This latter compound can be metabolized
(Figure 46).

   Pseudomonas cepacia AC1100 is a strain modified in the laboratory which utilizes 2,4,5-T as the sole source of carbon and energy
for growth (249). More than 97% of a 1 g/L solution was degraded
within 6 days with stoichiometric chloride release. In 2 hours, resting cells promoted 50% disappearance of 2,4,5-T although only 15%
chloride release was evident. Within 24 hours, there was complete
substrate disappearance with 94% chloride release. Resting cells also
mediated release of more than 80% of the chloride from 2,4,5-trichloro-
phenol, 2,3,4,6-tetrachlorophenol, and pentachlorophenol. In addition, resting cells were also shown by oxygen electrode determination to oxidize 2,4-D at relatively high rates but not 2,4-5-trichloro-
phenoxypropionic acid. Phenoxyacetic acid was oxidized at low rates

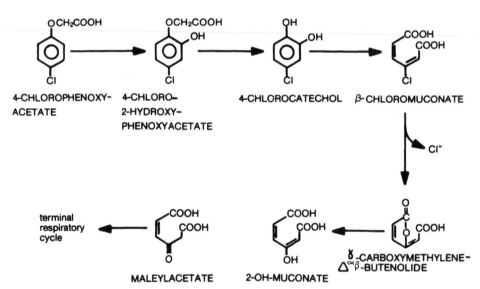

Figure 49 Pathway for 4-chlorophenoxyacetate metabolism by a soil
pseudomonad. (Adapted from Ref. 143.)

only. When inoculated into soil *P. cepacia* AC1100 mediated 95% chloride release of a 1 mg/g application of 2,4,5-T within 1 week (85). The optimum conditions were 25% moisture content at 30°C.

4-CHLOROPHENOXYACETIC ACID

The degradation of 4-chlorophenoxyacetic acid by a soil pseudomonad proceeds through 4-chloro-2-hydroxyphenoxyacetic acid to 4-chloro-catechol (Figure 49) (143). This organism also metabolizes 4-chloro-catechol to *cis,cis*-3-chloromuconic acid. While direct evidence for subsequent steps in the pathway was not obtained, the culture medium did contain a lactone which is analogous to that described for *Arthrobacter* sp. The pseudomonad was not induced to grow on 4-chlorophenol and this product was not found in the culture medium, indicating that this compound is not an intermediate in 4-chlorophenoxyacetic acid metabolism.

An unidentified Gram-negative organism isolated from soil also metabolizes 4-chlorophenoxyacetic acid through 4-chloro-2-hydroxyphenoxyacetic acid and 4-chlorocatechol, as determined by simultaneous adaptation experiments (146). The same technique was used to determine that *Achromobacter* sp. grown on 4-chlorophenoxyacetic acid immediately oxidizes 4-chloro-2-hydroxyphenoxyacetic acid, 4-chlorocatechol, and catechol, but not 4-chlorophenol (414). The first step of 4-chlorophenoxyacetic acid metabolism in these pathways is hydroxylation of the ring, followed by ether cleavage. This is in contrast to *Arthrobacter* sp. metabolism of several chlorophenoxy compounds in which cleavage of the ether linkage is the primary step yielding 4-chlorophenol (284).

OTHER PHENOXY HERBICIDES

The metabolism of phenoxy herbicides with longer aliphatic side chains has also been studied. Two mechanisms appear to mediate degradation of these compounds. The primary mechanism is β-oxidation, a mechanism common to plants (283). Evidence for β-oxidation comes from observations with cultures of *Flavobacterium* sp. which were grown on 4-(2,4-dichlorophenoxy)butyric acid (4-(2,4-D)B) and then tested for products arising from oxidation of higher carbon-number homologs. Phenols were detected in all cases but more so with compounds containing an odd number of carbons in the side chain (291). When the side chain contained an odd number of carbons, primarily 2,4-dichlorophenol was recovered, while 2,4-D was recovered from metabolism of compounds with an even number of carbons in the side chain. Extracts of these cultures also contained

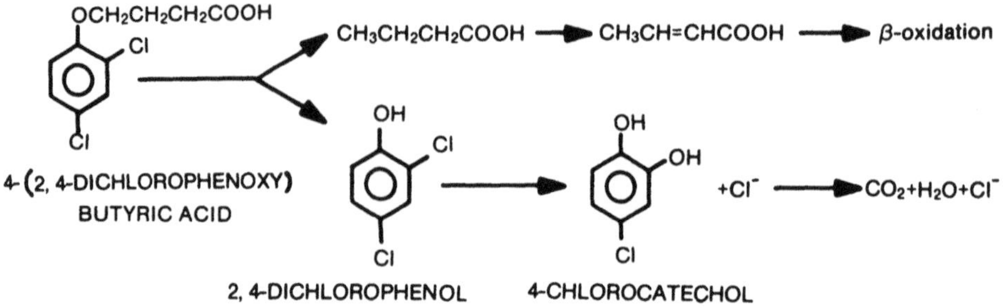

Figure 50   Pathway for the metabolism of 4-(2,4-dichlorophenoxy)-butyric acid by *Flavobacterium* sp. (Adapted from Ref. 292.)

free aliphatic acids which indicates ether cleavage, a second mechanism of phenoxy acid degradation similar to that shown for other herbicides (290,292). This organism when grown on 4-(2,4-D)B also oxidizes (as determined by manometric techniques) 3-(2,4-D)propionic acid, 2,4-dichlorophenol, and 4-chlorocatechol but not 2,4-D (60,292). The failure to oxidize 2,4-D argues against β-oxidation as the controlling mechanism for these degradations, as the product of β-oxidation of 4-(2,4-D)B would be 2,4-D (292). The enzymes involved in these oxidations are adaptive rather than constitutive. The oxidation of these substrates led to the proposal of a pathway for the degradation of 4(2,4-D)B through 2,4-dichlorophenol to 4-chlorocatechol with loss of one chloride ion and subsequent metabolism of 4-chlorocatechol by established pathways (Figure 50). The side chain is cleaved initially at the ether linkage and then is metabolized by β-oxidation (292).

Two species of *Nocardia*, *N. opaca* strain T16 and *Nocardia* sp. strain P2 have been shown to use β-oxidation for degradation of phenoxy acid homologs above the acetic congener (461). The corresponding chlorinated phenols are generated from 3-chlorophenoxypropionic and 4-chlorophenoxypropionic acids. The six-carbon homolog 2-chlorophenoxycaproic acid is metabolized to 2-chlorophenoxybutyric acid and 4-methyl-2-chlorophenozycaproic acid is metabolized to 4-methyl-2-chlorophenoxybutyric acid. Similarly, 2,4-dichlorophenoxycaproic acid is metabolized to 2,4-dichlorophenoxybutyric acid. Metabolism of 4-(4-chlorophenoxy)butyric acid to 3-hydroxy-4-(4-chlorophenoxy)butyric acid is followed by metabolic conversion to 4-chlorophenoxypropionic acid. A similar pathway is followed by 4-(3-chlorophenoxy)butyric acid (462). These studies with strain T16 showed that 3-hydroxy acid intermediates appear during the metabolism of all the o-arylobutyric acids.

An alternative mechanism has been noted in *N. coeliaca* (432). Although β-oxidation is operative in this organism, α-oxidation op-

erates in the metabolism of compounds with 10 or 11 side-chain carbons (phenoxydecanoic acid and phenoxyundecanoic acid). This
process, demonstrated for nonchlorinated molecules, involves two
enzymes: (a) a peroxidase catalyzing peroxidative decarboxylation
of the fatty acid to yield CO_2 and the fatty aldehyde with one less
carbon, and (b) a dehydrogenase catalyzing oxidation of the aldehyde to the corresponding acid.

The salt of phenoxy compound, sodium 2-(2,4-dichlorophenoxy)-
ethyl sulfate, is widely used in commercial formulations and is metabolized to 2-(2,4-dichlorophenoxy)ethanol by both *P. putida* FLA
and cell-free filtrates of *Bacillus cereus* var. *mycoides* (279,443).
2,4-D eventually appears in the *B. cereus* cell-free filtrates. Metabolism by *P. putida* does not result in production of 2,4-D. The
enzyme of *P. putida* which breaks the oxygen-sulfur bond does not
require prior activation (279). This is a novel alkylsulfatase, as
other microbial alkylsulfatases break the chain at the carbon-oxy-
gen bond.

FUNGAL METABOLISM OF PHENOXY HERBICIDES

Studies on fungal metabolism of phenoxy compounds have included
studies with *Aspergillus niger*. Hydroxylation is a major mechanism,
although not all vacant ring sites are hydroxylated. Thus, 2,4-
dichloro-5-hydroxyphenoxyacetic acid is the major metabolite of 2,4-
D metabolism (Figure 51). A minor metabolite, 2,5-dichloro-4-hydroxy-
phenoxyacetic acid, appears as the result of a novel hydroxyl-
chloride replacement and chloride shift (148). The latter compound
also is the only compound formed from metabolism of 2,5-D (148,149).
The hydroxyl-chloride shift is similar to that seen in 2,4-D metabolism by many plants, in which the major metabolite is 2,5-dichloro-
4-hydroxyphenoxyacetic acid and the minor metabolite is 2,3-dichloro-
4-hydroxyphenoxyacetic acid. Both metabolites require a hydroxyl-
chloride shift (283).

Hendersonula toruloidea metabolizes 2,4-D with production of
$^{14}CO_2$ (470). In 8 weeks, 28.8% of (carbon-1)-^{14}C 2,4-D and 2.8%
of ring-^{14}C 2,4-D were released. *Stachybotrys atra* produced only
3% of (carbon-1)-^{14}C 2,4-D as $^{14}CO_2$ after 8 weeks.

Phytophthora megasperma var *sojae* metabolized 10 mg/L 4-(2,4-
D)B with 45% disappearance in 21 days, but no production of 2,4-D
was noted (402). The organism also did not metabolize 2,4-D, suggesting that β-oxidation is not a primary mechanism in the metabolism of 4-(2,4-D)B.

MCPA is metabolized to 4-chloro-5-hydroxy-2-methylphenoxy-
acetic acid by *A. niger* (149). The metabolism of 2- or 4-chloro-
phenoxyacetic acid by fungi does not result in ring cleavage, in contrast to the activity of bacteria (150). Metabolism of 4-chlorophen-

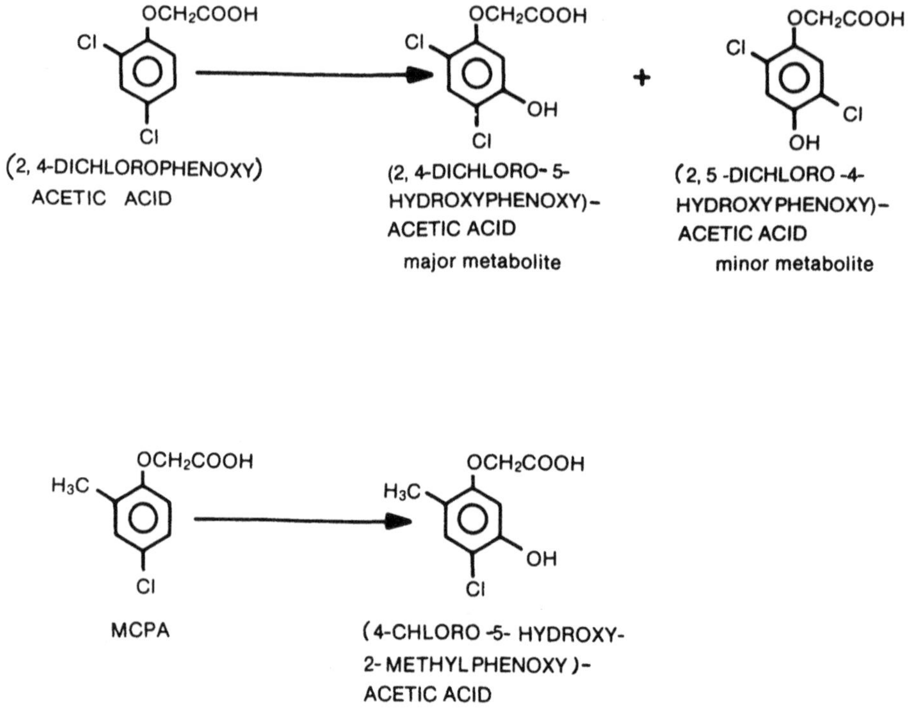

(2, 4-DICHLOROPHENOXY)
ACETIC ACID

(2, 4-DICHLORO- 5-
HYDROXYPHENOXY)-
ACETIC ACID
major metabolite

(2, 5 -DICHLORO -4-
HYDROXY PHENOXY)-
ACETIC ACID
minor metabolite

MCPA

(4-CHLORO -5- HYDROXY-
2- METHYL PHENOXY)-
ACETIC ACID

Figure 51 Metabolism of 2,4-D and MCPA by *A. niger.*

oxyacetic acid yields compounds hydroxylated in the 2- or 3-positions, the latter a novel product. Similarly, 2-chlorophenoxyacetic acid yields compounds hydroxylated in the 4- or 5-positions, the latter also a novel product. Minor products include 2-chloro-3-hydroxy and 2-chloro-6-hydroxy acids.

The microorganisms present in soil treated repeatedly with herbicides were isolated and identified (436). Bacteria capable of metabolizing 2,4-D include *Arthrobacter* sp., *Bacillus* sp., *Pseudomonas* sp., and *Sarcina* sp. Fungi include *Penicillium megasporum* and another *Penicillium* sp. Bacteria that can metabolize MCPA include *Arthrobacter* sp., *Corynebacterium* sp., and *Pseudomonas* sp., and fungi include *Fusarium culmorum*, *Mucor* sp., *Penicillium* sp., *Zygorhynchus moelleri*, and four *Verticillium* spp. Of these, two bacteria and five fungi also metabolize 2,4-D.

METABOLISM OF PHENOXY HERBICIDES IN SOILS

A sample of Philippine soil was treated with 2,4,5-T for 4 months, after which 2,4,5-trichlorophenol was recovered (377). A mixture of microorganisms was removed and incubated with 2,4,5-T. Loss of 10% of substrate was recorded with liberation of 8% of the initial radioactivity of the uniformly ring-labeled substrate as $^{14}CO_2$ in 25 days. The major metabolite was 2,4,5-trichlorophenol, which was readily metabolized with about 75% of the chloride in this metabolite liberated as free chloride. About 40% of this compound was released as $^{14}CO_2$ in 25 days. Products arising from incubation of the mixed culture include 3,5-dichlorocatechol, *cis,cis*-2,4-dichloromuconic acid, 2-chloro-4-carboxymethylene but-2-enolide, chlorosuccinic acid, succinic acid, and 4-chlorocatechol. These products, with the exception of 4-chlorocatechol, are all found in the 2,4-D degradative pathway subsequent to 3,5-dichlorocatechol (Figure 46).

The fate of uniformly ring-^{14}C labeled 2,4-D and 2,4,5-T was explored in six different soils (304). Metabolites of 2,4,5-T included 2,4,5-trichlorophenol and 2,4,5-trichloroanisole, while no metabolites were detected after 2,4-D incubation. About 20 to 35% of the substrates were recovered from the humic and fulvic acids and humin fractions, indicating formation of polymeric humic substances of 2,4-D and 2,4,5-T mediated by additional hydroxyl groups on the rings. Depending on the soil, up to 83% of 2,4-D and 71% of 2,4,5-T applied at 1 ppm concentration was converted to $^{14}CO_2$ in 150 days.

Diclorfop-methyl, (+)-methyl 2-[4-(2,4-dichlorophenoxy)phenoxy]-propionic acid, undergoes rapid hydrolysis of the ester bond in field soils (297). At 1 ppm the resultant diclorfop is rapidly metabolized in aerobic soils with isolation of two metabolites when the ^{14}C-label is the chlorinated ring and one metabolite when the label is in the nonchlorinated ring. The ubiquitous metabolite was identified as 4-(2,4-dichlorophenoxy)phenol. Other experiments indicated that intermediates include dichlorfop acid with subsequent decarboxylation to form phenyl ether (397). In 25 weeks, 25 to 35% of each type of labeled substrate was converted to $^{14}CO_2$ (397). In anaerobic soils diclorfop persists with no evolution of CO_2 and formation of only trace amounts of a metabolite.

The herbicide 2-(2,4-dichlorophenoxy)ethanol is often applied to soils in the inert form sodium 2-(2,4-dichlorophenoxy)ethyl sulfate (443). In sterile soils conversion to the active form occurs only at pH 3 to 4, while in nonsterile soils conversion takes place at pH 3 to 7, within 45 minutes after application. Thus, in soils with pH greater than 4, herbicide activation is thought to be biologically mediated.

The isopropyl, n-butyl, and isooctyl esters of 2,4,5-T, the n-butyl ester of 2,4-dichlorophenoxybutyric acid, and the isooctyl ester of 2,4-dichlorophenoxypropionic acid were applied to four Saskatchewan soils at 4 ppm concentration (400). In moist soils there was nearly complete conversion of the substrate to the free acids within 24 hours, with the exception of the isooctyl ester of 2,4-dichlorophenoxybutyric acid, which was completely converted in 72 hours. In air-dried soils there was very little loss of ester in the same time interval.

The isooctyl ester of dimethylamine salt of 2,4-D was applied to soils at a concentration of either 1.6 or 16 ppm (469). After 58 days 60 to 80% of the ring- or carboxyl-labeled substrate was released as $^{14}CO_2$, while 1% was recovered as 2,4-D. Eighty percent of the labeled substrate in the runoff water was recovered as $^{14}CO_2$ after 5 weeks, with an additional 3% more recovered in the next 5 weeks; the remaining material was not 2,4-D.

The primary effluent of municipal sewage was added to a nutrient medium containing 2,4-D (376). Within 7 days almost all the substrate disappeared. In a similar test phenoxyacetic acid disappeared within 12 days. Subsequent additions of either of these substrates resulted in metabolism without a lag period. No disappearance of 2,4,5-T was noted after 60 days.

Incubation of Maahas clay with medium containing these herbicides resulted in 90% disappearance of 2,4-D in 14 days on initial application, with 3 days required for 75% disappearance of additional applications of substrate (376). Phenoxyacetic acid required 16 days for initial disappearance, and subsequent applications were metabolized in 4 days. Evolution of $^{14}CO_2$ began from 7 to 60 days after herbicide application, and after 4 months from 5.2 to 34% was recovered depending on the soil.

After 12 weeks of incubation in sandy loam, 71 to 84% of either ring-^{14}C or (carbon-1)-^{14}C or (carbon-2)-^{14}C 2,4-D was released as $^{14}CO_2$ (470). The concentration of substrate in the soil was not given.

Several South Vietnamese soil and mud samples were treated with carboxyl-^{14}C labeled 2,4,5-T (65). Two samples were thought to be treated previously with a 50:50 mixture of 2,4-D and 2,4,5-T, while two samples were thought to be uncontaminated. At 1 ppm concentration, almost 70% was evolved as $^{14}CO_2$ in 49 days. At 15 ppm, three soils converted 70 to 80% of the substrate to $^{14}CO_2$ in 168 days, while one sample, thought to be previously uncontaminated, evolved more than 95% of the material as $^{14}CO_2$.

In moist Phillipine soils (upland conditions), 20 ppm 2,4-D disappeared more rapidly than when applied to flooded soils, but after 6 weeks the concentration of 2,4-D was similar in both moist and flooded soils (482). The same results were reported in the disappearance of 10 ppm 2,4,5-T from one of the soils. However, in another soil 2,4,5-T remained in flooded soils for a 4 week lag before undergoing rapid and complete disappearance in the next 4 weeks, while in the moist soils gradual disappearance was noted with about 40% remaining after 12 weeks. No disappearance of 2,4-D or 2,4,5-T was noted in sterile control samples after 12 weeks' incubation.

The persistence of 2,4-D and MCPA in soils following repeated applications was measured (436). Ten weeks were required for disappearance of 2,4-D upon first application. The herbicide disappeared after 7 weeks upon second application in the second year, and required only 4 weeks for disappearance after 18 years of repeated applications. MCPA required 20 weeks for disappearance in the first year, 10 weeks in the second year, and 7 weeks after 18 years. Soils pretreated with either herbicide showed accelerated 2,4-D disappearance after 18 years but not after 1 year of herbicide application. Enhancement of MCPA disappearance was noted after either one year or 18 years of pretreatment with either herbicide. The numbers of degradative bacteria or fungi were not significantly different after 0, 1, or 18 years of pretreatment.

A seed bioassay (mustard or cress) showed that 2,4-D disappears faster than MCPA, and 2,4,5-T much slower than either, from both a light clay soil and a sandy loam (414). The first application of 2,4-D to soil at 55 ppm concentration required 14 days for disappearance of herbicidal activity, while 7 days were required for disappearance of the second application at 120 ppm and 4 days upon third application of 200 ppm.

A soybean bioassay also indicated that 2,4,5-T lasted much longer than 2,4-D or MCPA (111). The rate of application had no effect at application rates from 5 to 20 lb/acre. The rate of disappearance of herbicide increased with increasing temperature and increasing moisture. Although MCPA is subject to degradation by photolysis, experiments with 1 mg/L MCPA incubated in rice water in the dark showed disappearance of 75% in 6 days, as opposed to 15% disappearance due to photolysis alone (405).

Application of bifenox, methyl 5-(2,4-dichlorophenoxy)-2-nitrobenzoate, to a greenhouse soil mix at a rate equivalent to 1.7 kg/ha showed that after 313 days 78% of the benzoate ring-^{14}C labeled material and 67% of phenoxy ring-^{14}C labeled material was bound to the soil (275). After 7 days following initial application very little additional bifenox disappeared, although only 20 to 26% was bound to the soils. The metabolites, which were identified by thin layer chromatography, included the acid of bifenox, 5-(2,4-dichlorophenoxy)-2-nitrobenzoic acid, nitrofen (2,4-dichlorophenoxy-4-nitrophenyl ether), 5-(2,4-dichlorophenoxy)anthranilic acid, and other unidentified compounds. These metabolites were also found as degradation products in plants grown in bifenox-treated soil.

CHLOROPHENYL HERBICIDES

Pseudomonas sp. strain CBS 3 utilizes 4-chlorophenylacetic acid as the sole source of carbon and energy (258). Initially formed metabo-

lites include 4-chloro-3-hydroxyphenylacetic acid, 3-chloro-4-hydroxy-
phenylacetic acid, and 4-chloro-2-hydroxyphenylacetic acid. This
strain, however, cannot grow on 3-chloro-4-hydroxy- or 4-chloro-3-
hydroxyphenylacetic acid (296). Metabolism of 4-chloro-2-hydroxy-
phenylacetic acid results in formation of 4-chloro-2,3-dihydroxy-
phenylacetic acid. This is not the primary pathway of substrate metabolism. Upon further incubation 3,4-dihydroxyphenylacetic acid
(homoprotocatechuate) appears, indicating direct removal and replacement of the chloride before ring cleavage. Homoprotocatechuate
is metabolized to homogentisic acid (2,5-dihydroxyphenylacetic acid)
and then to the *meta* cleavage product maleylacetoacetate resulting
from the action of homogentisate 1,2-dioxygenase. An *Arthrobacter*
sp. similarly produces 4-chloro-3-hydroxyphenylacetic acid from 4-
chlorophenylacetate, along with an additional unidentified metabolite
(110).

The herbicide chlorofenprop-methyl [2-chloro-3-(4-chlorophenyl)-
propionic acid methyl ester] is readily metabolized (264). In mixed
cultures of soil microorganisms, 4-chlorocinnamic acid and 4-chloro-
benzoic acid have been identified as metabolites. The latter product
has been recovered from soil amended with the herbicide. Two
strains, thought to be a *Flavobacterium* sp. and a *Brevibacterium*
sp., were isolated from the soil and found to convert 4-chlorocin-
namic acid to 4-chlorobenzoic acid. Two *Arthrobacter* spp. have been
shown to grow on 4-chlorobenzoic acid as the sole source of carbon
and energy. Thus, a consortia or a mixed soil population would be
able to mineralize chlorofenprop.

A chlorophenyl insecticide known as SD 8280 [2-chloro-1-(2,4-
dichlorophenyl)vinyl dimethyl phosphate] was studied with respect
to its degradation in soils (372). The major products were 2,4-di-
chlorobenzoic acid and 1-(2,4-dichlorophenyl)ethanol. Lesser amounts
of 2-chloro-1-(2,4-dichlorophenyl)vinyl methyl hydrogen phosphate
and 2',4'-dichloroacetophenone were also formed. Other products
were noted but not identified, although they were shown not to be
2,4-dichlorophenol or 2,4-dichlorobenzyl alcohol. None of the recovered metabolites represent alterations to the chlorinated ring of
the molecule.

SUMMARY

Most of the chlorophenoxy herbicides appear to be biodegradable to
CO_2 and free chloride under the right conditions. These results
have been shown for both pure cultures and in soils containing mixed
populations. Adaptation of the cultures to the substrate is required
and results in faster disappearance of the compound. The persistence of some compounds in some environments indicates that under

some conditions these herbicides could be considered to be recalcitrant compounds. Seasonal variations in herbicide degradation have also been noted (460).

Once a population has become adapted to metabolize a substrate, however, that capability persists for long periods of time. Thus, yearly applications of an herbicide are sufficient to maintain a degradative population in soil. A population adapted to metabolize a particular substrate is often also adapted to metabolize other related compounds. This has particular application where crop rotation is accompanied by usage of different herbicides.

10

Phenylamide and Miscellaneous Herbicides

The phenylamide herbicides include the groups of phenyl ureas, N-phenylcarbamates, and acylanilides. Each takes the general form R-NH-CO-X where R is a halogenated or nonhalogenated aromatic hydrocarbon (239,174). In urea herbicides, X is an amino group with methyl, alkyl, or methoxy substituents. Carbamates have the form whereby X is an alkoxy group. In acylanilide herbicides, X is an alkyl group. In many cases these herbicides are degraded to substituted anilines (305).

The urea herbicides are specific inhibitors of photosynthesis, but can have a selective effect because of their low water solubility and low mobility in soil. They can be applied to kill shallow-rooted weeds while having no effect on deeper-rooted plants of interest. In the late 1940s, a large number of substituted urea compounds were compared with 2,4-D for herbicidal activity and were found worthy of further development. Originally used as industrial weed killers, they have more recently been used in agricultural applications (174). The carbamates comprise a wide range of active compounds which, depending on the chemical substituents, are used for such purposes as herbicides, insecticides, and medicinals (186a). As insecticides, the carbamates inhibit the action of acetylcholinesterase, an enzyme required for proper neurotransmitter substance functioning. The degree of fit between the inhibitor and the enzyme lends selectivity to the activity of the carbamates. The herbicidal carbamates have also been used since the 1940s. Amide herbicides (substituted anilides) were developed in the 1950s and are used for such crops as corn and rice.

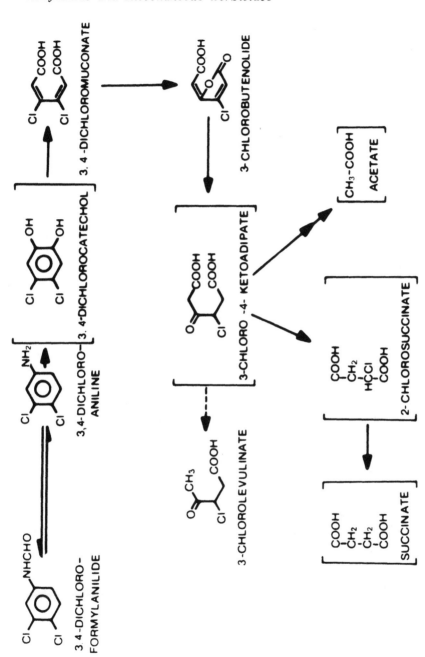

Figure 52 Pathways of 3,4-dichloroaniline metabolism by *P. putida*. (Adapted from Ref. 484.)

BACTERIAL METABOLISM OF CHLORINATED ANILINES

The degradation of such herbicides as monuron, diuron, linuron, and propanil results in formation of 3,4-dichloroaniline. A strain of *P. putida* has been isolated which mineralizes 3,4-dichloroaniline in the presence of aniline (483). The rate of mineralization was enhanced by increasing the concentration of aniline. In the presence of 500 mg/L propionanilide, as much as 50% of ring labeled chloroaniline (added at 10 to 60 mg/L concentration) was metabolized to $^{14}CO_2$ within 2 weeks, accompanied by some chloride release (484). A pathway for 3,4-dichloroaniline was proposed which involves *ortho* cleavage through 4,5-dichlorocatechol and further metabolism to succinic acid (Figure 52). This pathway is analogous to that shown for aniline. Dichloroaniline is also metabolized by a different mechanism to dichloroformylanilide (245,484).

Another *Pseudomonas* sp., strain G, mineralizes 3,4-dichloroaniline to CO_2 when grown in the presence of 4-chloroaniline (488). In 9 days, 15% of 0.5 mM dichaloroaniline was converted to CO_2.

Studies with 4-chloroaniline have shown that this metabolite is utilized as a sole source of carbon and nitrogen by *Pseudomonas* sp. strain G (487). After 10 days, 64% of a 2.5 ppm solution of the substrate was released as $^{14}CO_2$. Ammonium cation, rather than nitrate or nitrite anions, accumulates suggesting that the amino group is removed directly without oxidation. Other ^{14}C-labeled products accumulate but are not incorporated into the cell biomass. Resting cells of this strain grown on 4-chloroaniline also oxidize aniline, catechol, and 4-chlorocatechol but not 4-chloronitrobenzene or 4-chlorophenol, indicating dioxygenase attack on the molecule (487). The *meta*-cleavage metabolite 2-hydroxy-5-chloromuconic semialdehyde also accumulates in the medium. This strain also utilizes 2-chloroaniline and 3-chloroaniline as sole source of carbon and nitrogen. *P. multivorans* strain An 1, when grown with aniline present in the medium, converts 2-chloroaniline to 3-chlorocatechol, and 3-chloroaniline and 4-chloroaniline both to 4-chlorocatechol, which is metabolized subsequently to CO_2 and cell constituents (361).

A strain of *Alcaligenes faecalis* utilizes 3-chloroaniline or 4-chloroaniline under cooxidative conditions with sodium acetate or sodium pyruvate (419). Both chloroanilines are oxidized to 4-chlorocatechol, which undergoes *meta* cleavage to 5-chloro-2-hydroxymuconic semialdehyde and then to 2-chloro-4-oxalocrotonic acid. Similarly, an unidentified isolate which utilizes 4-chloroaniline as the sole carbon source demonstrated, via Warburg respirometry, oxidation of 4-chlorocatechol without a lag (56).

The metabolism of 4-chloroaniline by *Paracoccus* sp. under both aerobic and anaerobic conditions was investigated (43). Transformation was faster under anaerobic conditions with 100% of a 20 mg/L

solution converted within 2 days to a volatile product plus several
other metabolites. The volatile product was not identified but was
shown not to be CO_2. In the same period, 75% of the substrate was
utilized aerobically, although the aerobic population was larger than
the anaerobic population. The major product of aerobic metabolism
is 4-chloroacetanilide, although several other products are formed
as well. *Paracoccus* sp. also transforms 2-, 3-, 4-chloroaniline and
2,3-, 2,4-, 2,5-, and 3,4-dichloroaniline both aerobically and anaerobically, with more complete primary degradation occurring under anaerobic conditions.

A study of the anaerobic conversion of 4-chloroaniline by the
same organism showed pH-dependent formation of products (310).
In a medium containing nitrate, 80% of a 100 ppm solution was transformed within 48 hours to the condensation product 1,3-bis(*p*-chloro-
phenyl)triazine. However, sterile anaerobic solutions at pH of 5 to 6
also yielded this product, although no transformation took place at
pH 7. It was postulated that the conversion of nitrate to nitrite and
decrease in pH were the primary effects of bacterial metabolism, while
triazine formation was a nonbiological secondary effect. *Paracoccus* sp.
also formed small amounts of 4-chloroacetanilide during incubation with
4-chloroaniline.

A wide variety of both Gram-positive and Gram-negative bacteria
were isolated from soil which had been enriched with 4-chloroaniline
(134). The most active species was identified as *Bacillus firmus*.
This species could not use the substrate as the sole source of carbon, but when grown with ethanol transformed 4-chloroaniline to
4-chloroacetanilide as the main metabolite, with lesser amounts of 4-
chloropropionanilide also produced. Two other products were identified a 7-chloro-2-amino-3H-phenoxyazine-3-one and 7-chloro-2-amino-
3H-3-hydroxyphenoxyazine, and were postulated to result from spontaneous condensation of 4-chloroaniline with subsequent hydroxylation.
The acylation of the substrate described here is considered to be a
detoxification process in microorganisms (134).

Aniline-grown resting cells of *Rhodococcus* sp. An 117 convert
2-chloro- and 3-chloroaniline to 3-chloro- and 4-chlorocatechol, respectively. An additional product identified as 2-chloromuconic acid
results from metabolism of 2-chloroaniline. Cometabolism of 3-chloro-
aniline in the presence of $^{18}O_2$ resulted in production of another
product which was identified as the γ-lactone of 3-hydroxymuconic
acid, formed by incorporation of two molecules of oxygen (223). Appearance of this product was associated with disappearance of 4-
chlorocatechol. This suggests that dechlorination is associated with
lactonization of the 3-chloromuconic acid, a mechanism shown previously in pseudomonads. When fresh cultures were supplied with
benzoate plus 2- or 3-chloroaniline as the sole source of nitrogen,

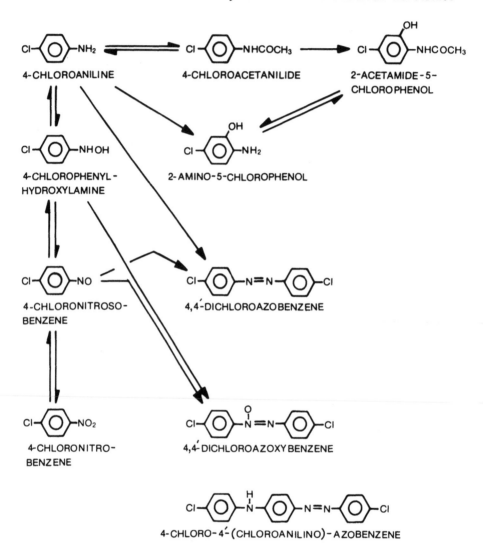

Figure 53 Metabolism of 4-chloroaniline by microorganisms. (Adapted from Refs. 49,155,242.)

growth of cells and disappearance of both substrates occurred (222). The growth yield was similar to that obtained when NH_4NO_3 was utilized as the sole nitrogen source. No growth occurred when 4-chloroaniline was supplied as the nitrogen source.

The fungicide 2,6-dichloro-4-nitroaniline is metabolized by many bacteria including *E. coli* and *P. cepacia* (442). The first step is reduction of the nitro group to an amine, forming 2,6-dichloro-p-phenylenediamine. This compound is then acetylated to produce 4-amino-3,5-dichloroacetanilide. The first reductive step occurs much faster under anaerobic conditions than under aerobic conditions.

FUNGAL METABOLISM OF CHLORINATED ANILINES

Fungi as well as bacteria produce peroxidases which are responsible for the polymerization of chloroanilines. Species with this capability include *Geotrichum candidum* L-3 and *Aspergillus* sp. (47,271). The substrate 3,4-dichloroaniline is converted to 3,3',4,4'-tetrachloroazobenzene by *G. candidum* L-3 and *Aspergillus* sp. (271) and is converted to 3,3',4,4'-tetrachloroazoxybenzene in *Fusarium oxysporum* cultures (243). However, this latter reaction varies with the culture conditions and the azoxy condensation product has not been found in soils.

The fungal metabolism of 4-chloroaniline follows several pathways (Figure 53). A major pathway is N-hydroxylation such as is demonstrated by *F. oxysporum* (242). This species metabolizes 4-chloroaniline to 4-chlorophenylhydroxylamine which is subsequently converted to 4-chloronitrosobenzene and 4-chloronitrobenzene. Condensation products which appear include both 4,4'-dichloroazobenzene and 4,4'-dichloroazoxybenzene. In addition, 4-chloroacetanilide appears as an acylation reaction and may undergo hydrolysis to yield 4-chloroaniline. Free chloride is produced as the result of some of these reactions.

The culture filtrate of *G. candidum*, as well as the purified fungal enzymes peroxidase and aniline oxidase, converts 4-chloroaniline to several condensation products including 4,4'-dichloroazobenzene and 4-chloro-4'-(4-chloroanilino)azobenzene (49). These reactions have also been demonstrated with horseradish peroxidase. *Streptomyces* sp. also formylates 4-chloroaniline with resultant production of 4-chloroformylanilide as well as 4-chloroacetanilide and at least two other metabolites (381).

Ring hydroxylation is a mechanism demonstrated by *F. oxysporum* which results in metabolism of 3-chloroaniline to 2-amino-4-chlorophenol and 4-chloroaniline to 2-amino-5-chlorophenol (Figure 53) (155). These molecules are hydroxylated in the *ortho* position. The aminophenols are relatively unstable and can undergo condensation and polymerization reactions, although 2-amino-4-chlorophenol has been detected in soil as well as in the pure culture studies reported above. The *ortho*-substituted substrate 2-chloroaniline is not hydroxylated by *F. oxysporum*.

METABOLISM OF CHLORINATED ANILINES IN SOILS

There is evidence that in soils 3,4-dichloroaniline is slowly mineralized. Radiocarbon-labeled humic-bound material is mineralized in
some soils ($^{14}CO_2$ production) at about the same rate as the average
soil organic matter polymer (384). The addition of aniline to soils
enhances mineralization of both free and humic-bound 3,4-dichloro-
aniline (483). This was attributed to selection by the aniline analogue
for chloroaniline-degradative populations as well as the induction by
aniline of the common metabolic pathway.

Application of 10 ppm ring-labeled ^{14}C-3,4-dichloroaniline to a
rice paddy ecosystem yielded a total recovery (extractable plus nonextractable) of almost 69% (219). Less than 4% of the material applied
to soil was recovered in the water or rice plants. CO_2 evolution was
not determined. Extraction of the soil yielded 3.4% of the original
material as dichloroaniline, 4% as tetrachloroazobenzene, and 44.7%
as polar material.

The degradation of aniline, 4-chloroaniline, and 3,4-dichloroaniline
in four different soils was compared (420). At 1 ppm application
rate, 16 to 26% aniline was converted to CO_2 after 10 weeks, while
after 16 weeks 12 to 27% chloroaniline and 4 to 12% dichloroaniline
were mineralized. In soils 4-chloroaniline can be converted to 4-
chlorophenylhydroxylamine via biological mechanisms (48). This compound then undergoes condensation with 4-chloroaniline to form 4,4'-
dichloroaozbenzene in a nonbiological reaction.

Condensation of two molecules of 3,4-dichloroaniline forms 3,3',
4,4'-tetrachloroazobenzene which is relatively persistent in the environment (51). This reaction has been demonstrated in Nixon sandy
loam (11,23,24) as well as in other soils (22,48,51,245). The conversion of chloroanilines to chloroazobenzenes has been shown to
occur by a peroxidase mechanism (25). A mixture of substituted anilines, hydrogen peroxide, and peroxidase resulted in the formation
of chloroazobenzene (51). Bacteria including *Bacillus* sp., *Arthrobacter* sp., and *Pseudomonas* sp. exhibit peroxidase activity (271).
A pathway has been proposed which includes transformation of the
hypothetical 3,4-dichloroanilidyl moieties to 3,4-dichlorophenylhydroxy-
lamine. Two of these molecules are condensed to the hypothetical 3,3',
4,4'-tetrachlorohydrazobenzene which is converted to 3,3',4,4'-tetra-
chloroazobenzene (51). However, studies in which dichloroaniline was
applied to herbicide-treated soils did not show formation of tetrachloroazobenzene, suggesting that dichloroaniline is not the prime precursor
for tetrachloroazobenzene (30). Also, production of tetrachloroazobenzene in soils incubated with propanil was not correlated with the
quantity of peroxidase-producing microorganisms recovered from the
soil (58). Cell-free peroxidase was found only rarely in the soil sam-

ples. Recovery of peroxidase increased upon amendment of the soils
with nutrient sources and additionally upon sonification of the samples,
which may have released cell-bound or intracellular enzymes. Addition
of proteose-peptone decreased recovery. However, this peroxidase
did not form tetrachloroazobenzene from dichloroaniline. Another condensation product detected as a humic-bound residue in soils treated
with propanil is 4-(3,4-dichloroanilino)-3,3',4'-trichloroazobenzene (22).

Warburg respirometry studies indicated that aniline-acclimated
activated sludge microflora oxidized 500 mg/L 2-chloroaniline at low
rates over a 192-hour incubation period (293). Oxidation of 4-chloro-
aniline occurred at slightly higher rates, and after a 100-hour lag
period rapid oxidation of 3-chloroaniline took place.

METABOLISM OF UREA HERBICIDES

A *Fusarium* sp. utilizes the larvacide diflubenzuron [1-(4-chloro-
phenyl)-3-(2,6-difluorobenzoyl)urea] as its sole source of carbon
and energy (389). The pathway was elucidated to include initial formation of 2,6-difluorobenzoic acid and 4-chlorophenylurea. The latter
compound is metabolized to 4-chloroaniline, and then 4-chloroacetani-
lide, followed by reductive dehalogenation to acetanilide and further
metabolism to cell constituents. Other fungi including *Cephalosporium*
sp., *Penicillium* sp., and *Rhodotorula* sp., although unable to utilize diflubenzuron as a sole carbon source, metabolized the compound
to 2,6-difluorobenzoic acid and 4-chlorophenylurea, indicating cleavage of the urea bridge.

Evidence that the urea herbicides are degraded in the environment accrues from studies which showed that such compounds as
3-(4-chlorophenyl)-1,1-dimethylurea (monuron) and 3-(3,4-dichloro-
phenyl)-1,1-dimethylurea (diuron), when applied at rates of 1 to 2
lb/acre annually in the eastern part of the United States, left no
residual phytotoxicity after 4 to 8 months (203). Higher application
rates required longer times for disappearance of phytotoxicity.

Potential mechanisms for disappearance include biological, leaching
volatilization and chemical decomposition (203). Leaching is considered to be significant only in porous soil or if there is a great
amount of rainfall. The low vapor pressure and aqueous solubilities
of these herbicides makes volatilization unlikely to be an important
mechanism. Photodecomposition may be a factor in dry areas in which
the herbicide remains on the soil surface, but these compounds are
stable to chemical decomposition in aqueous solutions. Biological
studies revealed that the rate of herbicide inactivation is greater
in nonsterile than in sterile soils. This was shown by the amount of
radiocarbon-labeled CO_2 evolved from soils amended with [14]C(methyl)-

labeled monuron, and a *Pseudomonas* sp. was isolated which oxidized this substrate in Warburg respirometry studies (203).

The herbicide N'-(4-chlorophenoxy)phenyl-N,N-dimethylurea is metabolized when placed in contact with soils such as sandy loam or humus soil (173). Sorption of the compound to the soils has an effect on the rate of degradation. Enriched bacterial cultures derived from the soil samples also metabolized the herbicide by successive demethylation to N'-(4-chlorophenoxy)phenyl-N-methylurea and subsequently to N'-(4-chlorophenoxy)phenylurea. No appreciable amounts of CO_2 were released. Fungal isolates of *Penicillium* sp. and *Aspergillus* sp. removed less than 50% of the carbonyl group but did not transform the compound further. The metabolism of this herbicide in plants also follows successive demethylation leading to CO_2 evolution.

Diuron is used for long-term weed control in peach orchards, and has been detected as long as 3 years after the last application in a field consisting of Fox loamy sand (248). The levels of 3,4-dichloroaniline were very low and decreased to undetectable levels in 3 years, while the potential condensation product tetrachloroazobenzene was not detected. The decomposition of diuron is enhanced by changes in environmental conditions that favor the growth of microorganisms (306). Thus increasing the temperature of incubation from 10 to 30°C or adding organic matter to the soils each increase the rate of diuron decomposition. The rate of herbicide inactivation is much greater than the rate of CO_2 evolution, and investigations showed that the loss of one methyl group decreases herbicidal activity by half while loss of both methyl groups completely inactivates the molecule. The pathway of diuron metabolism in soils was proposed to be 3-(3,4-dichlorophenyl)-1,1-dimethylurea to 3-(3,4-dichlorophenyl)-1-methylurea, then loss of the second methyl group to form 3-(3,4-dichlorophenyl)urea, followed by hydrolysis of the urea to form the aniline derivative 3,4-dichloroaniline, which accumulates as the major product (103).

Microbial enrichment cultures from pond water and pond sediment treated with diuron revealed a similar pathway of degradation (131, 132). Three additional unidentified products were also detected. The enrichment cultures included mixtures of fungi and bacteria as well as consortia of bacteria. Some of the mixed cultures converted more than 90% of the substrate to CO_2 within 3 weeks. Of 20 single isolates, however, only three could partially metabolize diuron after 4 weeks' incubation.

Under anaerobic conditions reductive ring dechlorination of diuron occurs (12). Enrichment cultures from pond water and sediment incubated anaerobically rapidly degraded diuron to 3-(3-chlorophenyl)-1,1-dimethylurea in stoichiometric amounts. This product was not degraded further and no other products were detected. Repeated additions of diuron resulted in rapid metabolism of the substrate.

Chlortoluron [N-(3-chloro-4-methylphenyl)-N'-dimethylurea] has a half-life in a variety of soils of 4 to 6 weeks (403). The major degradation product is monomethyl chlortoluron. However, the product of subsequent demthylation was never detected. This may indicate that cleavage of the molecule to form 3-chloro-4-methylaniline follows monomethyl chlortoluron formation. Although the substituted aniline was not detected, this product rapidly disappears from the soils when applied directly.

Methoxy phenyl urea herbicides have quite high herbicidal selectivity and additionally are not very persistent in soil after application (174). Metabolism of these compounds differs from that of the dimethyl phenyl urea herbicides.

Soils treated with 3-(4-chlorophenyl)-1-methoxy-1-methylurea (monolinuron) yielded a *Bacillus sphaericus* isolate which could co-metabolize ^{14}C(ureido)-monolinuron to $^{14}CO_2$ and 4-chloroaniline (452). Maximum degradation occurred after the end of logarithmic growth. The substituted aniline was not degraded by the organism but was lost from the culture through volatilization. Linuron [3-(3, 4-dichlorophenyl)-1-methoxy-1-methylurea] was metabolized to stoichiometric amounts of 3,4-dichloroaniline, while the dimethyl compounds monuron and diuron were not degraded. Cell-free extracts of *B. sphaericus* also transformed the methoxy herbicides to substituted anilines with the aliphatic portion of the molecule degraded to CO_2 plus another metabolite (453). The degradation product of linuron was identified as N,O-dimethylhydroxylamine (133). The cell-free extracts were less active against the dimethyl substrates monuron and diuron (453).

The soil fungus *Cunninghamella echinulata* Thaxter degrades linuron and monolinuron through stable hydroxymethyl intermediates (435). Linuron is metabolized to 3-(3,4-dichlorophenyl)-1-methoxy-1-hydroxymethylurea and subsequently 3-(3,4-dichlorophenyl)-1-methoxyurea and 3-(3,4-dichlorophenyl)-1-methylurea. Disappearance of compounds such as linuron from nonsterile but not from sterile soils has been noted (120). The degradation of monolinuron during waste composting was investigated (318). After three weeks of composting, N-methoxy-N'-4-chlorophenylurea was present in trace amounts and was the only metabolite detected.

METABOLISM OF CHLORINATED PHENYL CARBAMATE HERBICIDES

The phenyl carbamate herbicides (also known as carbanilates) are used to kill weeds on crop plants such as rice. Substituted anilines arise from degradation of these compounds, as the primary mechanism of degradation seems to be hydrolysis of the ester linkage (239).

The cell-free enzyme extract from a *Pseudomonas* sp. isolated from a soil enrichment culture was capable of converting several chlorophenylcarbamates to corresponding chlorinated anilines (244).

The compounds tested included CIPC, isopropyl-N-(3,4-dichloro-phenyl)carbamate, sec-butyl-N-(3,4-dichlorophenyl)carbamate, α-carboisopropoxyethyl-N-(3-chlorophenyl)carbamate, 2-chloroethyl-N-(3-chlorophenyl)carbamate, 2-(1-chloropropyl)-N-(3-chlorophenyl)-carbamate, 2-ethylhexyl-N-(3-chlorophenyl)carbamate and α-carbo-(2,4-dichlorophenoxyethoxy)ethyl-N-(3-chlorophenyl)carbamate. In contrast, the enzyme preparation had no activity against 3-(4-chloro-phenyl)-1,1-dimethylurea (monuron).

Penicillium jenseni was isolated from soil which had been treated with barban [(3-chlorophenyl)carbamic acid 4-chloro-2-butynyl ester] (472). The mold did not utilize barban as a carbon source, although trace amounts of 3-chloroaniline appeared after incubation. Incubation of the mycelia with barban resulted in production of large amounts of 3-chloroaniline, which was metabolized by the mycelia suspensions without production of free chloride.

A mixture of bacteria and fungi isolated from soil enriched with the nonchlorinated analogue isopropyl phenylcarbamate (propham) metabolized a wide range of herbicides to their corresponding chlor-inated aniline products (305). These included compounds ring-sub-stituted with 3-chloro-, 4-chloro-, 2,4-dichloro-, 3,4-dichloro-, and 2,4,5-trichloro- substituents. Microorganisms comprising the con-sortium included *Mycobacterium* sp., *Arthrobacter* sp., *Corynebac-terium* sp., *Fusarium* sp., *Nocardia* sp., *Streptomyces* sp., *Asper-gillus* sp., and *Penicillium* sp.

A wide variety of fungal isolates from treated soil metabolized swep [methyl N-(3,4-dichlorophenyl)carbamate] to 3,4-dichloroani-line with formation of trace amounts of 3,3',4,4'-tetrachloroazoben-zene (240). The fungi included *Aspergillus ustus, A. versicolor, Fusarium oxysporum, F. solani, Penicillium chrysogenum, P. nigu-losum* and *Trichoderma viride.* The isolates were most active on CIPC and only slightly active against the phenylureas diuron and 3-(3,4-dichlorophenyl)-1-methylurea. The rate of formation of 3-chloroaniline was the same as the rate of disappearance of CIPC due to metabolism by *Aspergillus fumigatus*, indicating that hydrol-ysis of the ester bond is the first step in degradation of this her-bicide (471).

Swep is metabolized to 3,4-dichloroaniline with formation of 3,3', 4,4'-tetrachloroazobenzene in soils such as Nixon sandy loam (11). Soil microorganisms obtained from muck soil metabolized 1240 mg/L isopropyl-N-(3-chlorophenyl)carbamate (CIPC, chlorpropham) to 3-chloroaniline with complete chloride release within 13 days (241). A similar pathway with complete dechlorination within 16 days was followed in the metabolism of 2-chloroethyl-N-(3-chlorophenyl)carba-mate (CEPC). The isolates lost degradative capability if they were maintained on nutrient agar for several days, but they could be readapted to use these herbicides as a sole source of carbon.

METABOLISM OF ACYL ANILIDE HERBICIDES

The substituted anilide herbicides are structurally related to the urea herbicides and the carbanilates, and like them are degraded to substituted anilines.

Strains of *Pseudomonas striata* and *Achromobacter* sp. metabolize N-(3-chloro-4-methylphenyl)-2-methylpentanamide (3'-chloro-2-methyl-p-valerotoluidide, solan) to 3-chloro-p-toluidine from which chloride is released quantitatively (Figure 54) (240). Azobenzene products were not detected in these experiments.

Fungi from several genera, including *Aspergillus* spp., *Fusarium* spp., *Penicillium* spp., and *Trichoderma* sp. also metabolize solan to 3-chloro-p-toluidine with chloride release (240). Cultures of *A. niger* metabolize solan to a product identified as 3'-chloro-4'-methyl-acetanilide (Figure 54) (455). Cell-free extracts converted solan to the substituted aniline, but in growing cultures this product was rapidly acetylated to effect detoxification and the free aniline could not be recovered.

Strains of *P. striata* and *Achromobacter* sp. metabolize both swep and N-(3,4-dichlorophenyl)propionamide (propanil) to 3,4-di-chloroaniline and small quantities of 3,3',4,4'-tetrachloroazobenzene (240). *Corynebacterium pseudodiphtheriticum* NCIB 10803 utilizes propanil as the sole source of carbon and energy for growth (186). The resulting products are 3,4-dichloroaniline, which accumulates in the medium, and the propionic acid moiety, which is utilized for cell growth.

A strain of *F. solani* was isolated which utilizes propanil as a sole source of carbon and energy (269). Dichloroaniline accumulates in the medium until it reaches toxic levels. The enzyme responsible for propanil hydrolysis to propionate and dichloroaniline was identified as an acylamidase (270). This enzyme is specific for molecules with a short chain, as it could not hydrolyze dicryl [N-(3,4-dichloro-phenyl)methacrylamide] or the six-carbon herbicide 2-methylpenta-namide. Other fungi, including *Aspergillus ustus*, *A. versicolor*, *Fusarium oxysporum*, *F. solani*, *Penicillium chrysogenum*, *P. nigu-losum*, and *Trichoderma viride*, also metabolize propanil to 3,4-di-chloroaniline with formation of trace amounts of 3,3',4,4'-tetrachloro-azobenzene (240).

The interaction of *Penicillium piscarium* and *Geotrichum candidum* incubated with propanil results in increased growth over either alone (50). *P. piscarium* contains an acylamidase which converts pro-panil to dichloroaniline. *G. candidum* cannot utilize propanil but contains a peroxidase which converts dichloroaniline to the tetrachloro-azobenzene. Each fungus reduces the toxic level of the other's by-product of metabolism, demonstrating a synergistic or mutualistic interaction.

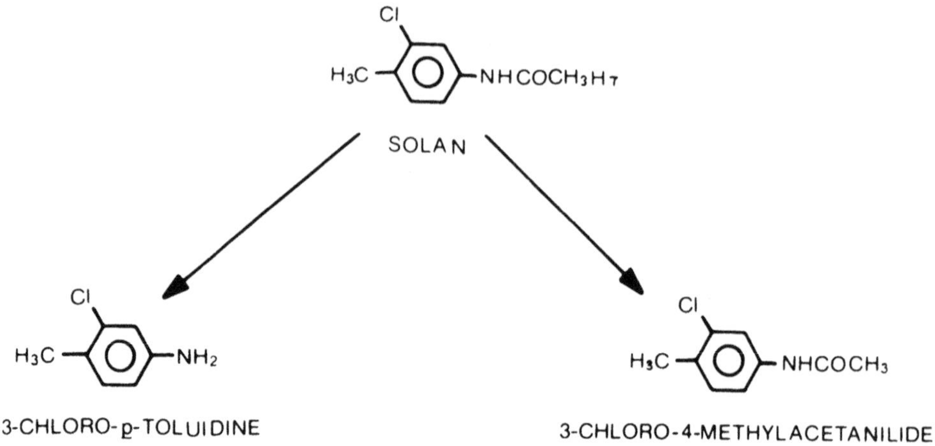

3-CHLORO-p-TOLUIDINE   3-CHLORO-4-METHYLACETANILIDE

Figure 54 Metabolism of solan by microorganisms. (Adapted from Refs. 240,455.)

The yeast *Pullularia pullulans* and two *Penicillium* spp. were isolated and found to utilize N-(3,4-dichlorophenyl)-2-methylpentanamide as a sole source of carbon and energy, metabolizing the herbicide to dichloroaniline and 2-methylvaleric acid (392). The enzyme was inducible and differed in level of activity and substrate specificity among the three species. Cell-free extracts of one of the *Penicillium* spp. hydrolyzed a wide variety of other acyl anilides as well, but had no activity against diuron or CIPC.

An unusual product, identified as N-(3,4-dichlorophenyl)-2-methyl-2,3-dihydroxypropionamide, was detected in the culture medium of a *Rhizopus japonicus* culture growing in the presence of dicryl (456). This product results from the double hydroxylation of the ethylene double bond of dicryl, and was the only metabolite detected. *R. japonicus* also hydroxylates the side chain of N-(3,4-dichlorophenyl)pentanamide to produce N-(3,4-dichlorophenyl)-3-hydroxy-2-methylpentanamide (457). This mechanism results in detoxification of the herbicide.

Since propanil is an inhibitor of photosynthesis it is a potential poison to the algae as well (473). A study of the effect of propanil on cyanobacteria indicated depression of photosynthesis but also showed conversion of propanil to the less toxic 3,4-dichloroaniline in both axenic (bacteria-free) and contaminated cultures. The species studied incorporated those found in flooded rice paddy fields and soil, and included *Anabaena cylindrica*, *A. variabilis*, *Nostoc muscorum*, *N. entophytum*, *Tolypothrix tenuis*, and *Gloeocapsa alpicola*.

In soils propanil is converted to 3,4-dichloroaniline and the condensation product 3,3',4,4'-tetrachloroazobenzene (30,87). Uniformly labeled ^{14}C-propanil applied to soils was transformed to 3,4-dichloroaniline and the multiple condensation product 4-(3,4-dichloroanilino)-3,3',4'-trichloroazobenzene which accumulated to 2% of the substrate (22). At high concentration (500 mg/L), the dichloroaniline was volatile, while at lower concentrations (5 to 10 mg/L) dichloroaniline and its condensation product were humic-bound. The aliphatic portion is degraded to CO_2. Studies with ^{14}C(carbonyl)-propanil revealed 70% conversion to $^{14}CO_2$ within 25 days, while soils amended with ^{14}C(ring)-propanil yielded only 3% $^{14}CO_2$ during the same time period (87). *Pseudomonas* sp. strain G also converts propanil to 3,4-dichloroaniline (488).

Propanil sprayed onto flooded rice plots was dissipated quickly and disappeared within 24 hours (112). The major metabolite was 3,4-dichloroaniline which sorbed to soils. Only a trace of 3,3',4,4'-tetrachloroazobenzene was detected, although dilution caused by the flooding may have precluded condensation of the dichloroaniline.

The amount of tetrachloroazobenzene formed in soils as a result of propanil or 3,4-dichloroaniline application was found to be highly variable (213). In a comparison of nine soils, those at a pH of 4.5 to 5.5 showed the most production. Tetrachloroazobenzene formation was not correlated with the organic matter content of the soils, and air-dried soil samples showed 87 to 99% reduction in product formation. More tetrachlorozobenzene was produced from direct application of dichloroaniline than from molar equivalent application of propanil. In a contradictory study, however, more tetrachloroazobenzene was produced from propanil than from the equivalent application of 3,4-dichloroaniline (58).

The degradation rates of herbicides in Nixon sandy loam have been correlated with the number of carbon atoms in the side chain (23). The four-carbon molecule dicryl is metabolized at a slower rate than propanil, and the six-carbon herbicide N-(3,4-dichlorophenyl)-2-methylpentanamide is the most persistent of the three. Each is converted to 3,4-dichloroaniline, 3,3',4,4'-tetrachloroazobenzene and another metabolite.

MISCELLANEOUS PESTICIDES

Chlordimeform [N'-(4-chloro-2-methylphenyl)-N,N-dimethylmethanimidamide] represents another class of formamide compounds used as insecticides. Chlordimeform is metabolized in sandy loam to several products within 90 days (220). These include 4-chloro-6-nitro-o-toluidine and 4-chloro-o-toluidine (Figure 55). The latter product is condensed to form 4,4'-dichloro-2,2-dimethylazobenzene. Other

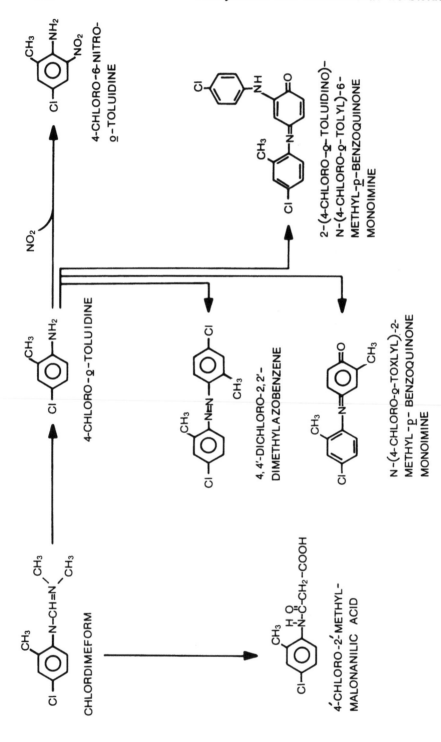

Figure 55 Metabolism of chlordimeform in soils. (Adapted from Refs. 220,378.)

coupling products which were detected include N-(4-chloro-o-tolyl)-2-methyl-p-benzoquinone monoimine and 2-(4-chloro-o-toluidino)-N-(4-chloro-o-tolyl)-6-methyl-p-benzoquinone monoimine. These are formed by a one-electron oxidation mediated by peroxidases, and appearance of these products is pH-dependent.

A mixed population of soil microorganisms mediated a novel conjugation of the aniline moiety of chlordimeform with malonic acid to form 4'-chloro-2'-methylmalonanilic acid (Figure 55) (378). This mechanism was previously found only in plants as a means for detoxification of certain D-amino acids.

The herbicide N-(1,1-dimethylpropynyl)-3,5-dichlorobenzamide underwent extensive metabolism of the side chain during 90 days' incubation in soils (481). However, no alteration of the chlorinated ring structure was noted. The cyanobacterium *Oscillatoria* sp. metabolized N'-(4-chloro-o-tolyl)-N,N-dimenthylformamidine with production of $^{14}CO_2$ from either tolyl-^{14}C- or ring-^{14}C-labeled substrate (34). Extensive nonbiological degradation of this compound occurs but no evolution of $^{14}CO_2$ in the absence of the algae was noted.

Techlofthalam [N-(2,3-dichlorophenyl)-3,4,5,6-tetrachlorophthalamic acid] is a bactericide used on rice plants (252). An analysis of its fate under flooded soil conditions analogous to those of rice paddy fields indicated that after 32 weeks most of the recoverable radiolabeled material was isolated as two or more products chlorinated in the tetrachlorophthalamic ring. Nine percent of the carboxyl-labeled material was converted to $^{14}CO_2$. Tetrachlofthalam was recovered as a minor metabolite. No further transformations occurred during the 32 weeks of the experiments.

Metabolism of chlomethoxynil (2,4-dichlorophenyl-3'-methoxy-4'-nitrophenyl ether) in flooded soil resulted in production of a number of compounds due to alteration of the molecule without loss of chloride (328). Cleavage of the ether bond results in production of 2,4-dichlorophenol.

Production of amino derivatives from other diphenyl ether herbicides was shown to be faster in flooded than in moist soils (329). A number of bacteria including those from the genera *Bacillus*, *Pseudomonas*, *Enterobacter* and *Escherichia* mediated this conversion, although disappearance of the herbicides was noted in sterile soils as well. The herbicides tested included nitrofen (2,4-dichlorophenyl 4'-nitrophenyl ether), 2,4,6-trichlorophenyl 4'-nitrophenyl ether, 2,4-dichloro-6-fluorophenyl 4'-nitrophenyl ether, and chlomethoxynil.

SUMMARY

The herbicides described here were formulated with degree of persistence after application as a prime consideration. All of these

compounds undergo primary degradation readily in soils and in a
variety of microbial cultures. The resulting products are chlorinated
anilines and other metabolites arising from the aliphatic side chain.
When the side chain is not chlorinated, it is metabolized to cell constituents and CO_2. Little research has been published regarding
the fate of chlorinated or other highly substituted side chain ma-
tabolites.

The bulk of the herbicides remain as chlorinated anilines. There
is some evidence for volatilization of these compounds in arid zones
if sorption to soils is delayed or if the concentration of chloroanilines is very high. However, the chlorinated anilines are readily
and strongly bound to soil humic substances. Monosubstituted
anilines, particularly 3-chloroaniline and 4-chloroaniline, are utilizable as sole carbon sources by some microorganisms and are metabolized in laboratory experiments. Dichloroaniline is toxic to
microorganisms at low concentrations and evidence for its degradation is indirect. It is not clear how sorption to soils affects degradation of the chloroanilines in the environment.

Chlorinated anilines are metabolized by dioxygenases to chlorocatechols. Organisms which convert the chlorinated anilines to
chlorocatechols and then also metabolize the chlorocatechols do so
by the *meta* pathway. Limited evidence suggests that the amine
group is removed directly without oxidation.

Under some conditions azobenzenes are formed from condensation
of chloroanilines via peroxidases. The concentration of chloroanilines
must be very high for this mechanism to be operative in soils.

Primary degradation of the urea herbicides occurs by successive
demethylation of the side chain. The side chain is then metabolized
leaving the chloroaniline moiety. Methoxy phenyl ureas are metabolized by a different mechanism, as microorganisms active against
these compounds have little activity against the dimethyl herbicides.

Reductive dechlorination to a limited extent of the aromatic
moiety has been reported. However, the effectiveness of this mechanism in the environment has not been investigated. Anaerobic utilization of 4-chloroaniline takes place readily, but the resultant
volatile product or products have not been identified.

The primary mechanism of degradation of the phenyl carbamate
herbicides is hydrolysis of the ester linkage to form chlorinated
anilines. These compounds are metabolized by a wide variety of
fungi and bacteria as noted above.

Hydrolysis of the acyl anilide herbicides results in formation of
the chlorinated anilines with utilization of the aliphatic side chain
for cell growth. An unusual mechanism employed by *R. japonicus*
results in hydrolysis of the side chains of dicryl and N-(3,4-dichloro-
phenyl)-2-methyl pentanamide. This modification results in detoxification of the molecules.

Studies with various species of algae have shown that these organisms metabolize acyl anilide compounds with production of chlorinated anilines.

Metabolism of these classes of herbicides is dependent upon factors which affect microbial activity, including such factors as pH and soil composition which affect the chemical sate of the compound as well. These compounds are relatively easily broken down to chlorinated anilines and side chain metabolites. The chlorinated anilines may be volatilized or bound to soils before or concomitantly with microbial metabolism. It is not clear what effects these competing processes have on biodegradation of these molecules. There have been few studies on the persistence or metabolism of compounds arising from the side chain, particularly those containing chloride ions.

11

Chlorinated Biphenyls

Polychlorinated biphenyls (PCBs) have been used widely in industrial applications because of their thermal stability, excellent dielectric (electrically insulating) properties, and resistance to oxidation, acids, bases, and other chemical agents. PCBs therefore have found use in capacitors and transformers as dielectric fluids, in hydraulic systems, gas turbines and vacuum pumps, and as fire retardants and plasticizers (215). In 1971, however, Monsanto Company, the sole U.S. producer, voluntarily restricted the use of PCBs to closed systems (capacitors, transformers, vacuum pumps, gas-transmission turbines) and discontinued production entirely in 1978 (281). These applications use complex mixtures of PCBs marketed under the trade names Aroclor (Monsanto Company, USA), Clophen (Germany), Phenoclor and Pyralene (France), Kaneclor and Santotherm (Japan), and Fenclor (Italy). Askarels are synthetic mixtures of chlorinated biphenyls and trichlorobenzenes.

The Aroclor products are denoted by a four-digit number in which the first two indicate the type of molecule ("12" indicates a chlorinated biphenyl) and the last two digits indicate the weight percent chlorine. Aroclor 1254 consists of chlorinated biphenyls with 54% by weight chlorine and on average five chlorines per molecule, although it has been reported to contain 69 different chlorinated biphenyl molecules (162). Similarly, Aroclor 1242 is 42% chlorine by weight and averages three chlorines per molecule. Aroclor 1016 also consists primarily of trichlorinated compounds but contains fewer penta- and hexachlorinated molecules than Aroclor 1242. There are 210 possible PCB compounds containing 0 to 10 chlorine atoms per biphenyl molecule. However, many of these have never been found in commercial PCB mixtures.

The PCBs have been released into the environment for many years and are a worldwide contaminant. They are lipophilic and sorb strongly to the lipids and fats of animals including fish, mussels, and birds. PCBs also undergo biological magnification in such common aquatic invertebrates as daphnids, mosquito larvae, stoneflies and crayfish. The concentration of PCBs in the invertebrates can be as high as 27,500 times that in water (162). These invertebrates subsequently are eaten by fish and birds, and bioaccumulation occurs at all levels of the food chain.

MICROBIAL METABOLISM OF PCBs

Most of the studies on microbial metabolism of PCBs have explored the biodegradability of the Aroclors in natural environments or in laboratories using pure strains or mixed cultures. These studies have shown that PCBs containing fewer than five chlorines per molecule are extensively degraded, while heavier molecules tend to persist in the environment (26, 91, 163, 167, 220a, 281, 394, 438, 474). These studies are corroborated by environmental analyses which indicate that PCBs found in weathered samples contain five or more chlorine atoms per molecule (162).

Few studies have attempted to elucidate the pathways of degradation of pure compounds of a chlorinated biphenyl. Several studies have shown that 4-chlorobiphenyl can be metabolized to 4-chlorobenzoic acid, indicating hydroxylation, ring cleavage, and degradation of the nonchlorinated ring of the molecule. This has been demonstrated with soil bacteria, a sewage effluent isolate identified as *Achromobacter* sp. pCB, an unidentified facultative anaerobe called strain B206, *Acinetobacter* sp. P6, and *Alcaligenes* sp. Y42 (1, 163, 164, 333, 426). *Acinetobacter* sp. P6 can use 4-chlorobiphenyl as the sole source of carbon for growth (164). Growth of both *Achromobacter* sp. strain B 218 and *Bacillus brevis* strain B 257 on 4-chlorobiphenyl as the sole source of carbon generates the same metabolites (298). The pathway of degradation involves formation of a 2,3-dihydroxy intermediate with *meta* cleavage to form eventually 4-chlorobenzoic acid. Other metabolites were isolated which represent successive oxidation and utilization of the aliphatic carbons from the cleaved ring (Figure 56).

The formation of chlorinated benzoic acids from chlorinated biphenyls is the most common route of PCB degradation. Both *Alcaligenes* sp. Y42 and *Acinetobacter* sp. P6 convert a large number of biphenyl compounds to the corresponding chlorinated benzoic acids (Table 5) (163, 168). For several compounds containing multiple chlorines with one on the second ring, loss of that chlorine occurs in the formation of the chlorobenzoic acid. Studies with ^{14}C-2,5,2'-trichlorobiphenyl confirmed the formation of ^{14}C-2,5-dichlorobenzoic acid and a yellow

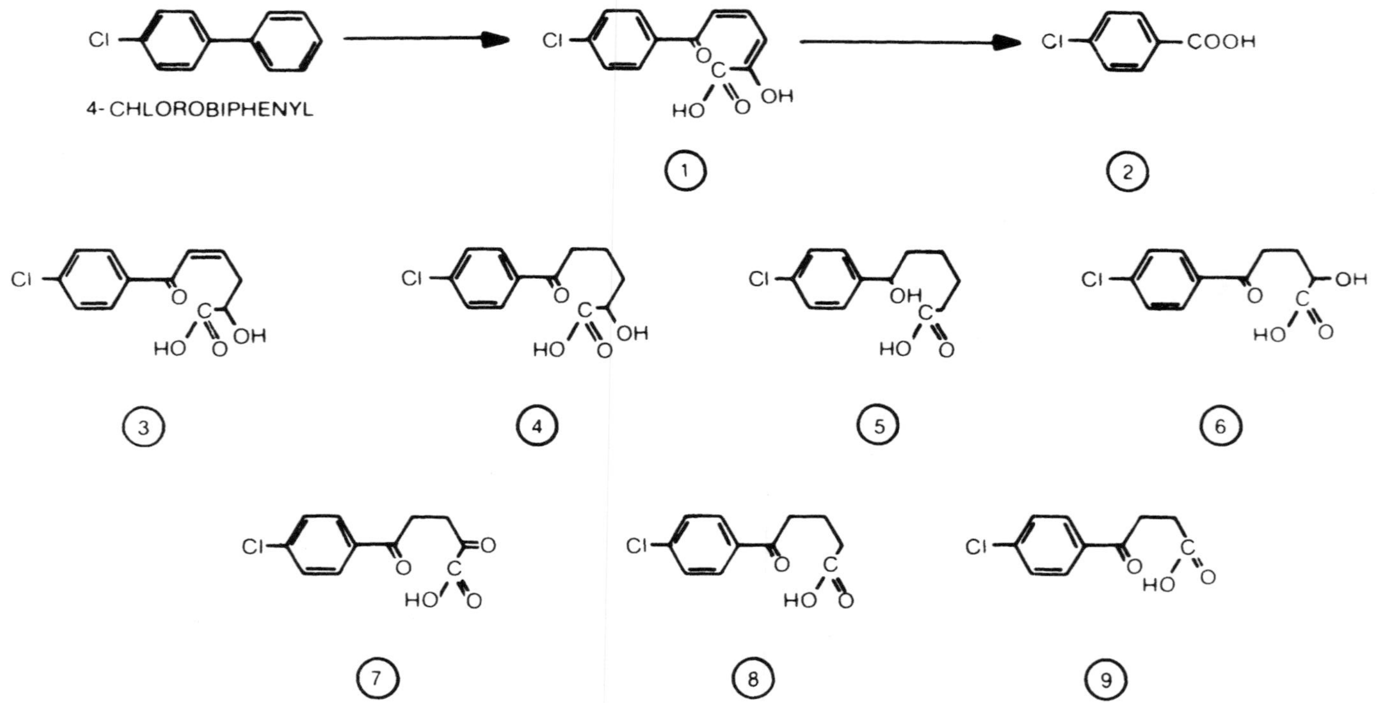

Figure 56 Microbial degradation of 4-chlorobiphenyl. Metabolites: (1) 2-hydroxy-6-oxo-6-(4'-chlorophenyl)-hexa-2,4-dienoic acid; (2) 4-chlorobenzoic acid; (3) 2-hydroxy-6-oxo-6-(4'-chlorophenyl)-4-hexenoic acid; (4) 6-oxo-6-(4'-chlorophenyl)-2-hydroxyhexanoic acid; (5) 6-hydroxy-6-(4'-chlorophenyl)hexanoic acid; (6) 2-hydroxy-5-oxo-5-(4'-chlorophenyl)pentanoic acid; (7) 2,5-dioxo-5-(4'-chlorophenyl)pentanoic acid; (8) 5-oxo-5-(4'-chlorophenyl)pentanoic acid; (9) 4-oxo-4-(4'-chlorophenyl)butanoic acid. (Adapted from Ref. 298.)

Table 5 Metabolism of Chlorinated Biphenyl Compounds by *Alcaligenes* Sp. Y42 and *Acinetobacter* Sp. P6

Substrate	Products
Substitutions on one ring	
2-Chlorobiphenyl	2-Chlorobenzoate
3-Chlorobiphenyl	3-Chlorobenzoate
4-Chlorobiphenyl	4-Chlorobenzoate
2,3-Dichlorobiphenyl	2,3-Dichlorobenzoate
2,4-Dichlorobiphenyl	2,4-Dichlorobenzoate
2,5-Dichlorobiphenyl	2,5-Dichlorobenzoate
2,6-Dichlorobiphenyl[a]	2,6-Dichlorodihydroxybiphenyl, 2,6-dichlorotrihydroxybiphenyl
3,4-Dichlorobiphenyl	3,4-Dichlorobenzoate
3,5-Dichlorobiphenyl	3,5-Dichlorobenzoate
2,3,4-Trichlorobiphenyl	2,3,4-Trichlorobenzoate
2,3,6-Trichlorobiphenyl[a]	2,3,6-Trichlorodihydroxybiphenyl, 2,3,6-dichlorotrihydroxybiphenyl
2,4,5-Trichlorobiphenyl	2,4,5-Trichlorobenzoate
2,4,6-Trichlorobiphenyl	2,4,6-Trichlorobenzoate[b], 2,4,6-trichlorodihydroxybiphenyl,[a] 2,4,6-trichlorotrihydroxybiphenyl[a]
2,3,4,5-Tetrachlorobiphenyl	2,3,4,5-Tetrachlorobenzoate
2,3,5,6-Tetrachlorobiphenyl	None
2,3,4,5,6-Pentachlorobiphenyl	None
Substitutions on both rings	
2,2'-Dichlorobiphenyl	2-Chlorobenzoate
2,4'-Dichlorobiphenyl	2-Chlorobenzoate
3,3'-Dichlorobiphenyl	3-Chlorobenzoate
4,4'-Dichlorobiphenyl	4-Chlorobenzoate
2,4,4'-Trichlorobiphenyl	2-Chlorobenzoate, 2,4-dichlorobenzoate
2,5,2'-Trichlorobiphenyl	2-Chlorobenzoate, 2,5-dichlorobenzoate

Table 5 Continued

Substrate	Products
2,5,3'-Trichlorobiphenyl	3-Chlorobenzoate, 2,5-dichloroben-zoate
2,5,4'-Trichlorobiphenyl	2-Chlorobenzoate, 2,5-dichloroben-zoate
3,4,2'-Trichlorobiphenyl	2-Chlorobenzoate
2,3,2',3'-Tetrachlorobiphenyl	2,3-Dichlorobenzoate, an unidentified dichloro compound
2,3,2',5'-Tetrachlorobiphenyl	Dichlorobenzoate, an unidentified dichloro compound
2,4,2',4'-Tetrachlorobiphenyl[a]	2,4-Dichlorobenzoate
2,4,2',5'-Tetrachlorobiphenyl[a]	Dichlorobenzoate
2,4,3',4'-Tetrachlorobiphenyl[a]	Dichlorobenzoate
2,5,2',5'-Tetrachlorobiphenyl[a]	2,5,2',5'-Tetrachlorodihydroxy-biphenyl
2,5,3',4'-Tetrachlorobiphenyl[a]	Dichlorobenzoate
2,6,2',6'-Tetrachlorobiphenyl[a]	None
3,4,3',4'-Tetrachlorobiphenyl[a]	3,4-Dichlorobenzoate
2,4,5,2',3'-Pentachlorobiphenyl[a]	2,4,5-Trichlorobenzoate, an unidentified trichloro compound
2,4,5,2',5'-Pentachlorobiphenyl[a]	2,4,5,2',3'-Tetrachlorodihydroxy-benzoate

[a]Metabolism by *Acinetobacter* sp. P6 only.
[b]Metabolism by *Alcaligenes* sp. Y42 only.
Source: Adapted from Refs. 163, 164, 168.

intermediate by resting cell suspensions of both *Alcaligenes* sp. Y42 grown on biphenyl and *Acinetobacter* sp. P6 grown on 4-chlorobiphenyl (164). The patterns of metabolism of PCBs by *Alcaligenes* sp. Y42 and *Acinetobacter* sp. P6 are similar. The general path of degradation proceeds through *meta* cleavage compounds to chlorobenzoic acids which accumulate during the metabolism of chlorobiphenyls. Metabolism of some chlorinated biphenyls is blocked after production of the dihydroxy intermediate (precursor to ring cleavage), while for other compounds the *meta*-cleavage intermediate accumulates.

Pseudomonas sp. strain 7509 grown on biphenyl metabolizes 2,4'-dichlorobiphenyl to two different monochlorobenzoates, indicating that both rings are capable of being attacked (26a). An intermediate metabolite was identified as 2-hydroxy-6-oxo-6-(chlorophenyl)chlorohexa-2,4-dienoic acid.

The degradation of 2,4,4'-trichlorobiphenyl by *Acinetobacter* sp. P6 grown on 4-chlorobiphenyl was studied in detail and a pathway was proposed involving *meta* cleavage after 2',3'-hydroxylation (Figure 57) (166). Hydroxylation occurs on the ring with the fewest chlorine substituents. The predominant metabolite is the *meta* cleavage product, although small amounts of dichlorobenzoic acid appear. The occurrence of a yellow *meta* cleavage product and subsequent production of chlorobenzoates in the degradation of other PCBs suggests that this may be a general pathway for most PCB metabolism in bacteria.

On the basis of these studies, several generalizations were made regarding the effect of the structure of the PCBs on microbial degradation (163, 164, 165, 168). Degradation decreases as the number of chlorines per molecule increases. Two chlorines on the *ortho* positions of a single ring (i.e., 2,6-) or on both rings (i.e., 2,2'-) inhibit degradation. PCBs with one unsubstituted ring are more readily metabolized than PCBs with the same number of chlorines on both rings. On PCBs with unequally substituted rings, the ring with fewer substitutions is preferentially cleaved. PCBs with a chlorine on the 4'-position are metabolized to stable *meta*-cleavage products.

The occurrence of nitro-containing metabolites in extracts of media containing the unidentified facultative anaerobe B 206 and 4-chlorobiphenyl was investigated (427). When ammonium sulfate is added to the culture medium as the nitrogen source, 4-chlorobiphenyl is metabolized to 2-hydroxy-4'-chlorobiphenyl and 4-hydroxy-4'-chlorobiphenyl. However, when the nitrogen source is sodium nitrate, the metabolites 2-hydroxy-nitro-4'-chlorobiphenyl and 4-hydroxy-nitro-4'-chlorobiphenyl appear. These metabolites were interpreted to result from a nonenzymatic reaction between an arene oxide intermediate and nitrate or nitrite anions. This organism subsequently accumulates 4-chlorobenzoic acid in the medium (426). The appearance of the monohydroxy intermediates suggests a rare monooxygenase mechanism for PCB degradation, although the phenylphenols may also be an artifact arising during the isolation procedure (71a).

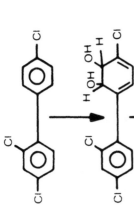

Figure 57 Pathway of 2,4,4'-trichlorobiphenyl metabolism by *Acinetobacter* sp. P6. (Adapted from Ref. 166.)

2. 4. 4'-TRICHLOROBIPHENYL

1-CHLORO-2. 3-DIHYDROXY-4-(2 4-DICHLOROPHENYL)-HEXA-4. 6-DIENE

2. 4. 4'-TRICHLORO-2'. 3'-DIHYDROXYBIPHENYL

3-CHLORO-2-HYDROXY-6-OXO-6-(2. 4-DICHLOROPHENYL) HEXA-2 4-DIENOIC ACID

2. 4-DICHLOROBENZOIC ACID

Acinetobacter sp. strain P6 resting cells were incubated for 4 hours with several Kaneclor PCB mixtures (167). Kaneclor KC200 (primarily dichlorobiphenyls) was metabolized to monochlorobenzoates. Kaneclor KC300 was metabolized to benzoates with one to three chlorines, dihydroxybiphenyls with two to four chlorines, ring *meta* cleavage products with two to three chlorines, and many other unidentified compounds with two chlorines. Kaneclor KC500 was scarcely metabolized, although some dihydroxy isomers were noted.

Studies were conducted utilizing *Alcaligenes* sp. strain BM 2 which was isolated on diphenylmethane and known to metabolize dichlorinated biphenyls (474). A mixture of di- and trichlorinated biphenyls at 0.05% concentration (32% by weight) was 80% metabolized in 1 day and completely metabolized in 3 days. At 0.25% concentration, 22% was metabolized in 1 day and 29% in 3 days. Under cometabolic conditions, a 100 mg/L PCB mixture of di-, tri-, and tetrachlorobiphenyls (41% by weight) was 70% metabolized in 2 days and 80% in 6 days. In a minimal medium with only a small quantity of carbon source, 30% was metabolized in 6 days. Most of the remaining substrate was tetrachlorobiphenyl. The metabolites included mono-, di-, and trichlorobenzoates, monohydroxychlorobiphenyl, 2-hydroxy-6-oxo-chlorophenylhexa-2,4-dienoic acid, chlorobenzoylpropionic acid, chlorophenylacetic acid, and 3-chlorophenyl-2-chloropropenoic acid (a substituted cinnamic acid).

Pseudomonas sp. strain 7509 also formed mono- and dichlorobenzoic acids during metabolism of Aroclor 1242 (26a). A number of non-chlorinated aromatic and aliphatic compounds were isolated after 2 months' incubation of Aroclor 1242 with several strains of bacteria isolated from lake water (233). No chlorinated metabolites or oxidized derivatives of PCBs were detected.

The disappearance of Aroclor products and individual PCB isomers during incubation with *Nocardia* sp. NCIB 10603 was monitored (26). The following isomers were 60 to 100% metabolized within 2 weeks: 2,4'-di-, 2,3-di-, 3,4-di-, 2,3,2'-tri-, 2,3,4'-tri-, and 3,4,3'-trichlorobiphenyl, while 2,5,4'-trichlorobiphenyl was 60% metabolized in 73 days. There was little transformation of 2,4,6-tri-, 2,4,2',4'-tetra-, or 2,4,6,2'-tetrachlorobiphenyl in 9 days, and no degradation of 4,4'-dichlorobiphenyl was detected after 121 days. However, 4,4'-dichlorobiphenyl was 50% metabolized in 2 days when present as a component of Aroclor 1242. During 52 days' incubation Aroclor 1242 was 88% metabolized and in 100 days was 95% metabolized. Aroclor 1016 was 96% metabolized in 52 days.

There are two reports in the literature concerning the degradation of PCBs by fungi. *Rhizopus japonicus* converts 4-chlorobiphenyl to 4-chloro-4'-hydroxybiphenyl and 4,4'-dichlorobiphenyl to an unidentified hydroxylated metabolite (454). *Cunninghamella echinulata* Thaxter metabolizes 2,5-dichloro-4'-isopropylbiphenyl by oxidation

of the isopropyl group to form 2,5-dichloro-4'-biphenylcarboxylic
acid and by hydroxylation of the chlorine-substituted phenyl group
(440).

METABOLISM OF PCBs BY MIXED MICROBIAL CULTURES

A mixed microbial population in lake water metabolized 2-chlorobi-
phenyl to 2-chlorobenzoic acid and chlorobenzoylformic acid (394).
In contrast, 2,4'-dichlorobiphenyl was not metabolized after 8 months'
incubation.

A mixed microbial culture derived from river sediments was able
to metabolize 4-chlorobiphenyl rapidly, with 99% removal in 30 days
(265). Acclimated and nonacclimated cultures showed similar results.
There was transitory formation of a metabolite thought to be 2-hy-
droxy-6-oxo-6-(4-chlorophenyl)hexa-2,4-dienoic acid as well as production of 4-chlorobenzoic acid. The substrate was ^{14}C-labeled on
the chlorinated ring only and production of $^{14}CO_2$ was noted after
4-chlorobenzoic acid formation. The substrates 2-chloro- and 3-
chlorobiphenyl were also metabolized, but 2-chlorobenzoic acid was
not degraded further.

A marine mixed microbial community metabolized all three monochlorinated biphenyls (52). Metabolites were not identified.

A river water die-away test demonstrated 50% removal of 1 to 100
mg/L 2-chloro-, 3-chloro-, or 4-chlorobiphenyl within 2 to 5 days
(17). The compound was uniformly ^{14}C-labeled in the chlorinated
ring and up to 50% of the label appeared as monochlorobenzoic acid
and subsequently as $^{14}CO_2$. No degradation of 2,2',4,4'-tetrachloro-
biphenyl was noted after 98 days' incubation.

A mixture of bacteria was isolated by enrichment culture with
garden soil using benzene as the substrate (18). Several PCB isomers were incubated with the mixed culture and benzene for up to
6 weeks and the medium subsequently tested for presence of chlorobenzoate metabolites. Neither 2-chloro- nor 2,2'-dichlorobiphenyl
was metabolized to chlorobenzoic acids, although 4-chlorobiphenyl
formed 4-chlorobenzoic acid. The substrate 2,4,4'-trichlorobiphenyl
formed copious amounts of 4-chloro- and 2,4-dichlorobenzoic acid in
the ratio 5:2. Loss of one chloride was noted in the formation of 4-
chlorobenzoic acid from 2,4'-dichlorobiphenyl, 2,5-dichlorobenzoic
acid from 2,2',5-trichlorobiphenyl, 4-chlorobenzoic acid and 3,4-
dichlorobenzoic acid from 3,4,4'-trichlorobiphenyl, 4-chlorobenzoic
acid and 2,4-dichlorobenzoic acid from 2,4,4'-trichlorobiphenyl, and
2,3,4,5-tetrachlorobenzoic acid from both 2,3,3',4,5-pentachloro-
biphenyl and 2,3,4,4',5-pentachlorobiphenyl.

A soil plot was treated with 1 ppm 2,2'-dichlorobiphenyl (317).
After one growing season almost half the remaining material was unchanged substrate. About 9% were soluble metabolites and almost

42% were unextractable residues. After 1 year, 74% of the remaining material were unextractable residues. The metabolites included monohydroxy derivatives of the substrate as well as other products.

Metabolism of 4,4'-dichlorobiphenyl by a mixed microbial culture obtained from the filtrate of an activated sludge sample resulted in formation of 4,4'-dichloro-2,3-hydroxybiphenyl and 4-chlorobenzoic acid (439). Metabolism of this substrate is repressed by the presence of alternative carbon sources. Under similar experimental conditions, including an incubation time of 14 days, there was no metabolism of 2,4,5'-tri-, 2,2',5,5'-tetra-, 2,2',3,4,5'-penta-, 2,2',3,4,5,5'-hexa-, and decachlorobiphenyl.

Metabolism of 2,4,2',5'-tetrachlorobiphenyl and 2,5,2'-trichloro-biphenyl occurred in seawater with production of a compound thought to be a lactone acid (68a). No degradation was noted during the incubation of the tetrachlorobiphenyl with anaerobic marsh mud during a 45-day incubation period.

A microbial consortium obtained from activated sludge metabolized the isopropyl group of 4-chloro-4'-isopropylbiphenyl to a hydroxyl substituent, forming 4-chloro-4'-hydroxybiphenyl, followed by formation of 4-chlorobenzoic acid (440). Intermediates of the isopropyl metabolism pathway were identified. Addition of glucose as an alternate carbon source repressed metabolism of the chlorinated substrate.

Biphenyl enrichment of both uncontaminated soils and soils contaminated with PCBs resulted in isolation of mixed cultures which were incubated with Aroclor 1242 (91). Cometabolism with sodium acetate enhanced metabolism of all the PCBs including the higher chlorinated molecules, although this phenomenon may be due to the increased biomass resulting from growth on the simpler carbon source. Extensive degradation of all the lower chlorinated isomers was noted with up to 68% metabolism of the tetrachlorinated biphenyls in 15 days.

A related study was conducted to determine the influence of inoculum concentration on the aerobic bio-oxidation of 3,3'-dichlorobenzidine which is used in the manufacture of azo dyes (57). The effluent from a domestic sewage treatment plant was used as the inoculum and the substrate concentration was 20 mg/L. When present as the sole carbon source, the substrate was not metabolized. However, the presence of yeast extract in the medium promoted extensive disappearance of the substrate within 28 days. In another experiment, 2 mg/L 3,3'-dichlorobenzidine was added to lake or reservoir water and incubated 14 days (10). Neither metabolites nor $^{14}CO_2$ was recovered after incubation. Increasing disappearance of substrate with time was correlated with increasing biomass, which served as a sorbent for the substrate. The supernatant fluid from settled activated sludge material served as the inoculum for flasks containing 3,3'-dichloro-benzidine and additional carbon sources. After 4 repeated weekly subcultures to flasks of fresh media, no metabolites of the substrate were recovered.

SUMMARY

The limited number of studies on the degradation of specific chlorinated biphenyl compounds by pure strains of bacteria has served to establish some general features of PCB metabolism. A few strains of bacteria have been shown to mineralize some chlorinated biphenyls. In most cases bacteria with the capability of degrading one ring of a chlorinated biphenyl compound are unable to degrade the resulting chlorinated benzoates. These compounds accumulate in pure cultures. Evidence exists for the complete mineralization of chlorinated benzoates by other strains of bacteria (discussed in Chapter 5 on chlorobenzoates), and mixed cultures of bacteria have been shown to mineralize PCBs with four or fewer chlorines per molecule. More heavily substituted PCBs appear to resist degradation and accumulate even in environments where less chlorinated PCBs are degraded.

The mechanism of hydroxylation of PCBs by bacteria has not yet been elucidated, nor have the enzymes mediating the steps in the proposed pathways been isolated. Two pathways have been proposed, the first analogous to the pathway of degradation of biphenyl. Initial hydroxylation occurs in the 2,3-position of the less substituted ring, followed by *meta* cleavage and subsequent degradation of the aliphatic portion of the molecule to form substituted benzoic acids. Chlorines on the aliphatic carbons are lost during this process. However, this may not be the mechanism for degradation of PCBs substituted in all the *ortho* positions. A second pathway based on presence of a monooxygenase in bacteria has been proposed after discovery of 4-hydroxy-4'-chlorobiphenyl in extracts of bacterial cultures incubated with 4-chlorobiphenyl. More evidence corroborating this mechanism needs to be obtained to determine how widespread this pathway is. Limited evidence on fungal metabolism of PCBs indicates activity of a monooxygenase in a manner similar to that shown in biphenyl metabolism.

12

DDT and Related Compounds

DDT, 1,1,1-trichloro-2,2-bis(p-chlorophenyl)ethane, received its abbreviation from the trivial name dichlorodiphenyltrichloroethane. This chlorinated aromatic is one of the most persistent pesticides in the environment. Since it is lipophilic, it readily accumulates in microorganisms and invertebrates and undergoes biomagnification as fish and birds higher in the food chain ingest DDT-contaminated organisms (79, 229, 263, 311). Widespread effects of DDT poisoning include eggshell thinning and birth defects. Reviews of the literature pertaining to DDT metabolism in microbial systems have been published in 1976 and 1980 (138, 228).

BACTERIAL METABOLISM OF DDT

Due to the widespread incidence of toxicity demonstrated after DDT ingestion, the intestinal flora of various animals became the focus for studies of the metabolism of DDT. DDT is converted directly to DDD (1,1-dichloro-2,2-bis[p-chlorophenyl]ethane, also referred to as di-chlorodiphenyldichloroethane) (Figure 58). The reaction involves removal of a chloride ion from the aliphatic portion of the molecule. This is a reductive dechlorination reaction requiring anaerobic conditions, in contrast to the oxidative pathways of metabolism of most other pesticides.

The involvement of intestinal bacteria in DDT metabolism by mammals has been demonstrated in experiments which showed that rats converted DDT to DDD when fed by stomach tube, but not when injected intraperitoneally (308). The coliform bacteria *Escherichia coli* and *Enterobacter aerogenes* isolated from rat feces demonstrate

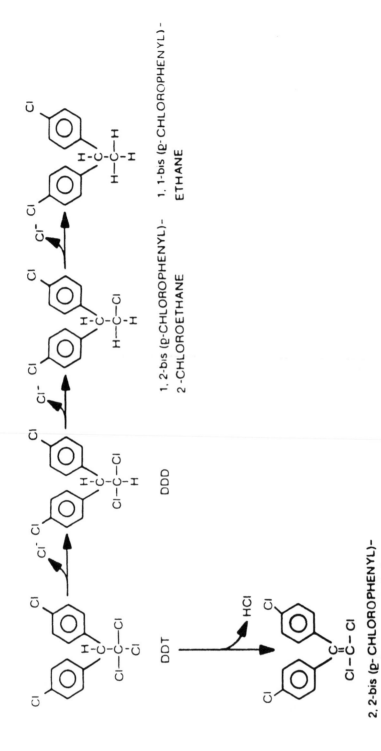

Figure 58 Reductive dechlorination of DDT and dehydrochlorination of DDT.

154

this reaction as well. Microorganisms isolated from rat intestines which convert DDT to DDD include *Clostridium perfringens*, *Streptococcus* sp., *Bacteroides* sp., *E. coli* and other coliforms, yeasts, and to a lesser extent *Lactobacillus* sp. (55). *Klebsiella pneumoniae* also converts DDT to DDD (464). DDE [1,1-bis(p-chlorophenyl)-2-chloroethylene] can be detected after 20 hours at concentrations from 5 to 10% in cultures of *Streptococcus* sp., *Bacteroides* sp., *Pseudomonas* sp., and *Lactobacillus* sp. (the latter after a 72-hour incubation) (55). However, DDE has been shown to be produced nonenzymatically as well as enzymatically (464).

Proteus vulgaris, isolated from the intestines of DDT-resistant mice, converts DDT to DDD and subsequently to 1,1-bis(p-chlorophenyl)-2-chloroethane and 1,1-bis(p-chlorophenyl)ethane, representing three successive reductive dechlorinations (Figure 58) (20, 21). DDE also appears in the medium. The excreta of stable flies became the source of three bacteria, *E. coli*, *Serratia marcescens*, and a third unidentified strain, which convert DDT to DDD (90%) and DDE (5%) after 24 to 72 hours anaerobically but not aerobically (415). Bacterica from bovine rumen fluid convert ^{14}C-DDT to ^{14}C-DDD (311). This same reaction was noted for DDT incubation with water from Clear Lake, California and with reduced iron porphyrins (hemoglobin or hematin) (311). The isomer o,p'-DDT which constitutes to 15 to 20% of technical grade DDT is converted by rumen microorganisms to o,p'-DDD at the same rate as p,p'-DDT (161). This conversion occurs was well in *E. aerogenes* both aerobically and anaerobically (309).

The direct conversion of DDT to DDD without the DDE intermediate was confirmed in *E. aerogenes* using deuterated DDT (355). The deuterium atom present at the 2-position in DDT is retained in the product, indicating that the chlorine is replaced by hydrogen (or hydrogen ion) directly without the intermediary species. The membranes of *E. coli* were shown to be the site of reductive dechlorination of this species. The process required flavine-adenine dinucleotide (FAD) and anerobic conditions (160).

Cell-free extracts of *E. aerogenes* also convert DDT to DDD. This activity is due to reduced Iron(II) cytochrome oxidase (464). More complete degradation of DDT occurs in both whole-cell preparations and cell-free extracts of this organism (Figure 59) (463, 465). Metabolism follows the pathway DDT → DDD → DDMU → DDMS → DDNU → DDA → DPM → DBH → DBP, where the abbreviations represent the compounds as follows: (DDMU) 1-chloro-2,2-bis(p-chlorophenyl)ethylene; (DDMS) 1-chloro-2,2-bis(p-chlorophenyl)ethane; (DDNU) unsym-bis(p-chlorophenyl)ethylene; (DDA) 2,2-bis(p-chlorophenyl)acetate or more commonly dichlorodiphenylacetate; (DPM) dichlorodiphenylmethane; (DBH) dichlorobenzhydrol; and (DBP) dichlorobenzophenone. The enzymatic conversion of DDT to DDE is a dead-

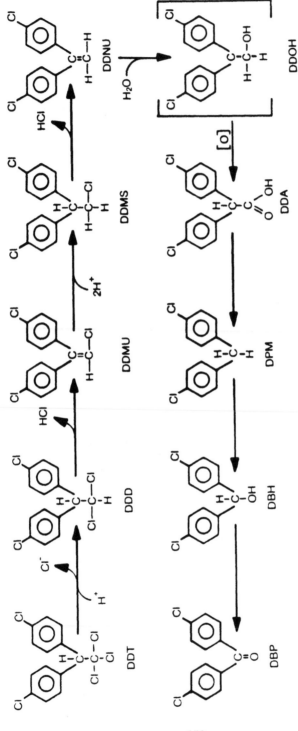

Figure 59 Metabolism of DDT by bacteria. DDT = 1,1,1,-trichloro-2,2-bis (p-chlorophenyl)ethane; DDD = 1,1-dichloro-2,2-bis(p-chlorophenyl)ethane; DDMU = 1-chloro-2,2-bis(p-chlorophenyl)ethylene; DDMS= 1-chloro-2,2,-bis(p-chlorophenyl)ethane; DDNU =unsym-bis(p-chlorophenyl)ethylene; DDOH = 2,2-bis(p-chlorophenyl)ethanol (hypothetical metabolite); DDA = dichlorodiphenylacetate; DPM = dichlorodiphenylmethane; DBH = dichlorobenzhydrol; DBP = dichlorobenzophenone. (Adapted from Refs. 463, 465.)

end side reaction. DDA is the end product of vertebrate metabolism. The conversion of DDA to DBP does not require anaerobic conditions. This pathway has also been demonstrated in anaerobic cultures of *E. coli* (268).

A single study reports that under aerobic conditions, cultures of *Bacillus cereus* metabolize DDT by this pathway within 7 days, although the use of screw-cap flasks in these experiments may have allowed some anaerobiosis to develop. Cultures of *E. coli* incubated aerobically for 24 hours with intermittent shaking converted DDT to DDD (75%) and DDE (25%) (247).

A *Hydrogenomonas* sp. isolated from sewage and grown on diphenylmethane metabolized dichlorophenylmethane to p-chlorophenylacetic acid, indicating ring fission under aerobic conditions (156a). Cell suspensions effected the same transformation (156b). Similarly, cell suspensions incubated with 1,1-diphenyl 2,2,2-trichloroethane promoted the formation of the ring cleavage product 2-phenyl-3,3,3-trichloropropionic acid (156b). Under anaerobic conditions, cell-free extracts of *Hydrogenomonas* sp. metabolized ^{14}C-DDT to DDD, DBP, DDMS, DDMU and DDE (350a). The reaction mixture containing these metabolites was subsequently exposed to aerobic conditions along with a fresh culture of *Hydrogenomonas* sp. A new metabolite identified as p-chlorophenylacetic acid was formed as a result of ring cleavage. This acid was further metabolized by an *Arthrobacter* sp. to p-chlorophenylglycolaldehyde.

Other environmental isolates also have the ability to convert DDT to DDD. Viable cells of *Bacillus megaterium* convert DDT to DDD (201). Three hundred bacterial strains from Lake Michigan each converted DDT to DDD and many converted ^{14}C-DDD to ^{14}C-DDNS (1-bis(p-chlorophenyl)ethane) (302). Fresh water/sediment and sewage ecosystems emended with ^{14}C-DDT promoted formation of DDD, DBP and other metabolites (350a). Chlorophenylacetic acid added to these ecosystems was also metabolized. Bacteria isolated from marine and brackish water and sediment converted ^{14}C-DDT to water-soluble metabolites (232). Forty-seven of one hundred isolates effected 5 to 10% conversion while an additional 38 isolates converted less than 5% of the starting material. Twenty-five isolates did not produce water-soluble metabolites, indirectly indicating that the presence of those metabolites in the other cultures was biologically mediated. Twenty-three of twenty-six plant pathogenic and saprophytic strains of bacteria, representing nine genera, converted DDT to DDD anaerobically (230). Eighteen bacterial strains, mostly *Pseudomonas* spp. which previously had been shown to metabolize dieldrin, also degraded DDT to DDD (349). In addition, 14 of these isolates produced DDA and 10 produced a dicofollike compound.

Pseudomonas aeruginosa 640× isolated from DDT-polluted soil of the Crimean region was used to construct 2 derivatives (185). Strain BS816 carries a plasmid encoding the genes which degrade naphthalene and salicylate by *ortho* cleavage, and strain BS827 carries a plasmid which effects *meta* degradation. Both plasmids were obtained from strains of *P. putida*. The parent *P. aeruginosa* and the two derivatives all metabolize DDT with the formation of the same metabolites. Strain BS816, carrying the plasmid coding for *ortho* cleavage, degrades DDT most extensively, converting 89% of the DDT to DDD, 1,1-dichloro-2,2-bis(*p*-chlorophenyl)ethylene (DDDE), phenylpropionic acid (PPA) and phenylacetic acid (PAA).

An organism identified as a *Pseudomonas* sp. was isolated by enrichment culture for its ability to use diphenylethane as a sole source of carbon and energy for growth (157). The ability of this organism to grow on several other metabolites of DDT was tested. Diphenylethane is converted to 2-phenylpropionic acid and metabolized further. Diphenylmethane is metabolized with intermediate production of phenylacetic acid and 1-(*p*-chlorophenyl)-1-phenylethane is metabolized with the production of 2-(*p*-chlorophenyl)propionic acid as the only metabolite, indicating cleavage of the unsubstituted ring. The DDE analog 1-(*p*-chlorophenyl)-1-phenylethene is metabolized to 2-(*p*-chlorophenyl)-2-propenoic acid. The substrate 1-(*p*-chlorophenyl)-1-phenylethanol is converted to 2-(*p*-chlorophenyl)-2-hydroxypropionic acid which is slowly metabolized further. During metabolism of this substrate minor amounts of the nonchlorinated metabolite 2-hydroxy-2-phenylpropionic acid are produced, indicating attack on the chlorinated ring. Each of the above substrates served as the sole source of carbon and energy for growth. The compound 2,2-diphenylethanol is not metabolized. Analogs of DDT which have chlorine substituents on both rings are not metabolized, and 1,1-diphenyl-2,2,2-trichloroethane is not metabolized. However, when diphenylethane is available in the medium, this organism cometabolizes bis(*p*-chlorophenyl)methane to *p*-chloropehnylacetic acid which accumulates (158). With the same cosubstrate, the organism cometabolizes 1,1-bis(*p*-chlorophenyl)ethane to 2-(*p*-chlorophenyl)-propionic acid with transient appearance of two hydroxylated metabolites, 1-(*p*-chloro-*o*-hydroxyphenyl)-1-(*p*-chlorophenyl)ethane and 1-(*p*-chloro-*m*-hydroxyphenyl)-1-(*p*-chlorophenyl)ethane. Accumulation of toxic chlorinated carboxylic acids may have inhibited metabolism of the substrate. Compounds which are recalcitrant to cometabolic activity have substitutions in the ethane or ethene sections of their structures which may cause steric hindrance.

A large sampling study of oceanic and near-shore environments established that 35 of 95 isolates degraded DDT to many of the metabolic products previously identified, with DDD the major metabolite (350). The only environmental samples which failed to mediate DDT degradation were the oceanic water samples.

FUNGAL METABOLISM OF DDT

The earliest studies involving microbial metabolism of DDT were conducted using commercial yeast cakes (*Saccharomyces cerevisiae*) (234). Reductive dechlorination was demonstrated by the appearance in culture media of ^{14}C-DDD from ^{14}C-DDT labeled in the phenyl group. DDE was shown not to be a necessary intermediate metabolite.

A study of 8 fungi incubated with DDT for 6 days did not reveal degradation (80). However, in the same study six of nine actinomycetes were found to convert DDT to DDD. These 6 actinomycetes are *Nocardia erythropolis*, *Streptomyces aureofaciens*, *S. viridochromogenes*, *S. cinnamoneus*, and with lesser efficiency *S. albus* and *S. antibioticus*. All cultures in this study were incubated aerobically with shaking. Another study of microbial cultures with the capability to degrade dieldrin showed that two *Trichoderma viride* strains could degrade DDT to DDD, a "dicofollike" metabolite, and DDA (349). The dicofol-like compound was subsequently identified as 1-bis(*p*-chlorophenyl)ethane (302). Earlier studies with several strains of *T. viride* established differences in the metabolites produced by each strain (300).

Shake cultures of *Mucor alternans* in nutrient media containing ^{14}C-DDT produced three hexane-soluble and two water-soluble metabolites within 2 to 4 days (7). The total activity recovered was about equally divided between the two phases. The major metabolite is water soluble. These metabolites were unidentified since the results obtained from thin layer chromatography were different from those for DDE, DDD, DDA, DBP, dicofol, or 1,1-bis(*p*-chlorophenyl)-ethane. *M. alternans* converted 15% of the DDT starting material to three unidentified water-soluble products in another experiment as well (232). A comparison of these products with 1,1-bis(*p*-chlorophenyl)acetic acid (DDA), PCPA, DBP, DBH and 2-chlorosuccinic acid failed to reveal their identities. Attempts to reproduce this metabolic activity in the natural environment were unsuccessful, as the addition of *M. alternans* spores to DDT-treated soil failed to promote any degradation after 11 weeks' incubation (7).

A sequential experiment was developed to study the interactive effects of bacteria and fungi (156). *Hydrogenomonas* sp. was grown in a medium containing dichlorodiphenylmethane or *p*-chlorophenylacetic acid, both metabolic products of DDT. *Hydrogenomonas* sp. cannot liberate free chloride from metabolism of these compounds. The culture supernatant fluid was extracted and added to a basal salts solution, which became the growth medium for a culture of *Fusarium* sp. Growth occurred under anaerobic conditions, and chloride was detected in the medium, indicating that the products of *Hydrogenomonas* sp. metabolism were degraded to CO_2, H_2O, and HCl by *Fusarium* sp. The ability to perform this mineralization decreased if the two microbial populations were incubated together.

In another study of the interactive effect of other fungi on DDT degradation by *M. alternans*, the addition of other fungal cultures or the cell-free spent media from some cultures repressed the formation of water-soluble metabolites (6). Other fungi, including *Aspergillus flavus, A. fumigatus, A. niger, Fusarium oxysporum, Penicillium notatum, Rhizopus arrhizus,* and *Trichoderma viride,* failed to produce water soluble metabolites of DDT. However, water-soluble ^{14}C-products appeared after incubation of ^{14}C-DDT with the excretory products retained in culture media after growth of all of the above fungi including *M. alternans,* with the exception of *R. arrhizus.* This discrepancy in the appearance of water soluble metabolites may be attributed to sorption of degradation products by the mycelia or to further degradation by cells to metabolites that are not water soluble.

The path of DDT metabolism by *Fusarium oxysporum* has been established and follows the route DDT → DDD → DDMU → DDHO → DDOH → DDA → DBP, with DDE formed from DDT (135, 136, 136a). DDHO is the aldehyde intermediate which is rapidly converted to DDOH and DDA. This path is similar to that described for bacteria. The enzymes involved in DDT metabolism which have been isolated include DDT dehydrochlorinase and those that decompose DDMU, DDA and DDOH (265a). DDT inhibits the fungal esterase while DDD strongly activates the same enzyme (159). The net effect is enzyme activation which results in detoxification of the molecule.

FUNGAL METABOLISM OF OTHER COMPOUNDS

The acaricide chlorobenzilate (ethyl 4,4'-dichlorobenzilate) ^{14}C-labeled in the aliphatic moiety was cometabolized by *Rhodotorula gracilis* with glucose as an additional carbon source (312, 313). Production of ^{14}CO$_2$ was correlated with culture growth. Metabolites included 4,4'-dichlorobenzilic acid, dichlorobenzophenone, and other unidentified products. The same results were obtained in the metabolism of chloropropylate (isopropyl 4,4'-dichlorobenzilate). Alteration of the chlorinated rings was not noted.

PERSISTENCE AND DEGRADATION OF DDT IN THE ENVIRONMENT

DDT is converted to DDD by anaerobic but not aerobic sludge microorganisms (202). The same results were found using Pawnee silt loam treated with ^{14}C-DDT. Under anaerobic conditions DDD was recovered while under aerobic conditions DDE was the only metabolite (190). Another study of anaerobic soil treated with DDT reported formation of DDD and traces of other metabolites including DDE, DDA, dicofol, 2-chlorobenzoate, DBP, and DBM (189a). Conversion of DDT to DDD in flooded soil was faster when more organic matter was present (69).

The conversion of DDT to DDD has been demonstrated in sterile as well as nonsterile environments (59) and has been related to the redox potential (Eh) of the soil. Sewage sludge samples sterilized in a variety of ways all resulted in conversion of DDT to DDD if the Eh was sufficiently low (489). The rate of DDT degradation is highest in soils with the lowest redox potentials, in the range of −90 to −250 mV (184). Studies of DDT degradation in a variety of anaerobic and aerobic environments, using various carbon sources and various soils, have shown differences in the amount of degradation and the efficiency of substrate and product recovery which preclude extrapolation from laboratory to the environment or even from one experiment to another (346, 347). Little conversion was found in moist anaerobic soil with Eh of +350 mV or in flooded anaerobic soil with an Eh that dropped from +400 to +200 mV. Flooded stirred soil (Eh = 0 mV) also showed little degradation. However, stirred anaerobic soil treated with lime (Eh = −250 mV) and glass beads inoculated with muck (Eh = −250 to −300 mV) mediated greater than 95% conversion of DDT (346).

It appears that at low Eh, DDT undergoes an irreversible redox type of reaction with transient formation of a free radical before conversion to DDD (184). The reaction is thought to be mediated by reduced iron porphyrins, with cell metabolism not being necessary (489).

Numerous studies have been conducted with regard to the persistence of DDT and its metabolites in environments treated with the insecticide (115, 121, 278, 324, 344, 429, 444, 485). All of the studies have reported residues of DDT, DDE, and DDD in the environment at the time of sampling. The longest period of time between last application of DDT and sampling was 17 years, and on the basis of these data half-life numbers for DDT of 2.5 to 35 years have been reported (324). In general, the amount of these residues that remains in the soil 1 to 2 years after application is similar to the amount recovered 9 years or longer after the first sampling.

SUMMARY

Studies on metabolism of DDT by bacteria and fungi have shown that reductive dechlorination of the nonaromatic portion of the molecule is the necessary primary step. The pathway of metabolism of DDT, first described for *E. aerogenes* but subsequently confirmed in other bacteria and in fungi, describes a series of steps requiring anaerobiosis (DDT to DDA) followed by a series of steps that may require aerobic conditions (DDA to DBP). Although some studies report aerobic conversion of DDT to metabolites, the oxygen tension was not rigorously defined in these experiments. All of these metabolites retain the chloride ions on both aromatic rings of the molecule.

The effect of a concerted attack by several microbial species has been demonstrated in a two-stage experiment in which *Hydrogenomonas* sp. was grown in a medium containing either of the DDT metabolites dichlorodiphenylmethane or p-chlorophenylacetic acid. The resulting filtered media became the growth substrate for cultures of the fungus *Fusarium* sp., which anaerobically liberated free chloride, indicating mineralization to CO_2, H_2O, and HCl. The nature of this pathway and the enzymes involved have not yet been elucidated.

Other studies which have shown differing efficiencies of DDT metabolism between pure culture and consortia or sewage/soil studies, indicate that degradation of this compound is highly dependent on environmental factors including coexistence of other organisms capable of metabolizing the compound to at least DDD. Of particular importance is the redox potential of the environment. Only under highly reducing conditions can the necessary first step of conversion of DDT to DDD be achieved. This reaction does not require microbial mediation. Further steps in DDT degradation may require environmental conditions and microbial activities which have not yet been elucidated. DDT, DDD, and DDE are highly persistent in all environments treated with DDT.

13

Chlorinated Dioxins and Dibenzofurans

The chlorinated dibenzo-p-dioxins and dibenzofurans are produced as byproducts during the formation of many other chemicals, including 2,4,5-T, hexachlorophene, pentachlorophenol and other chlorinated phenols, and polychlorinated biphenyls (360). Chlorinated dioxins have been found in the fly ash and flue gases from municipal generators in Switzerland, presumably due to pyrolysis of chlorophenol salts, and the formation of chlorinated furans has been tied to the pyrolysis of polychlorinated biphenyls and polychlorinated diphenyl ethers. These compounds are used as heat exchange fluids and as hydraulic liquids. From 3 to 25% of the polychlorinated biphenyls burned may be converted to chlorinated dibenzofurans (360). There is no known technical use for the chlorinated dibenzo-p-dioxins, of which 75 congeners can exist, and the chlorinated dibenzofurans, of which there are 135 theoretical congeners (360). The positional isomers of the dioxins vary greatly in their acute toxicity and biological activity, and the most potent isomer, 2,3,7.8-tetrachlorodibenzo-p-dioxin (TCDD), is considered the most potent low-molecular-weight toxin known (mean lethal dose in guinea pigs 0.6 µg/kg body weight) (387).

Interest in these compounds was generated after an epidemic of "chick edema factor" in 1957 due to 1,2,3,7,8,9-hexachlorodibenzo-p-dioxin that caused the death of millions of broiler chickens, and an accident in a chemical plant in 1976 in Seveso, Italy that released a cloud of toxic materials including TCDD to the surrounding environment (259, 356). TCDD has also been shown to cause chick edema factor (387). The extreme toxicity of the compound of major interest, TCDD, has focused most research on this isomer.

MICROBIAL METABOLISM OF DIOXINS AND FURANS

To date none of the chlorinated or nonchlorinated dioxins or furans
have served as a sole source of carbon or energy for growth by any
microorganism in a wide range of screening and enrichment experiments (214, 259, 260, 301). *Pseudomonas* sp. NCIB 9816, which can
utilize naphthalene as a sole carbon source, can cometabolize the nonchlorinated molecule dibenzo-*p*-dioxin when salicylic acid is present
in the growth medium (259). Studies to determine the products of
cometabolism were conducted with a mutant, *Pseudomonas* sp. NCIB
9816 strain 11, which oxidizes naphthalene only to *cis*-1,2-dihydroxy-
1,2-dihydronaphthalene. Dibenzo-*p*-dioxin is cometabolized to 2
neutral products, identified as *cis*-1,2-dihydroxy-1,2-dihydrodibenzo-
p-dioxin and 2-hydroxydibenzo-*p*-dioxin. When the first product
is incubated aerobically or anaerobically with cell extracts of the
parent organism in a medium containing NAD^+, a third product is
formed and was identified as 1,2-dihydroxydibenzo-*p*-dioxin (Figure
60). This metabolite completely inhibits or inactivates the enzyme
1,2-dihydroxynaphthalene oxygenase, which in the parent splits the
naphthalene ring in the analogous pathway.

Similarly, a *Beijerinckia* sp. grown on dibenzo-*p*-dioxin and succinic acid produces 1,2-dihydroxydibenzo-*p*-dioxin (260). Cell growth
is inhibited after 4 hours. Cell extracts incubated with 1,2-dihydroxy-
dibenzo-*p*-dioxin show a brief initial rate of oxidation followed by a
decline to the nonenzymatic rate. The rate of oxidation was determined polarographically using an oxygen electrode to measure oxygen
consumption. Two oxygenases were isolated from the cell extract,
2,3-dihydroxybiphenyl oxygenase which also oxidizes 1,2-dihydroxy-
dibenzo-*p*-dioxin, and catechol oxygenase which has no activity
against the dioxin metabolite. Both these oxygenases are inhibited
when incubated with cell extracts.

This *Beijerinckia* sp. utilizes biphenyl as a sole carbon and energy
source. When grown on succinic acid plus biphenyl, resting cells
oxidize 1-chloro- and 2-chlorodibenzo-*p*-dioxin at a high rate. Di-
benzo-*p*-dioxin and the isomers 2,3-, 2,7-, and 2,8-dichlorodibenzo-
p-dioxin are oxidized at a lower rate, followed by 1,2,4-trichloro-
dibenzo-*p*-dioxin (260).

A mutant strain of this species called *Beijerinckia* sp. B8/36 was
isolated which metabolizes several aromatic hydrocarbons to *cis*-
dihydrodiols (260). When grown on succinate plus dibenzo-*p*-dioxin,
a neutral product is formed which was identified as *cis*-dihydroxy-
1,2-dihydrodibenzo-*p*-dioxin and has identical characteristics to the
product of *Pseudomonas* sp. NCIB 9816 metabolism. Also appearing
in the medium is the metabolite 2-hydroxydibenzo-*p*-dioxin. The
mutant *Beijerinckia* sp. B8/36 cometabolizes 1-chloro- and 2-chloro-
dibenzo-*p*-dioxin to neutral products which appear to be *cis* dihydrodiols, but no products appear after cometabolism with 2,3-dichloro-
or 2,7-dichlorodibenzo-*p*-dioxin.

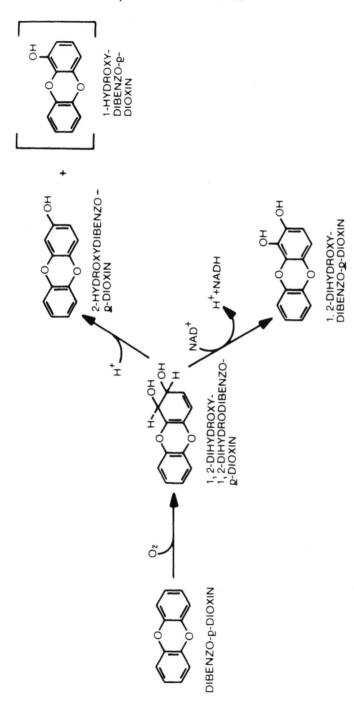

Figure 60 Oxidation of dibenzo-p-dioxin by *Pseudomonas* sp. NCIB 9816. Bracketed compound is hypothetical intermediate. (Adapted from Ref. 259.)

An unidentified bacterium was isolated from contaminated Seveso soil and incubated aerobically in a complex nutrient medium containing ^{14}C-TCDD (214, 351, 352). After 54 weeks two polar metabolites appeared. One was isolated in very small quantities and was not identified. The other was found to be a hydroxylated derivative. This microbial metabolite also appears in a culture of *P. testosteroni* strain G1036 after incubation for 36 weeks and in a culture composed of a mixture of six bacteria from Seveso soil. The metabolite was postulated to be 1-hydroxy-2,3,7,8-TCDD, assuming no chlorine rearrangement took place.

TCDD was metabolized by *Bacillus megaterium* to several polar metabolites (358). The most active cultures were incubated with 5 µg/L TCDD introduced in an ethyl acetate carrier, a solvent which increases cell permeability. When ethyl acetate was the carrier and the amount of soybean extract in the medium was reduced, as much as 55% of the dioxin was recovered as polar metabolites. TCDD was also converted in small amounts to a polar metabolite in farm soil which had been incubated for 2 months. The quantity of this metabolite did not increase with time after 2 months. Other soils similarly incubated failed to produce any metabolites. Two strains of bacteria which converted TCDD to polar metabolites were isolated from the farm soil samples.

A large screening study examined 100 bacterial isolates for ability to metabolize TCDD (301). These strains all had shown previous ability to metabolize persistent pesticides, but only five showed some ability to metabolize TCDD as determined by thin-layer chromatography. The product or products were not identified.

Degradation of TCDD by an extracellular laccase (*p*-diphenol:- oxygen oxidoreductase) produced by the fungus *Polyporus versicolor* was investigated (68). Crude enzyme extracts incubated with TCDD under a variety of conditions failed to modify the substrate.

There has been one report on the microbial metabolism of dibenzo- furan (77). A comparison was made of the cooxidation of this compound by *Cunninghamella elegans, Beijerinckia* sp. and *Beijerinckia* sp. B8/36 discussed previously with regard to dibenzo-*p*-dioxin metabolism. The mutant strain oxidizes dibenzofuran to a mixture of 1,2-dihydroxy-1,2-dihydrodibenzofuran and the unstable 2,3- dihydroxy-2,3-dihydrodibenzofuran which under acidic conditions dehydrates to a mixture of 2-hydroxy- and 3-hydroxydibenzofuran (Figure 61). The fungal culture forms a much more stable 2,3-di- hydrodiol which yields 2-hydroxydibenzofuran and 3-hydroxydibenzo- furan only when heated with acid, although the ratios of the 2 products are similar after bacterial or fungal metabolism. These results are consistent with the unstable bacterial metabolite being of the *cis* configuration and the stable fungal metabolite arising from an epoxide to form a *trans* configuration. The fungal culture also forms

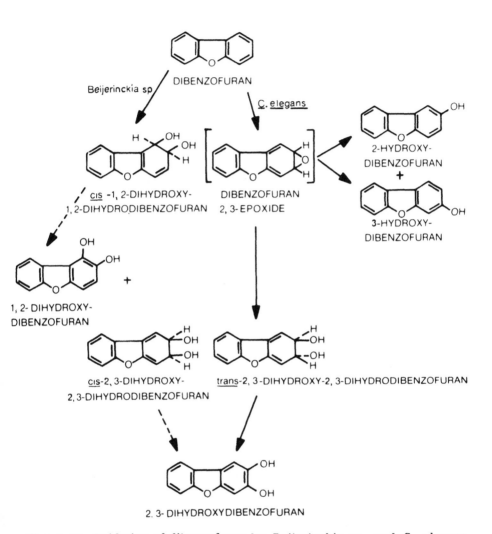

Figure 61 Oxidation of dibenzofuran by *Beijerinckia* sp. and *C. elegans*. Bracketed compound is hypothetical intermediate. Dashed lines indicate postulated reactions. (Adapted from Ref. 77.)

a small amount of 2,3-dihydroxydibenzofuran, and the parent *Beijerinckia* sp. in the presence of NAD$^+$ forms 1,2-dihydroxy- and 2,3-dihydroxydibenzofuran.

DIOXIN PERSISTENCE AND DEGRADATION IN SOILS

Evaluation of the persistence of TCDD in soils is complicated by the strong sorptive properties of TCDD, making recovery for analysis difficult (301, 352). In addition, artifacts may arise during the exhaustive extraction and analytical procedures involved. TCDD incubated with lake sediment for less than 1 hour and then extracted with solvents and analyzed by thin layer chromatography and ^{14}C-radioactivity showed 6 to 7% conversion to metabolites, indicating the generation of artifacts during the procedure or the presence of impurities (458). Analysis of a commercial ^{14}C-TCDD preparation revealed the presence of 7% contaminants, including TriCDD, some anisole isomers, and some other components (352). Upon incubation, TCDD became less easily extractable while the other components were in comparison readily extractable, leading to artificial enrichment of the isomer during the analysis. This could cause misinterpretation of experimental results. Formation of metabolites would be expected to increase with time unless precluded by a toxicity threshold or by further metabolism.

Finally, since TCDD is only present in the environment as a contaminant of other chemicals, the analytical procedure must be able to measure TCDD at levels lower than a few parts per million (197). Recovery of ^{14}C-TCDD as measured by combustion from soils receiving 1.78 ppm TCDD were 52% after 1 year from Hagerstown silty clay loam containing 2.5% organic matter, and 67% from Lakeland loamy sand containing 0.9% organic matter (246). At an application rate of 17.8 ppm, 89% was recovered from Hagerstown loam and 73% from Lakeland sand. No metabolites were detected. Little ^{14}CO$_2$ was evolved from TCDD-treated soils during 10 weeks' incubation. Extracts of soil treated with 2,7-dichlorodibenzo-*p*-dioxin (DCDD) contained a major metabolite in addition to the parent substrate (246). The metabolite was not identified. About 5% of the added radioactivity in a 0.7 ppm application of DCDD to soils was evolved after 10 weeks. TCDD incubated in lake water for 589 days was not altered (458). However, a lake water and sediment system incubated for the same length of time produced metabolites amounting to 1 to 4% of the original substrate. These products were polar (water soluble) and some were extractable and some were nonextractable in chloroform. The addition of nutrients enhanced formation of metabolites.

SUMMARY

The chlorinated and nonchlorinated dioxins and furans have not yet been shown to be utilizable as a sole carbon source for growth and energy. The parent nonchlorinated dibenzo-p-dioxin can be hydroxylated by several bacteria. The hydroxylated products accumulate and are recalcitrant to further oxidation. Some of the mono-, di-, and trichlorinated isomers are also oxidized by some bacteria, although at lower rates. TCDD is also hydroxylated by a few species of bacteria. Not all of the metabolites arising from TCDD metabolism have been identified. The dibenzofurans have been less well studied. although it has been demonstrated that the nonchlorinated substrate can be hydroxylated by both bacteria and fungi. Studies of the metabolism of these compounds have been hampered by the difficulty of extraction and product analysis, and by the extreme toxicity of TCDD, the isomer of greatest interest.

14

Biodegradation of Chlorinated Aromatic Compounds in Scaled-up Biological Treatment Processes

INTRODUCTION

The previous sections have demonstrated that a wide variety of chlorinated aromatic compounds are subject to biodegradation by a diversity of pure and mixed bacterial cultures. The significance of this information relates to the perceived potential for both environmental and wastewater biodegradation and/or detoxification of chlorinated aromatic pollutants. The extrapolation of such laboratory-derived results to environmental degradation and waste treatment is imprecise due to optimization, acclimation and high cell density cultures employed in most biochemical and physiological studies, which are rarely if ever met in real world biodegradation scenarios. In addition, real world complexity of the environment matrix in which biodegradation occurs frequently necessitates the use of imperfect measures of biodegradation that cannot readily be correlated with those used to assess biodegradation in a laboratory environment.

Much of the work assessing biodegradation potential has been done in small-scale laboratory glassware with the hope that adequate comparisons might be drawn to full-scale environmental or treatment systems. Many factors may vary between the small-scale lab systems and the full-scale systems. A summary of factors that influence organic biodegradability is presented in Tables 6, 7, and 8 (385). Small-scale laboratory tests can assess the importance of many of these factors in well-designed, controlled experiments. However, many issues of importance in relating small-scale test results to full-scale process performance depend on interactions of the various individual factors and the rates of material and biomass changes. These often are influenced by the physical design of the system and mass and energy transfer considerations. The turbulence and mixing potential of the system are also of major importance.

Table 6 Chemical Factors Influencing Organic Biodegradability

Chemical factors	Consequences
Substrate structural considerations	
Molecular weight or size	Limited active transport
Polymeric nature	Extracellular metabolism required
Aromaticity	Oxygen-requiring enzymes (in aerobic environment)
Halogen substitution	Lack of dehalogenating enzymes
Solubility	Competitive partitioning
Toxicity	Enzyme inhibition, cell damage
Xenobiotic origin	Evolution of new degradative pathways
Environmental factors	
Dissolved oxygen	O_2-sensitive and O_2-requiring enzymes
Temperature	Mesophilic temperature optimum
pH	Narrow pH optimum
Dissolved carbon	Organic/pollutant complexes are concentration dependent for growth
Particulates, surfaces	Sorptive competition for substrate
Light	Photochemical enhancement
Nutrient and trace elements	Limitations on growth and enzyme synthesis

Source: Ref. 385.

Table 7 Biological Factors Influencing Organic Biodegradability

Biological factors	Consequences
Enzyme ubiquity	Low frequency of degradative species
Enzyme specificity	Analogous substrates not metabolized
Plasmid encoded enzymes	Low frequency of degradative species
Enzyme regulation	Repression of catabolic enzyme synthesis Required acclimation or induction
Competition	Extinction or low density populations
Habitat selection	Lack of establishment of degradative populations
Population regulation	Low population density of degradative organisms

Source: Ref. 385.

Table 8 Experimental Factors Influencing Organic Biodegradability

Experimental factors	Problems encountered
Analytical method	
Substrate disappearance	Competing abiotic processes
Biotransformation	Complex analysis
Mineralization	Incomplete biochemical pathways
Scale up/down	Comparability among reactor designs and effects on kinetics
Feedstock complexity	
Chemically/biologically defined	Poor simulation and predictability
Complex waste/wastewaters	Difficult interpretation

Source: Ref. 385.

For these reasons, environmental scientists and engineers have turned to larger-scale experiments to simulate the design and physical features of the full-scale system, whether a treatment or an environmental system. In these systems, material and energy kinetics can be measured, competitive abiotic processes can be studied, and these results can be linked to mathematical models to describe the system and allow scale-up to full-scale systems with more certainty.

This chapter focuses on scale-up studies to determine the biodegradability of chlorinated aromatics reported in the literature. Much information is available on continuous wastewater treatment systems. This was emphasized in this chapter over the environmental microcosm work, since the focus of this work is on treatability as opposed to persistence in the environment. Little information was found on scaled-up studies in the soil matrix.

PENTACHLOROPHENOL

Pentachlorophenol has been studied in several scaled-up systems. These include studies in aerobic "fiber-wall" reactor systems (140) as well as more conventionally-designed laboratory activated sludge systems (37, 129, 315).

A laboratory-scale test with a continuous, aerobic, "fiber-wall" reactor was used to study the biooxidation of PCP in a synthetic and in an authentic wood-preserving wastewater (140). In he synthetic case, the concentration of PCP in the feed was 20 mg/L, the COD was 300 mg/L, the acclimation period was 15 days, and the operational period was 30 days. Reagent grade, commercial grade, and improved commercial grade pentachlorophenol were used. The improved commercial grade had fewer impurities, including chlorodioxins. Table 9 shows a summary of the results of these experiments. Other concentrations were also tested. These data show a general inhibition of the disappearance of PCP in the commercial preparation relative to the reagent and improved commercial grade preparations. Presumably, this is related to impurities present, possibly chlorodioxins. Actual waste rivaled the synthetic tests in PCP degradation performance. No proof of mineralization or estimation of other fate mechanisms was given in these tests.

A series of laboratory-scale continuous-stirred tank reactors (CSTR) was used to determine the aerobic biodegradation of PCP in wastewater treatment applications (315). The testing protocol included a phase where the inoculum was acclimated to PCP over 90 days from initial concentrations of 1 mg/L to 20 mg/L of PCP in a "fiber-wall" reactor. This sludge was then introduced into continuous-stirred tank reactors with no sludge or cell recycle. The hydraulic residence times (HRT) and the mean cell residence times (MCRT) were, therefore, equal and ranged from 3.2 to 18.3 days. Data

Table 9 The Effect of Pentachlorophenol Purity on Disappearance in Continuous Systems

Parameter	Reagent grade[a]	Commercial grade[b]	Improved commercial grade[c]	Actual waste
Feed PCP (mg/L)	20	20	20	17.8
Feed COD (mg/L)	515	515	515	1336
Hydraulic residence time (hr)	6	6	6	6
Effluent PCP (mg/L)	0.53—0.7	1.9	0.3	0.2
Effluent COD (mg/L)	15.8—45.6	29.1	52.8	216
Activated sludge initial pentachlorophenate degrading capacity (mg PCP g cell^{-1} hr^{-1})	0.4—0.49	0.11	0.4	

[a] 98% pure PCP.
[b] Blend from four manufacturers, 75 to 85% pure.
[c] Same PCP concentration as commercial grade, "substantially reduced" chlorodioxins [Dow Chemical Co., Improved Commercial Grade Penta (XD-8108.00L)].
Source: Ref. 140.

collected included disappearance data on COD and PCP as well as reactor suspended solids concentrations. Proof of mineralization of PCP by use of ^{14}C-PCP was employed in related batch studies and fate information was collected on sorption and stripping mechanisms in adjunct batch experiments. No confirmatory specific analysis of radiolabeled intermediates in the effluent or in the biomass are offered.

Results from this study include a first-order kinetic rate constant, (the maximum specific growth rate, μ_m, divided by the Monod saturation constant, K_s) of 0.0017 L-μg^{-1} d^{-1} with a minimum attainable PCP CSTR reactor concentration of 27 μg/L. Aqueous phase concentrations of PCP ranged between 51-293 μg/L in the reactor. PCP had little effect on the removal of other COD in this study. The batch fate testing indicated that neither stripping nor sorption were significant PCP removal mechanisms and PCP was mineralized with some carbon incorporated into the cellular material.

A synthetic waste containing PCP was treated in a continuous laboratory-scale activated sludge system consisting of a 6.25 L mixed liquor vessel and a 1.66 L external clarifier (129). Air was added at the rate of 6.25 L/minute through a sintered glass sparger. Sludge was wasted at 15 min intervals throughout the experiment. HRTs ranged from 8.9 to 10.4 hours and the MCRT was maintained at 6.2 days. Parameters measured included total solids, sludge volume index, PCP and reducing sugar concentrations in the clarifier, and clarifier effluent turbidity.

Screened wastewater threatment plant sludge was acclimated to PCP using a fill and draw reactor. An *Arthrobacter* sp. strain (ATCC 33790) was also added to another acclimation reactor with the effect of a lag period reduction to 1 to 3 days as opposed to over 6 days for the unamended sludge.

Steady-state operation of the activated sludge reactor achieved reductions in concentration from 40 mg/L PCP to about 1 mg/L. Unsteady-state transient conditions were studied by increasing the feed to 120 mg/L. Shock loading effects were analyzed with the use of kinetic models. Systems where the *Arthrobacter* sp. was added continuously showed a considerably improved transient response to the shock load than did the acclimated systems without addition of the strain.

Continuous laboratory-scale activated sludge units, consisting of 11 L aeration reactors (air sparged in at ~ 2 L/minute) and 4 L external clarifiers with partial sludge recycle were challenged with 8.6 mg/L of PCP (37). The feed stream consisted of a pulp mill foul condensate with substantial amounts of non-PCP carbon, largely as methanol. Mean cell residence times were 4.9 and 9.3 days and hydraulic residence times were 25.4 and 24.0 hours, respectively. PCP was added as $U-^{14}C$-PCP and label analysis and specific PCP analysis were performed on the mixed liquor supernate as well as waste biomass and offgas. Other operating conditions are reported as well.

Unlike the other studies, the emphasis of this study was to develop information on PCP sorption on biomass and, therefore, a biomass not acclimated to PCP was used. BOD_5 and TOC removals from the aqueous phase were 63 to 69% and 86 to 96%, while PCP removals were 11.6 to 7.0%, depending on the reactor MCRT. Essentially, all of the PCP removal was either sorbed or soluble in the aqueous component of the waste sludge. No significant biological transformation or mineralization was evident and stripping of PCP was below the detection limit. Good accountability was found for both labeled and unlabeled PCP.

Batch sorption tests were also incorporated and comparison of the batch data and the continuous runs suggests that the data fits a Langmuir-type isotherm with an apparent saturation of the biomass at an aqueous phase concentration of 2 mg/L PCP. The PCP data collected at concentrations less than the apparent saturation concentration agree well with a proposed sorption equilibrium equation based on the PCP octanol-water partition coefficient. An equation proposed for estimation of stripping based on the Henry's law constant for associated PCP was found to exaggerate the amount stripped. However, use of the Henry's law constant for the pentachlorophenate form was expected to show better agreement. No such Henry's law constant is available.

Consistent with information in earlier chapters, PCP has been found to biotransform in scaled-up systems using acclimated biomass or systems amended with known PCP degraders. No scaled-up studies have attempted to elucidate transformation products other than $^{14}CO_2$, but complete mineralization is strongly indicated in at least one study. Inhibition of other compounds (potentially chlorodioxins) on the biomass has been noted and, therefore, the waste matrix in which the PCP resides may be important in determining the extent and rate of biodegradation. Conventionally designed activated sludge systems operating with biomass not acclimated to PCP or experiencing transient shock loads either fail to achieve biotransformation of PCP or do so at greatly reduced rates. In these cases, PCP removal from the aqueous phases is poor and sorption to biomass (or other suspended solids) is expected to be a significant removal mechanism.

In summary, effective continuous PCP degradation appears to require a biomass with specific PCP degradative capability (PCP degrader subpopulation) as evidenced by the importance of acclimation and a steady and transient-free feed. Systems allowing longer cell residence times may also have an advantage over designs incorporating short cell residence time, although evidence here is not conclusive.

CHLORINATED BIPHENYLS

Several studies have occurred in which various mixtures and/or specific congeners of chlorinated biphenyls (PCBs) have been tested for biodegradability in larger, continuous experiments. These have included tests on consortia taken from operating wastewater treatment plants (200, 236, 438), as well as tests using specific organisms (280). Still other research has been completed focusing on the fates of PCBs in environmental microcosms (272).

PCB commercial mixtures (Aroclors 1221, 1016, 1242, 1254, and MCS 1043) were tested for biodegradation using a laboratory-scale activated sludge test (438). Because of the test protocol, evaluation of the HRT and MCRT was not possible. The sludge inoculum was obtained from a municipal treatment plant and was acclimated for several weeks on a synthetic feed composed of glucose, nutrient broth, and KH_2PO_4. PCBs were not included in the acclimation feedstream. Initial MLSS concentrations were adjusted to 2500 mg/L. PCBs dissolved in ethanol were injected into the reactors. Disappearance of PCBs from the mixed liquor was measured using a hexane extraction followed by specific GC-EC or UV analysis. PCB spikes into the mixed liquor were used to measure analytical recoveries of PCBs from the liquor containing biomass and supernate.

Sorption on biomass was checked by sonic homogenization and extraction of the mixed liquor from an Aroclor 1016 run and extraction in the standard manner. Comparison with results without sonic homogenization showed similar PCB recoveries. No sorption checks were made with higher chlorinated PCBs, nor was proof of mineralization or biotransformation reported. Stripping was checked with Aroclor 1221, MCS 1043, and Aroclor 1016 by use of hexane offgas scrubbers connected to the reactors. Stripping rates of 4.2, 6.1, and 3.6%, respectively, were reported for the above PCB mixtures.

Results of these tests, equating disappearance with degradation, are presented in Table 10. A 48-hour cycle with an addition rate of 1 mg PCB over this interval was used.

A laboratory study of activated sludge challenged with carbon radiolabeled 2,5,4'-trichlorobiphenyl and 2,4,6,2',4'-pentachlorobiphenyl was performed to determine the fates of these compounds in biological processes (200). Municipal sewage sludge was placed in an aerated glass column and the offgas was scrubbed in hexane and toluene to recover stripped ^{14}C-PCB. No acclimation of the sludge to PCBs was reported. Initial concentrations were 0.178 mg/kg and 0.231 mg/kg trichlorobiphenyl and pentachlorobiphenyl, respectively.

Metabolic byproducts of the trichlorobiphenyl were found in the sludge while none were found in the aqueous phase. One percent of

Table 10 Disappearance of Commercial Chlorobiphenyl Mixtures

Mixture	Percent chlorine of mixture	Percent disappearance during test
Biphenyl	0	100
Aroclor 1221	21	81 ± 6
MCS 1043	30	56 ± 16
Aroclor 1016	41	33 ± 14
Aroclor 1242	42	26 ± 16
Aroclor 1254	54	15 ± 38

Source: Ref. 438.

the ^{14}C-trichlorobiphenyl was estimated to undergo biodegradation.
The degradation products were found to be less volatile than the parent compound and, therefore, accumulated in the biomass. No evidence
of degradation of the pentachlorobiphenyl was found in this study.

A laboratory-scale semicontinuous test using a "fill-and-draw"
technique (236) was performed using municipal sludge acclimated over
3 months on 1, 5, and 10 μg/L of Kanechlor 500 (a PCB product with
its main component being pentachlorobiphenyl). The units were
aerated for 12 hours and then settled for 0.5 hour, after which time
the clear supernate and sludge in excess of 25% of the reactor volume
was wasted. The reactor was recharged with synthetic feed containing the PCBs, glucose, sodium glutamate, and inorganic nutrients.

The BOD of the feed was 320 mg/L. Specific PCB analysis in the
sludge included centrifugation of the sludge, digestion of the solids
with methanol-KOH, and subsequent hexane extraction. The aqueous
phase was extracted with hexane and extracts were combined, water
washed, dried with anhydrous Na_2SO_4, cleaned up with a Florisil
column, concentrated, and analyzed by GC-EC. COD measurements
and respirometric measurements were also taken during the study.

In batch respirometric tests, oxygen uptake of biomass with 1
and 5 μg/L PCBs was stimulated relative to the control without PCBs.
Semicontinuous reactors fed up to 10 μg/L PCBs experienced high
BOD removal efficiencies (98.6 to 99.1%). Major removals of PCBs
in the semicontinuous reactors were found at all feed concentrations
over the 12-hour aeration period. Equilibrium concentrations were
achieved during the first hour of the period. Table 11 shows the
distribution of PCBs in the semicontinuous reactors for sludge acclimated at 1, 5, and 10 μg/L PCB.

Table 11 Distribution of Kanechlor 500 in Activated Sludge Semicontinuous Systems

Sludge PCB acclimation concentration (μg/L)	Initial PCB concentration (μg/L)	Removal in wasted sludge (%)	Remaining in effluent (%)	Unaccounted (%)
1	0.16	45.6	31.3	23.1
5	0.48	75.9	15.0	9.1
10	0.85	81.5	12.1	6.4

Source: Ref. 236.

The majority of the PCBs charged to the system were removed with the wasted sludge. The authors tested stripping of PCBs in their semicontinuous reactors at air flow rates of 0.1 L/minute of air per liter of mixed liquor and found significant disappearance (65%) after 20 hours' aeration. They conclude that stripping could account for the PCB losses experienced in the semi-continuous runs. Finally, challenge of a municipal anaerobic digester sludge with 31 μg/L of PCBs (wet weight) incubated at 38° C for 40 days showed no disappearance of the starting PCB material. Resistance to anaerobic biotransformation was concluded.

A laboratory-scale continuous aerobic study of the degradation of Aroclor 1221 by a *Pseudomonas* sp. strain 7509 culture is reported (280). Inocula were acclimated on a feed in which Aroclor 1221 was the sole carbon source. A fermentation vessel was used as the reactor and raw sewage with a BOD_5 of 140 to 170 mg/L, fortified with 20 mg/L each of nitrogen (as NH_4Cl) and phosphorus as (KH_2PO_4 + K_2HPO_4), was amended with Aroclor 1221 at concentrations of 50 and 100 mg/L. The waste was fed at rates between 13 and 91 ml/hour to the 14 L reactor in which the agitator rate and the dissolved oxygen concentration were monitored. The HRT ranged from 4.6 to 32 days and was equal to the MCRT.

The broth (or mixed liquor) was sampled periodically and acidified prior to hexane extraction. GC-FID was used for disappearance analysis. No distinction can be made between the compounds in the aqueous and biomass compartments of the mixed liquor since the solids were not separated prior to analysis. Neither stripping nor sorption was measured and no proof of transformation or mineralization was reported.

At high HRTs (16 to 32 days), all of the Aroclor 1221 fed disappeared. At lower HRTs (4.6 to 10.7 days), some of the specific congeners began to appear in the mixed liquor indicating, according to the author, a preference of the organisms for certain congeners. The implication is that under nonstressed conditions, biphenyl, 2-chlorobiphenyl, and 4-chlorobiphenyl will degrade readily while 2,2'-dichlorobiphenyl, 2,4'-dichlorobiphenyl, and 4,4'-dichlorobiphenyl are more recalcitrant. 2,4'-Dichlorobiphenyl was found to build up as an indicator of the lower biodegradation rates. Aroclors 1016 and 1254 were also tested with resulting accumulation of all of the components. Switching from continuous to batch operation indicated disappearance of many of those congeners. Specific congeners were not identified in the Aroclor 1016 and 1254 runs.

Fates of 2,2',4,5'-tetrachloro-, 2,2',4,4',5,5'-hexachloro-, and 2,2',3,3',4,4',5,6'-octachlorobiphenyls were introduced into a model system that included sediment, water, and air compartments (272). The model systems were fitted with a special gas bubbling device to enable investigation of removal by jet drop entrainment, and were operated in the dark. Anaerobic, sterile (bactericidal $HgCl_2$), and aerobic biotic systems with and without macroinvertebrates were studied. The macroinvertebrates were grown in separate vessels and then added to the model systems. The equivalent of 7500 chironomids/m^2 (*Chironomus plumosus* and 25,000 tubificids/m^2 (*Tubifex tubifex*) were added to the system. Gas flow to the system was estimated to be from 0.00005 to 0.003 ml/min of gas per ml of water. Thus, air flow to liquid volume ratio was from 30 to 2000 times less than that found in diffused air biological treatment systems. ^{14}C-PCB was used and no specific compound analysis was performed on the parent compound or possible metabolites or conversion products. Congeners recovered in each experiment averaged 68%, 40%, and 31% for tetra-, hexa-, and octachlorobiphenyls, respectively. Distribution of the PCBs in the various tests are found in Tables 12, 13, and 14.

Table 12 Distribution of 2,2',4,5'-Tetrachlorobiphenyl in a Sediment-Water-Air Model System

Compartment	Aerobic with macroinverte-brates	Aerobic without macroinverte-brates	Sterile	Anaerobic
Sediment	97.7[a]	98.3[a]	99.8[a]	99.3[a]
Dissolved in water	0.03	0.06	0.02	0.03
Particles in water	0.06	0.03		
Macroinvertebrates	0.55			
Glass walls (particle adhesion)	0.20			0.02
Glass walls (extractable)	0.10	0.02		
Air filters	0.03	0.01		0.02
Vessel stoppers	0.06	0.08		0.04
Surface micro-layers	0.03	0.01		
Jet-drop impactors	0.93	1.48	0.18	0.04

[a]All values as percent of recovered compound.
Source: Ref. 272.

Table 13 Distribution of 2,2',4,4',5,5'-Hexachlorobiphenyl in a
Sediment-Water-Air Model System

Compartment	Aerobic with macroinverte-brates	Aerobic without macroinverte-brates	Sterile	Anaerobic
Sediment	96.1[a]	99.9[a]	98.9[a]	99.9[a]
Dissolved in water	0.06			0.02
Particles in water	0.14			
Macroinverte-brates	2.73			
Glass walls (particle adhesion)	0.48	0.01		0.02
Glass walls (extractable)	0.02		0.17	0.02
Air filters	0.01		0.02	
Vessel stoppers			0.02	
Surface microlayers	0.04	0.03		0.03
Jet-drop impactors	0.36	0.03	0.10	0.03

[a]All values as percent of recovered compound.
Source: Ref. 272.

Although this experiment was intended to simulate a natural eco-system and clearly differs from experiments on engineered systems, some conclusions are suggested that may relate to engineered treatment systems. First, although the vast majority of the congeners partition with the sediments (90 to 99.9%), the presence of biomass (especially the macroinvertebrates) was a determinative factor in the partitioning relative to dispersion of sediment particles and jet-drop entrainment. In an engineered system, the turbulence of the system may override the turbulence from the macroinvertebrates with the result of much greater suspended material potentially being available for adhesion to walls and other surfaces. Of potentially greater importance is increased jet-drop entrainment, because of higher suspended solids concentrations and substantially higher air flow to water volume ratios in the engineered system. Truly aerosolized jet drops may be collected in scrubbers or filters and therefore may be included

Table 14 Distribution of 2,2',3,3',4,4',5,6'-Octachlorobiphenyl in a
Sediment-Water-Air Model System

Compartment	Aerobic with macroinverte-brates	Aerobic without macroinverte-brates	Sterile	Anaerobic
Sediment	90.0[a]	99.6[a]	99.8[a]	99.4[a]
Dissolved in water	0.10	0.03	0.06	0.45
Particles in water	1.0	0.02	0.09	0.01
Macroinverte-brates	3.00			
Glass walls (particle adhesion)	4.67	0.07		
Glass walls (extractable)	0.07			0.01
Air filters	0.01			
Vessel stoppers	0.3			0.05
Surface micro-layers	0.07	0.03		0.11
Jet-drop impactors	0.99	0.24		0.04

[a]All values as percent of recovered compound.
Source: Ref. 272.

in measures of stripping potential, even though the substrate is
actually not a vapor but a mechanically-carried liquid-solid droplet.
Unfortunately, no literature addressing this potentially important
removal mechanism in engineered systems is evident.

A single scaled-up study has been reported to date that conclusively demonstrates PCB biodegradation. In this study, metabolic
byproducts accounting for about 1% of the trichlorobiphenyl were
recovered. No other studies present conclusive evidence for biodegradation. This is largely due to use of parent compound disappearance
data for analysis and lack of proof for biotransformation or mineralization. No work is evident using labeled congeners to support an
argument for enzymatic processes. Several studies have shown major
potential for PCB disappearance relative to stripping and to sorption
on biomass or other solids (sediments).

Analytical difficulty in extractive recovery of the higher chlorinated congeners from sediments suggests that in the absence of methods to digest the biomass or to determine specific recoveries of the higher congeners in the biomass, extractive analysis of the biomass or the total mixed liquor (biomass and supernate) may not recover substantial portions of the parent compounds strongly sorbed to the cellular or solids matrix. Spiking with a PCB mixture and complete extractive recovery is not conclusive proof of recovery of PCBs from unknown biomass samples unless no saturation effect of the biomass with sorbed PCBs exists and sufficient time is allowed for the PCBs to achieve equilibrium partitioning with the biomass. Absence of either condition would lead to more of the spiked sampled residing in the aqueous compartment and misleadingly high apparent recovery values. In other words, special caution must be taken when equating PCB-spiked mixed liquor or biomass recovery with actual sample PCB recovery.

The strongest evidence of biodegradation, notwithstanding the above comments, is the study by Liu (280) in which ratios of the specific congeners in Aroclor 1221 change depending on the operating conditions. It is difficult to see how sorption or stripping mechanism would vary as widely as reported for the same congeners under different operating conditions unless enzymatic processes are at work. Also the PCB mixture studied is comprised of congeners that, in other microbial tests (reported in an earlier chapter), have been shown to undergo biotransformation.

In general, further scaled-up studies on PCBs are needed that include analytical protocols to offer proof of biotransformation or mineralization, quantify the sorption and stripping mechanisms, and study other potentially important mechanisms such as jet-drop entrainment in engineered systems, before engineered biological processes may be considered for PCB treatment.

DICHLOROPHENOL

A laboratory-scale activated sludge reactor with sludge recycle (liquid volume of aeration tank, 3 L) was used to study the simultaneous biodegradation of 2,4-dichlorophenol (DCP) and phenol (33). The phenol concentrations ranged from 14.9 to 45.7 mg/L and the DCP concentrations ranged from 52.4 to 121 mg/L. A 1:1 carbon ratio of both substrates was desired. The sludge was previously acclimated to phenol and then was acclimated to DCP by gradual replacement of the phenol in the feed with DCP. The acclimation process required 70 days to complete. The reactors were operated with HRT between 2.5 and 6.25 hours and MCRT of between 1.75 and 10.7 days.

MLVSS concentrations ranged between 46 and 299 mg/L. Analysis focused on substrate disappearance with no determination of stripping or sorption. Chloride analysis as a test of mineralization and

COD determinations did not account for the phenol or DCP disappearance, implying incomplete mineralization of the substrates.

Biokinetic rate constants for the runs yielded disappearance rate constants of 0.00098 L mg^{-1} hr^{-1} for phenol and 0.045 hr^{-1} for DCP. The Monod half-saturation constant for DCP was 63 mg/L. Yield coefficients for phenol and DCP were 0.67 mg VSS/mg phenol and 0.39 mg VSS/mg DCP, respectively. A combined biomass decay coefficient of 0.014 hr^{-1} was presented.

A mathematical rationale leading to deterministic estimates of fates from biological wastewater treatment processes has been proposed (38). For continuous, complete-mix, activated sludge units where the biological disappearance of the parent compound is described by a rate equation first order in substrate concentration, and where sorption is occurring at concentrations below any potential biomass saturation concentration, the following equations are proposed for first estimates of the percent substrate stripped, sorbed, and wasted in the waste biomass, and biotransformed to another compound:

$$REM_s = \frac{A}{1 + A + S + B} \tag{2}$$

$$REM_{st} = \frac{S}{1 + A + S + B} \tag{3}$$

$$REM_b = \frac{B}{1 + A + S + B} \tag{4}$$

$$REM_e = \frac{1}{1 + A + S + B} \tag{5}$$

$$A = HRT \; \frac{X \, K_{ow} \, f_L}{1000 \, p_L \, MCRT} \tag{6}$$

$$S = HRT \; K^{st} = \frac{Q_{air}}{V} \, 3.71 \times 10^{-3} \, (H_c)^{1.045} \tag{7}$$

$$B = HRT \; K^b \tag{8}$$

where:

REM_s, REM_{st}, REM_b, REM_e are the percent removals of the substrate from the system by the sorption, stripping, biological transformation, and effluent fate mechanisms, respectively,

HRT is the hydraulic residence time of the activated sludge system (hr),

X is the concentration of biomass as MLSS (mg/L),

K_{ow} is the substrate octanol-water partition coefficient (concentration in octanol/concentration in water),

f_L is the fraction of lipids or lipophilic compounds in the biomass (weight fraction),

p_L is the mean density of the lipophilic biomass compounds (g/L),

MCRT is the mean cell residence time for the biomass in the systems (hr),

Q_{air}/V is the ratio of the air flow rate into the system to the system hydraulic volume (min^{-1}),

K^b is a biological disappearance rate constant, first order in substrate concentration (hr^{-1}).

Equations 6 to 8 are discussed individually in separate papers (37, 385, 437).

If assumptions are made such that the equations described above are applicable to the DCP experiment discussed earlier (complete-mix system, $Q_{air}/V = 0.1$ min^{-1}, H_c of DCP = 13.4 torr L mol^{-1}, K_{ow} of DCP = 1202, etc.), then the data reported can be used to calculate K^b for each experimental series and the mechanism removals can be estimated. Table 15 presents the results of this analysis. It may be concluded from this analysis that stripping can be a significant removal mechanism, especially in the instance where the K^b is relatively slow. In run 4, 16% of the DCP removed from the system (other than in the effluent) was stripped. As the biotransformation rate increased, the stripping potential was vastly reduced to less than 0.1% of the total DCP removed (other than in the effluent) (Run 1). Sorption of DCP to the waste sludge taken from the system is not a significant fate mechanism for DCP. However, depending on the level of extra-cellular water wasted with the biomass (waste sludge solids concentration), more DCP could be lost in the waste sludge than that shown in Table 15.

Finally, the first-order biotransformation rate constants, K^b, are similar to values reported by Beltrame et al. (33) derived using more conventional empirical biokinetic rate constant methods ($K^b = 0.045 \pm 0.005$ hr^{-1}). Calculated rate constants from Table 10 show wider variance and indicate strong relationships between biological rates and the MCRT.

TRICHLOROCARBANILIDE

Trichlorocarbanilide (TCC) was studied in both laboratory-scale batch flask tests and continuous activated sludge systems (184a). TCC with ^{14}C label on the 4-chloraniline (p-chloroaniline or PCA) ring as well as the dichloroaniline (DCA) ring was used.

Sludge was obtained from a municipal wastewater treatment plant. In the activated sludge units, the biomass concentration and air flow rates were controlled at 4000 mg/L as MLSS and 0.05 standard $ft^3/$ hour of CO_2-free air. Offgas was trapped in an amine solution to

Table 15 Fate Estimates of 2,4-Dichlorophenol Removal from a Laboratory-Activated Sludge System Using Proposed Equations

Run	MCRT (hr)	HRT (hr)	Calculated K^b (hr^{-1})	Percent removal of DCP by			
				Sorption	Stripping	Biotransformation	Effluent
1[a]	257	25	0.075	0.008	0.034	63.0	33.6
2[a]	42	6.25	0.043	0.67	2.0	21.6	76.8
3[b]	109	6.25	0.070	0.25	1.8	29.8	68.1
4[b]	20	6.25	0.020	0.78	2.2	10.8	86.2

[a]Runs with phenol and 2,4-dichlorophenol at a ratio of 1 to 1 carbon from each substrate.
[b]Runs with glucose and 2,4-dichlorophenol at a ratio of 1 to 1 carbon from each substrate.
Source: Data taken from Ref. 33.

recover $^{14}CO_2$. The ^{14}C content of the sludge was determined by a combustion $-^{14}CO_2$ recovery method; 200 µg/L of TCC were added to the feed stream.

Batch flask tests showed that ~90% of the theoretical $^{14}CO_2$ evolved from incubation with both raw sewage and activated sludge after 12 weeks. A lag of 2 weeks was noted in the raw sewage before $^{14}CO_2$ evolution. Concentrations of 200 µg/L of TCC resulted in nearly 100% theoretical $^{14}CO_2$ evolution at 12 weeks while concentrations of both 20 µg/L and 2000 µg/L TCC resulted in less mineralization at this incubation time.

Sorption of TCC to activated sludge was determined by contacting activated sludge at concentrations from 0 to 2000 mg/L with 20 to 200 µg/L of TCC. Sorption was calculated based on the disappearance from the aqueous phase over 2 hours. A Freundlich isotherm was generated showing TCC-on-sludge concentrations of 2 and 0.1 µg/mg at aqueous concentrations of 78 and 8 µg/L, respectively. Continuous operation of the activated sludge unit at a 10-hour HRT for 90 days showed 20 to 30% of theoretical $^{14}CO_2$ evolution for the dichloraniline ring while from 40 to 60% of theoretical $^{14}CO_2$ was evolved from the chloroaniline ring. Table 16 highlights the fate of the radioactivity in the continuous tests.

Although no specific analysis of the sludge was undertaken to determine the chemical composition of ^{14}C found in the sludge (i.e., parent compound, metabolites, cellular material), such analysis was

Table 16 Fate of Trichlorocarbanilide in Laboratory-Scale Activated Sludge Systems

Substrate	Stream	% of feed	% recovered
^{14}C-PCA-TCC[a]	Effluent	3.2	
	$^{14}CO_2$	56.1	93.4
	Activated sludge	34.1	
^{14}C-DCA-TCC[b]	Effluent	30.3	
	$^{14}CO_2$	25.9	91.4
	Activated sludge	35.2	

[a] Trichlorocarbanilide with a labeled 4-chloroaniline ring.
[b] Trichlorocarbanilide with a labeled dichloroaniline ring.
Source: Ref. 184a.

done on the effluent. Chloroaniline, dichloroaniline, aniline condensation products, and unknowns were found in addition to the TCC parent compound.

This study convincingly supports conclusions related to the biotransformation and mineralization of TCC. These findings are consistent with those stated in an earlier chapter on chloroaniline herbicides even though this molecule's structure varies somewhat. In addition, this study is among the earliest work found that considered major fate mechanisms and offers proof of biotransformation or mineralization in a scaled-up biological wastewater system treating chloroaromatic compounds. Unfortunately, lack of information on the MCRT limits the calculation of biokinetic rate constants and the direct extrapolation of these results to other design configurations.

DICHLOROBENZENE

A pilot scale activated sludge system was operated on a side-stream of sewage from the city of Zurich, Switzerland (303). 1,4-Dichlorobenzene was present at all times in the system feed and it was used as an indicator compound to determine the nonbiological removal mechanisms of stripping and sorption on the biomass. The major assumption made in this study was that 1,4-dichlorobenzene was conserved and was not biotransformed at the operating conditions of the study. The aeration vessel volume was 11.3 to 15 m^3 and the HRTs ranged from 2.5 to 6.5 hours with MCRTs ranging from 74 to 182 hours. Sorption and stripping mechanisms were quantified but proof of biotransformation (or the absence of) was not provided.

The fates of DCB were reported to be 72% stripped, <3% sorbed on wasted sludge, 10% in the effluent, and 15% unaccounted. Data presented in an earlier section indicated that chlorobenzenes (except hexachlorobenzene) can be mineralized but there is a lack of knowledge on the biochemical pathways. The DCB not accounted for in this study may be undergoing biotransformation but the variance in the material balance related to analytical shortcomings precludes a definite statement. A study discussed later in this chapter offers evidence of DCB biotransformation.

COMBINED STUDIES ON SEVERAL CLASSES OF CHLOROAROMATICS

Laboratory and Pilot Studies

Several studies have occurred using a given experimental protocol on a variety of chloroaromatic compounds. Table 17 shows the classes of chloroaromatics studied for each of these compounds.

Benzoic acid, 2-chlorobenzoic acid, 3-chlorobenzoic acid, 4-chlorobenzoic acid, 2,4-dichlorobenzoic acid, 2,5-dichlorobenzoic acid,

Table 17 Classes of Chloroaromatics Studied in Several Experimental Studies

Compound	Reference					
	61	127	250	251	391	416
Pentachlorophenol	X		X	X		X
Chlorinated biphenyls						
Aroclor 1242	X					
Aroclor 1254	X					
Chlorophenols						
2-Chlorophenol	X					
2,4-Dichlorophenol	X	X	X	X		X
2,4,6-Trichlorophenol	X					
Chlorobenzenes						
Chlorobenzene	X					
1,2-Dichlorobenzene	X	X	X	X		X
1,3-Dichlorobenzene	X	X				
1,4-Dichlorobenzene	X	X				
1,2,4-Trichlorobenzene	X	X				
Hexachlorobenzene	X					
Chlorobenzoates						
2-,3-, and 4-Chlorobenzoate					X	
2,4-, 2,5-, 2,6-, and 3,5-Dichlorobenzoate					X	
Chlorophenoxy compounds						
2,4-Dichlorophenoxyacetic acid					X	

2,6-dichlorobenzoic acid, 3,5-dichlorobenzoic acid, phenoxyacetic acid, and 2,4-dichlorophenoxyacetic acid were studied in laboratory-scale continuous flow reactors resembling chemostats (391). The HRT and MCRT were equal in these reactors. The reactors were completely mixed by aeration and used suspended biomass. The lag for acclimation of municipal biomass to the specific substrates was determined. Kinetic disappearance data were collected on the specific substrates and on dissolved organic carbon (DOC) as well. Batch testing also was performed to determine kinetic rate constants for comparison with the continuous tests. Proof of mineralization of the specific substrates was determined by measurement of chloride ion release.

The lag for acclimation of the initial sludge for the monochloro-benzoic acids was in the range of 10 to 20 days. The biomass began to show acclimation to 3,5-dichlorobenzoic acid at about 20 days but the acclimation process was continued through 100 days. 2,5-Dichloro-benzoic acid became acclimated abruptly at 100 days; 2,4- and 2,6-dichlorobenzoic acids did not show acclimation during this testing protocol.

The author argues that long-term acclimation for some of the compounds is evidence for genetic changes in the organisms as opposed to enzyme induction or population effects. Maximum specific growth rates, μ_m, the Monod half-saturation constant, K_s, and the yield coefficient, Y, are shown in Table 18.

Good agreement was found between μ_m and K_s for the continuous flow reactors and associated batch tests. The MCRT of the systems

Table 18 Biokinetic Results for 2,4-D and Chlorobenzoates

Compound	μ_m (day^{-1})	K_s (mg/L)	Y (mg/mg)
2,4-Dichlorophenoxyacetate	2.3	5.4	0.14
2-Chlorobenzoate	1.0	2.4	0.22
3-Chlorobenzoate	0.6	2.0	0.14
4-Chlorobenzoate	1.2	1.1	0.25
2,5-Dichlorobenzoate	0.6	1.5	0.16
3,5-Dichlorobenzoate	0.05	25.3	

μ_m = maximum specific growth rate. K_s = monod half-saturation constant. Y = yield coefficient.
Source: Ref. 391.

were found to be strongly related to the effluent concentrations of the specific substrates. MCRTs of 3 to 15 days were required to achieve a 0.5 mg/L effluent concentration of the monochlorobenzoic acids while 6 to 50 days were needed to achieve 0.25 mg/L effluent concentrations.

The substrates were mineralized to CO_2 and cellular material but no fate measurements were made on stripping or sorption mechanisms. Performance of systems with glucose and the 2,4-dichlorophenoxyacetic acid indicated no effect of glucose on 2,4-D disappearance but lower glucose disappearance rates related to 2,4-D presence. However, no strong inhibition or toxic effects of the substrates on the biomass were seen at feed concentrations of 50 to 200 mg/L. The effluent concentration is, of course, much less than the feed concentrations.

A major study on a variety of organics was undertaken (250, 251, 416). Continuous laboratory-scale activated sludge units were challenged with chlorophenols, chlorobenzenes, and pentachlorophenol as well as a number of other organics. These compounds were added to a synthetic "base mix" containing ethylene glycol, ethyl alcohol, acetic acid, glutamic acid, glucose, phenol, and various inorganic nutrients. The chloroaromatics were added such that the BOD_5 achieved was $\sim$250 mg/L. HRT was held to about 8 hours and the MCRT ranged between 43 and 146 hours. Stripping was measured by trapping the organics from the offgas on a solid sorbent and all specific compound analysis generally followed EPA analytical protocols. Methodology for sorption quantitation was not clear. No proof of mineralization or biotransformation was offered. Acclimation to the compounds was allowed for 4 weeks before 2 months of continuous data collection. The reactor volume was 3 L for aeration and 3.3 L for an internal clarifier. Air flow to the reactors ranged from 2 to 3 L/minute. Table 19 presents data on the fate of specific chloroaromatics.

Table 19 Fate of Several Chloroaromatics in a Laboratory-Scale Activated Sludge System

Compound	Percent removed by		
	Stripping	Sorption	Biotransformation
2,4-Dichlorophenol			95.2
Pentachlorophenol		0.58	97.3
1,2-Dichlorobenzene	21.7		78.2

Source: Ref. 250, 252, 416.

These studies utilized an air flow-to-liquid volume ratio greater than
1 min^{-1} and the resultant data probably exaggerate the stripping
mechanism. Disappearance attributed to biotransformation for DCP
and PCP are in agreement with earlier studies reviewed in this chapter.
Biotransformation of 1,2-dichlorobenzene is in conflict with a study
discussed earlier but is consistent with general predictions on the biodegradability of chlorobenzenes.

Full-Scale Studies

Many full-scale biodegradation studies on chloroaromatics are reported.
However, only a few have attempted to quantify the abiotic fates of
the compounds. Also, variation in the waste composition and flow and
in other physical variables often makes it difficult to interpret the
results. Two of the most significant examples of full-scale plant
studies are discussed here.

A study to determine the fates of priority pollutants for 50 pub-
licy owned wastewater treatment plants has generated input-output
data on several chlorinated aromatics including 2-chlorophenol, 2,4-
dichlorophenol, 2,4,6-trichlorophenol, pentachlorophenol, chlorobenzene,
1,2-dichlorobenzene, 1,3-dichlorobenzene, 1,4-dichlorobenzene,
1,2,4-trichlorobenzene, hexachlorobenzene, Aroclor 1242, and Aroclor
1254 (61). This work was designed to allow statistical analysis on the
occurrence and fates of priority pollutants in the system feedstream,
intermediate process streams, system effluent, and waste sludge
streams. The material that disappeared in the process was reported
but since no air sampling was undertaken, no quantitative attempt
was made to distinguish between stripping and biotransformation
removal mechanisms. EPA analytical protocols were used and often
the compounds were grouped into the "volatile," "acid extractable,"
and "base neutral" groupings arising from the analytical workup.
Sampling periods were for approximately 6 days with 24-hour compos-
ite samples, and compounds were often at concentrations near the
analytical detection limit. Since the influent was often variable and
the sampling period was of the same order of magnitude as the plant's
MCRT, material balance data and conclusions must be viewed with
caution.

A related study was performed on a single publicly-owned wastewater treatment plant for a 30-day period (127). Over this period,
variations in influent flow and substrate concentrations could be more
precisely quantified and conclusions regarding compound fate could
be made. Table 20 summarizes the removal data of 2,4-dichlorophenol,
1,3-dichlorobenzene, 1,4-dichlorobenzene, and 1,2,4-trichlorobenzene
found in this study. The removal calculation sums the substrate en-
tering the system from all streams and the substrate leaving in all
streams. Thus, sorption and effluent removal mechanisms are not

Table 20 Mass Removal of Chloroaromatic Compounds in a Full-Scale
Wastewater Treatment Plant

	Percent removal[a] in		
Compound	Primary treatment	Secondary treatment[b]	Overall treatment
2,4-Dichlorophenol	2	46	47
1,3-Dichlorobenzene	14	30	40
1,4-Dichlorobenzene	0	88	88
1,2,4-Trichlorobenzene	12	79	82

[a]Calculation includes (total mass accounted for in minus total mass
accounted for out) /total mass accounted for in. Thus, compound in
the aqueous effluent is combined with that found in the waste solids.
Removal mechanisms here are biotransformation, stripping, and other
abiotic mechanisms excluding sorption.
[b]Based on activated sludge units alone.
Source: Ref. 127.

considered "removed" whereas biotransformation, stripping, and other
abiotic mechanisms, excluding sorption, are considered "removed."

   Full-scale plant data require special planning and careful imple-
mentation to yield satisfactory data on biotic and abiotic removal
mechanisms. Conclusive material balances often are impeded because
of feed variability and low compound concentrations. Use of labeled
compounds is expensive at large scale and questions related to the
environmental release of labeled compounds exist. Therefore, the capability of collecting data leading to proof of biotransformation or mineralization is limited.

SUMMARY

In order to conclusively establish biodegradation of chlorinated aromatic compounds in larger scale systems and to collect data that is of use
in extrapolation and systems design, several factors must be included
in the experimental design. These include: (a) measurement or prediction of abiotic fate processes, (b) proof of biotransformation or
mineralization, and (c) suitable measurement and reporting of important process variables relating to the calculation of biokinetic rates.

Failure to include these factors leads to inconclusive results on compound fates or the inability to use the data for predictions on other (even similarly designed) systems.

Calculation of biokinetic rate constants based only on "removal" leads to wide variances in the rate constants for compounds with major abiotic fate tendencies and precludes reliable scale-up and more direct comparison between systems. Only a few scale-up studies are available on chlorinated aromatics in general, and only a subset of these contain data suitable for drawing conclusions relative to biodegradation and which allow comparison between systems. At times, coupling the results from several studies may allow enlightened judgment regarding biodegradation, but there is no substitute for a single well-designed study. In general, much additional work is needed to generate reliable scale-up data for biodegradation of chloroaromatic compounds.

15

Overview of Microbiological Decomposition of Chlorinated Aromatic Compounds

Most of the studies reviewed here have explored the metabolism of a single compound by a single organism. A few have reported the metabolites formed in soils or by contrived microbial consortia. Taken together, however, these studies indicate the potential ultimate fate of the chloroaromatic compounds.

The chlorophenoxy herbicides and the chlorobenzenes can be metabolized to chlorophenols (Figure 62). Chlorophenols may in turn be metabolized to chloroanisoles, but the most common route of biodegradation is to chlorocatechols. Phenylamide herbicides and other compounds with nitrogen-containing substituents are metabolized to chloroanilines. The chloroanilines form a variety of products including chlorocatechols (Figure 63). Other products represent alterations of the aliphatic moiety.

Chlorocatechols, in turn, may be metabolized by several different mechanisms to nonchlorinated ring cleavage products (Figure 64). There are two main pathways. One results from *meta* cleavage to form a chlorohydroxymuconic semialdehyde which, after loss of the chloride, forms pyruvate and an aldehyde. The second pathway involves *ortho* cleavage to form β-ketoadipate via chloromuconate. The products of this pathway that are incorporated into cell constituents are succinate and acetyl-CoA.

Chlorobenzoic acids may be metabolized by three routes (Figure 64), the first through protocatechuic acid (a substituted catechol) to 3-ketoadipic acid. The second route metabolizes chlorobenzoic acids through chlorosalicylic acid to maleylpyruvic acid. Chloronaphthalenes are also metabolized through chlorosalicylic acid. Anaerobic metabolism of chlorobenzoic acids involves reductive dechlorination to

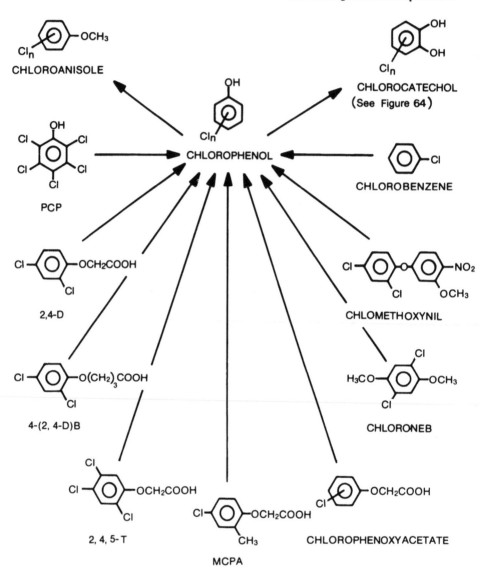

Figure 62 Chlorinated aromatic compounds metabolized to chlorophenols. This figure presents possible pathways extrapolated from various studies. In actual environmental systems a given transformation may be inhibited by a number of factors. Terminal compounds shown may be recalcitrant or insufficient research may exist on which to base a conclusion. Refer to text for further discussion.

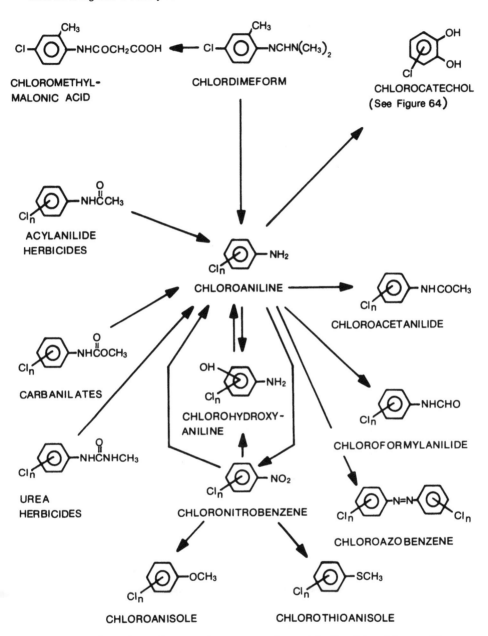

Figure 63 Chlorinated aromatic compounds metabolized to chloroanilines. This figure presents possible pathways extrapolated from various studies. In actual environmental systems a given transformation may be inhibited by a number of factors. Terminal compounds shown may be recalcitrant or insufficient research may exist on which to base a conclusion. Refer to text for further discussion.

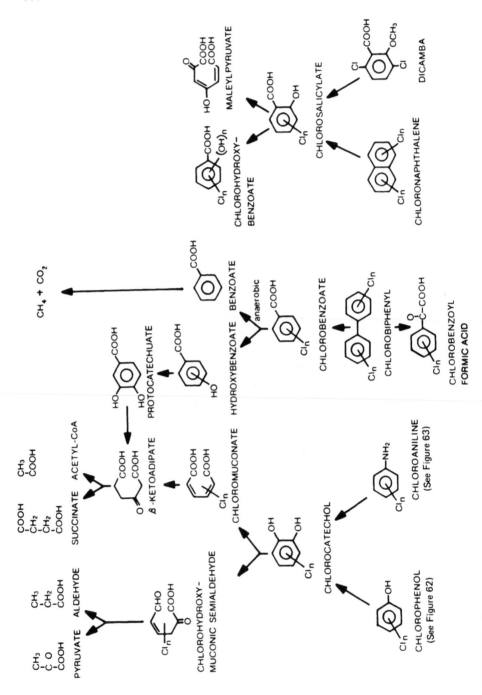

benzoate followed by formation of CH_4 and CO_2. The ultimate products of each of these metabolic pathways are either available for incorporation into cellular material or represent ultimate mineralization.

These pathways all represent a general biochemical potential. Whether or not a specific compound is actually metabolized to nonchlorinated products depends on many factors. The compounds which are most readily metabolized are the lower chlorinated forms. However, some compounds with only one chlorine, for instance 2-chloroaniline, are not readily metabolized. Thus both the position and the number of substituents are important in determining the biodegradability of a molecule. It is not only the number of chlorines but also the general form of the molecule itself which governs biodegradability. Compounds such as chlorodioxins and DDT are not shown on these figures because they appear to be highly resistant to substantive microbial attack. The exact physicochemical features of any given molecule that govern its biodegradability remain to be elucidated.

Even chlorinated molecules for which complete pathways of mineralization have been developed are not metabolized in all systems. Although a few compounds are amenable to anaerobic biodegradation, most require the availability of molecular oxygen for ring cleavage. Other environmental parameters which may place restrictions on biological activity include pH, temperature, and moisture. Upon exposure to the environment, the chemical state of the substrate may be altered to a form resistant to microbial attack.

The chemical itself may be degradable but the system may lack other nutrients necessary for microbial activity. Alternatively, other chemicals present may be preferred substrates, preventing metabolism of the substrate of interest. Another compound may also act to repress the activity of enzymes required for substrate metabolism. Accumulation of toxic metabolites may also repress further metabolic activity.

Ample evidence exists that some chemicals require the activities of several different groups of microorganisms for complete mineralization. Such consortia may not be found in the system containing the substrate. The interactions of these microorganisms pose additional constraints regarding production and utilization of potentially toxic metabolites as well as competition for nutrients and growth factors.

Figure 64 Metabolism of chlorocatechols, chlorobenzoates, and chlorosalicylates. This figure presents possible pathways extrapolated from various studies. In actual environmental systems a given transformation may be inhibited by a number of factors. Terminal compounds shown may be recalcitrant or insufficient research may exist on which to base a conclusion. Refer to test for further discussion.

Finally, the dynamics of pollutant appearance in the system is of critical importance. Most microorganisms require a period of acclimation to the substrate before metabolism occurs. During this period, the substrate level must be high enough to promote acclimation without being toxic or inhibitory. Prior exposure to the compound helps to shorten the acclimation period. Such exposure to other pollutants may also predispose the microbial population to adaptation to the substrate of interest. However, such acclimation may result in destruction of microorganisms capable of substrate utilization in favor of a population adapted to a different substrate, or in virtually complete destruction of the microbial flora.

This review of microbiological decomposition of chlorinated aromatic compounds indicates that while certain metabolic pathway generalizations exist, as reflected in Figures 62, 63, and 64, biodegradation of the compounds shown is dependent on many other variables and cannot be assumed in any given biotic system. Data on many of these variables are needed to allow prediction of the metabolic fate of chlorinated aromatic compounds.

16

Conclusions

1. The biodegradability potential of a given compound depends
on many factors. Chemical determinants include the ionic state of the
compound, the number, types, and position of substituents, and the
general form of the molecule. Environmental parameters affecting
microbial metabolism include pH, temperature, redox state, moisture,
reactor configuration, kinetics, and system turbulence considerations,
and interference by competitive processes such as sorption, stripping,
and photodegradation. Other factors involved in microbial growth
and metabolism include availability of nutrients and growth factors,
concentration of substrate, competitive interference by other substrates, and formation of toxic metabolites.

2. Many microorganisms require a period of acclimation before
biodegradation occurs. Once a population is acclimated, it sometimes
retains its predisposition to metabolize the substrate, and subsequent
additions of substrate are metabolized after a shorter lag or no lag
period.

3. Some compounds are mineralized by consortia of microorganisms,
in cases where no single species has been shown to be capable of that
process.

4. Fungi and bacteria metabolize most compounds by different
biochemical pathways.

5. In general, chlorophenoxy herbicides and chlorobenzenes
can be metabolized to chlorophenols and then to chlorocatechols.
Phenylamide herbicides and other compounds with nitrogen-containing
substituents are decomposed to chloroanilines, which can be oxidized
to chlorocatechols and a variety of products. Chlorocatechols may be
converted by a variety of different mechanisms to nonchlorinated ring
cleavage products. Chlorobenzoic acids may be transformed by three

different routes to ring cleavage products: (1) through a substituted catechol, (2) through chlorosalicylic acid, or (3) anaerobically through benzoic acid to methane and carbon dioxide.

6. The above conclusions represent general pathways for substrate dissimilation. Prediction of the biodegradability of a particular compound on which no data exist, based on data about similar types of compounds, can be loosely made. For instance, the degree of chlorination affects the rate and extent of metabolism of PCBs and other compounds.

7. Experiments in the laboratory or in a particular environment cannot be readily extrapolated to other environments, because of the need to consider the parameters affecting biodegradability potential listed above.

8. Small-scale lab tests are essential in order to perform properly controlled experiments to assess the effects of many of the above-mentioned biodegradation factors. Experimental design necessary to achieve the highly controlled results desired in small-scale testing nearly always requires compromise of factors needed to apply results to larger-scale systems.

9. Scale-up testing in larger-scale systems designed to simulate the desired full-scale application (wastewater treatment system, land farming, environmental scenario, etc.) is required to collect data (often kinetic data) allowing the prediction of process performance and comparison among full-scale systems.

10. Many of the available scaled-up studies focus on substrate disappearance or removal from the given feedstream without quantification of abiotic fates or adequate regard for calculation of biokinetic rate constants.

11. Additional emphasis should be given to establishing consistent scale-up methodology and to implementing additional work for generation of reliable scale-up data on the biological treatment of chlorinated aromatic compounds. Factors that should be included in this methodology include (1) quantification of abiotic fate processes, (2) experimental proof of biotransformation and/or mineralization of the substrate, and (3) measurement and reporting of important process variables related to the biokinetic rates.

Symbols

a	empirical constant in Langmuir isotherm
b	empirical constant in Langmuir isotherm
C_a	concentration of substrate in the mixed liquor
C_l	concentration of substrate in the liquid phase
C_s	concentration of substrate in the solid phase
f_L	weight fraction of lipid-like compounds in the biomass
K^b	a biological disappearance rate constant, first order in substrate concentration
K_f	empirical constant for the Freundlich isotherm equation
K_{ow}	octanol-water partition coefficient
K_s	Monod half-saturation constant
K^{st}	first order stripping rate constant
μ	specific growth rate
μ_m	maximum specific growth rate
n	empirical constant in Freundlich isotherm equation
Q_{air}/V	ratio of the air flow rate to the hydraulic reactor volume
REM_b	percent removal of the substrate from the system by the biotransformation fate mechanism
REM_e	percent removal of the substrate from the system in the effluent

REM_s percent removal of the substrate from the system by sorp-
              tion on biomass

REM_{st} percent removal of the substrate from the system by the
              stripping fate mechanism

X   concentration of biomass in the mixed liquor

Y   yield coefficient

Glossary

Acidophile Organism that grows very well at an acidic pH.

Active transport Process in which substrate entry into a cell is coupled to an energy-requiring process.

Adenine Purine base unit of a nucleoside.

Adenosine triphosphate (ATP) Ribonucleoside 5'-triphosphate that serves as a phosphate-group donor in the cell's energy cycle.

Allosteric enzyme Enzyme that contains a regulatory site.

Anaerobic respiration Oxidative process similar to aerobic respiration but which utilizes nitrate or another inorganic compound as the terminal electron acceptor.

Autoradiograph Image formed on film resulting from exposure to radioactive emission.

Autotroph Organism that obtains its energy from the oxidation of inorganic compounds, or sunlight, and fixes CO_2.

Bacteriophage Virus that attacks bacteria.

Barophilic Requiring high barometric pressures for survival.

Barotolerant Ability to survive at a wide range of barometric pressures.

Basophile Organism that grows at alkaline pH.

Binary fission Process of cell replication whereby a single cell divides into two.

Biochemical oxygen demand (BOD) Oxygen consumed during a standardized test in which a solution containing the compound of interest is inoculated with biomass and incubated for a fixed time.

Budding Process of cell replication whereby a protruberance from a cell grows into another cell.

Capsule Extracellular polysaccharide which accumulates around the cell and functions in cell attachment, defense, and protection.

Carboxylase Enzyme that catalyzes the ATP-dependent addition of carbon dioxide to the acceptor substrate.

Catabolite repression Inhibition of enzyme activity by binding of a control protein to the operator site.

Cell wall Outer portion of bacterial cell that confers upon the cell its shape and integrity.

Chemical oxygen demand (COD) Oxygen consumed during a standardized test in which a chemical oxidant is added to a solution containing the compound of interest and the oxidation reaction is allowed to go to completion.

Chemostat An apparatus used for the continuous culture of microbial populations in a steady state in which the growth rate is maintained by the substrate dilution rate.

Chemotaxis Movement in response to a specific chemical.

Chemotroph *See* Autotroph.

Chimera Molecule consisting of a replicon and another fragment of DNA.

Chromosome A single large molecule of DNA that contains many genes, and maintains all the required inheritable information of a cell.

Codon A group of three adjacent nucleotides that codes for an amino acid.

Cometabolism Process by which a substrate is metabolized by a cell while the cell utilizes another substrate as its energy source.

Competency State in which a cell is able to undergo transformation.

Competitive inhibitor Chemical which has a similar structure to an enzyme substrate and therefore binds to the enzyme, but which does not activate the enzyme.

Complete-mix reactor A reactor in which, at any instant in time, the concentration of constituents is the same at any point in the reactor. The effluent concentration of a constituent is therefore also the same as the concentration in the reactor.

Conjugation Process of DNA transfer from one bacterial cell to another by direct cell to cell contact.

Conjugative plasmid Plasmid that contains genes for bacterial conjugation.

Consortium A mixture of different mutualistic microbial populations.

Continuous stirred tank reactor (CSTR) A reactor in which mechanical agitation is used to generate a complete-mix condition where, at any instant in time, the concentration of constituents is the same at any point in the reactor. The effluent concentration of a constituent is therefore also the same as the concentration in the reactor.

Copy number The number of copies of a single plasmid present within a single cell.

Cosmid Constructed vector derived from a bacterial virus used for cloning large fragments of DNA.

Cytochrome P-450 Protein that serves as an electron carrier in enzymatic hydroxylation reactions and can also transfer electrons to oxygen.

Cytoplasmic membrane The limiting boundary of the cell protoplasm, composed of protein and phospholipid, that functions in substrate transport, osmotic regulation, cell wall synthesis, oxidative metabolism, and energy production.

Cytosine Pyrimidine base unit of a nucleoside.

Death phase Period in which reduction in population occurs due to cell death.

Decarboxylase Enzyme that catalyzes decarboxylation of the substrate.

Dehydrogenase Enzyme that mediates the loss of a hydrogen ion from a substrate with the acceptor being other than molecular oxygen.

Denaturation Process of separating double-stranded DNA into single strands.

Deoxyribonuclease Enzyme that catalyzes random cleavage of double-stranded or single-stranded DNA.

Deoxyribonucleic acid (DNA) A polynucleotide consisting of deoxyribonucleotide units that serves as the carrier of genetic information.

Deterministic Use of knowledge of the causes of a process to arrive at a prediction of its performance.

Dioxygenase Enzyme that catalyzes the addition of two atoms of molecular oxygen to a molecule.

Enzyme Protein that both lowers the energy of activation of and directs the metabolic pathway taken by chemical reactions in an organism.

Eukaryote Organism characterized by having diploid chromosome(s) surrounded by a membrane.

Exogenous DNA DNA molecule that is not an integral part of the cell genome.

Exonuclease Enzyme that removes single nucleotides from the end of a DNA molecule.

Exponential phase (log phase) Growth phase during which cells divide by binary fission.

Extracellular Outside of the outermost layer of a cell (cell membrane or cell wall).

Extracellular water The water in biomass "solids" residing between cells. Compounds residing in extracellular water are not included in measures of passive or active cellular uptake.

Facultative anaerobe Organism that grows in the presence or absence of air.

Feedback inhibition Inhibition of an allosteric enzyme early in a metabolic pathway by a later product of the pathway.

Fermentation pathway Metabolic pathway in which organic compounds serve as both the electron donor and the electron acceptor.

Fiber wall reactor A reactor in which the biomass is contained within a fibrous inner cavity. The aqueous solution and its dissolved constituent are thus permitted to pass through the fiber wall but the biomass (fixed film and suspended) is contained within the inner cavity. This type of reactor obviates the need for biomass separation by settling and recycle to the system.

Fill-and-draw reactor A mode of operating a reactor in which reactants and products are added or removed over discrete time intervals but the reactions are allowed to proceed continuously. Since concentrations of constituents change cyclically during the test, results of this type of test may only approximate true continuous and steady-state tests.

Flagellum An appendage attached to a cell that functions in motility.

Freundlich isotherm An isotherm equation relating the equilibrium partitioning of a compound between liquid and solid compartments (or phases). The equation is of the form

$$C_s = K_f C_l^{1/n}$$

where C_s and C_l are concentrations of the compound in solid and liquid compartments, respectively, and K_f and n are empirical constants.

Gene A DNA segment that codes for a single polypeptide chain or RNA molecule.

Genetic recombination Process of combining DNA from different sources into a molecule.

Genome The entire group of genes of a cell.

Gram negative Term given to bacteria that lose the primary stain (crystal violet) of the Gram-staining procedure upon exposure to alcohol or other decolorizing agent.

Gram positive Term given to bacteria that retain the primary stain (crystal violet) of the Gram-staining procedure upon exposure to alcohol or other decolorizing agent.

Gram stain Differential staining procedure in which crystal violet, Gram's iodine, decolorizing agent such as alcohol, and safranin are sequentially applied to bacterial cells. Most bacteria can be divided into two groups based on whether they retain or lose the primary stain (crystal violet) during the procedure. The response of bacterial cells to this procedure has been linked to differences in the cell wall composition.

Guanine Purine base unit of a nucleoside.

Halophile Organism that requires high salt concentrations for growth.

Henry's law constant (H_c) A constant that describes the equilibrium partitioning of a compound between liquid and gas compartments (or phases) at a given condition, where the concentration of the compound in the liquid compartment is sufficiently low. Many units are possible with this constant and caution should be taken in its use.

Heterotroph Organism that requires an organic form of carbon for energy.

Holdfast Appendage of some bacteria consisting of a fine stalk which may possess adhesive material; functions in attachment.

Hybridization Process of double strand joining two different single strand polynucleotides.

Hydraulic residence time (HRT) The time that the bulk aqueous phase resides in a continuous reactor volume. This may be calculated as the ratio of the system's hydraulic volume and the flow rate of the aqueous feed stream, assuming no change in fluid density during the reaction.

Hydrolase Enzyme that mediates hydrolytic reactions.

Hydrolysis Cleavage of a chemical bond by reaction with water.

Hydrophilic Water-loving; refers to polar molecules that associate with water.

Hydrophobic Water-hating; refers to nonpolar molecules that are insoluble in water.

Hypha A fungal filament.

Inducer A molecule that induces the synthesis of an enzyme.

Insertion sequence Inverted repeated DNA sequences allowing random integration into a DNA sequence.

In situ In its original location.

Intron An intervening sequence in a gene that is transcribed but excised before translation of the gene.

In vitro In a test tube or beaker ("in glass").

In vivo In a living organism.

Irreversible inhibitor Chemical that destroys or binds to a functional group on an enzyme, thereby preventing its catalytic activity.

Isomerase Enzyme that catalyzes a change in the atomic configuration of a molecule without a change in the number or kind of atoms.

Jet drop entrainment A mechanism in which small particles may be launched into a gaseous phase based on the collapsing of bubbles at the gas-liquid surface. Small droplets of fluid originating at the base of the bubble are formed and accelerated as the top of the bubble breaks and the bottom of the bubble merges with the gas-liquid surface. These drops may be aerosolized and thus carried with the gas phase.

3-Ketoadipate pathway Aerobic pathway of aromatic compound dissimilation in which the end products are the tricarboxylic acid cycle intermediates succinate and acetyl-CoA.

Kinase Enzyme that catalyzes transfer of a phosphate group from ATP or other nucleoside triphosphate to the substrate.

Lag phase Growth phase during which adaptation to the environment occurs and no increase in cell number is seen.

Langmuir isotherm An isotherm equation relating the equilibrium partitioning of a component between liquid and solid compartments (or phases). The equation is of the form:

$$C_s = \frac{ab\ C_L}{1 + b\ C_L}$$

Where C_S and C_L are concentrations of the compound in solid and liquid compartments, respectively, and a and b are empirical constants. This isotherm may be derived from considerations of mass transport onto surfaces with limited capacity to sorb the compound.

Ligase Enzyme that catalyzes the formation of a product resulting from the condensation of two different molecules, coupled with the cleavage of a pyrophosphate linkage in ATP.

Lipophilic Refers to molecules that associate with lipids.

Lithotroph Bacteria that use inorganic compounds as substrates for respiratory metabolism.

Log phase See exponential phase.

Lyase Enzyme that catalyzes the addition of a chemical group to the double bond of a substrate or the removal of a chemical group to form a double bond.

Lyse Breaking apart of the cell wall.

Lysogeny Infection of a bacterial cell by a virus during which the viral genome becomes integrated into the cell genome, is repressed, and is replicated with the cell genome.

Lytic cycle Process by which a bacteriophage infects a cell, replicates, and is released into the environment.

Macroinvertebrates Group of organisms that lack a backbone; in this context refers to species that live in the water.

Maximum specific growth rate (μ_m) The specific growth rate of biomass on a substrate that is limiting growth as defined by the Monod equation:

$$\mu = \mu_m \frac{C_a}{K_s + C_a}$$

where μ is the specific growth rate, μ_m is the maximum growth rate measured, K_s is the Monod half-saturation constant ($K_s = C_a$ at a growth rate of $\mu_m/2$), and C_a is the concentration of the growth-limiting substrate.

Mean cell residence time (MCRT) The mean of the cellular time distribution descriptive for the reactor configuration under consideration. This equals the hydraulic residence time in complete-mix, suspended growth reactors that have no biomass separation and recycle. This is a design variable that may be determined in suspended growth reactor systems that have provision for biomass separation and recycle and is determined by biokinetic rate constants and physical design in fixed film systems.

Mesocosm A constructed laboratory representation of an environment including atmospheric, hydrospheric, and geospheric parts with associated flora and fauna.

Mesophile Organism that grows best at temperatures from 15° C to about 45° C.

Messenger RNA (mRNA) RNA molecule that serves to carry the genetic message from the DNA to the ribosome.

Meta **position** Position on an aromatic molecule separated from the point of reference by one carbon position.

Methylase Enzyme that adds a methyl group to particular nucleotides.

Microaerophile Organism that has a narrow range of tolerance for its gaseous environment and requires either a reduced air environment or, in some cases, an increased proportion of carbon dioxide.

Microcosm A small-scale version of a mesocosm. *See* Mesocosm.

Mineralize Convert a molecule to inorganic ions and molecules.

Mitochondria Organelles in a eukaryotic cell which are the sites of oxidative metabolism.

Mixed liquor The mixture of the aqueous phase and suspended biomass in the aeration reactor of a biological treatment process.

Mixed liquor suspended solids (MLSS) A measure of biomass in suspended biological processes where solids are separated from the mixed liquor, are dried, and weights are determined gravimetrically. Standard methods exist for this test.

Mixed liquor volatile suspended solids (MLVSS) A measure of biomass in suspended biological processes where solids are separated from the mixed liquor, dried, heated to remove volatile organics, and weights are determined gravimetrically. Standard methods exist for the test.

Monod half-saturation constant (K_s) A constant defined in the Monod equation. See maximum specific growth rate.

Monooxygenase Enzyme that catalyzes the addition of one atom of molecular oxygen to a molecule.

Mutant Cell in which the genome has undergone mutation.

Mutase Enzyme that catalyzes transfer of a functional group between two positions on the same molecule.

Mutation Alteration of the genetic message.

NAD$^+$, NADH Nicotinamide adenine dinucleotide, a coenzyme which functions in oxidation-reduction reactions as hydrogen and electron carriers.

NADP$^+$, NADPH Nicotinamide adenine dinucleotide phosphate. Same function as NAD; see NAD.

Neutrophile Organism that grows best at a neutral pH.

Nick A breakage in one strand of a double-stranded DNA molecule.

NIH shift Migration of a hydrogen atom from one carbon to the adjacent carbon on an aromatic molecule.

Noncompetitive inhibitor Chemical that binds to an enzyme in an area other than the binding site, thereby altering and inactivating the catalytic site.

Nucleoid The area in a prokaryotic cell that contains the chromosome and is not bounded by a membrane.

Nucleoside A compound composed of a purine or pyrimidine base covalently linked to a pentose sugar.

Nucleotide A nucleoside with a phosphate group attached to one of the pentose hydroxyl groups.

Nucleus The membrane-bound organelle in a eukaryotic cell that contains the chromosome.

Obligate aerobe Organism that grows in the presence of air and uses aerobic respiration to obtain energy.

Obligate anaerobe Organism that grows only in the absence of air.

Octanol-water partition coefficient (K_{ow}) A constant the describes the equilibrium partitioning of a compound between equal volumes of n-octanol and water at a given temperature. Partitioning between other immiscible fluids and between other compartments can be mathematically related to this constant.

Operator region Regulatory site of a gene.

Organelle A discrete portion of a cell, with a specific function.

Organotroph Bacteria that use organic compounds as substrates for respiratory metabolism.

Origin of replication Sequence of DNA required for replication of the molecule.

Ortho **position** Position on an aromatic molecule adjacent to the point of reference.

Osmotic shock Sudden change in the solute concentration of the environment surrounding a cell.

Oxidase Enzyme that catalyzes loss of a hydrogen ion with molecular oxygen as the acceptor.

Oxidoreductase Enzyme that mediates alterations of the CH-OH group of a substrate and requires NAD^+ or $NADP^+$ as the hydrogen acceptor.

Para **position** Position on an aromatic molecule separated from the point of reference by two carbon positions, effectively opposite the point of reference.

Parts per billion (ppb) ($\mu g/L$)

Parts per million (ppm) ($\mu g/ml$ or mg/L)

Passive transport Process by which substrates enter a cell by free diffusion dependent on the difference in substrate concentration inside and outside of the cell.

Pathogenic Capable of causing disease.

Periplasmic space Area between the cytoplasmic membrane and the cell wall.

Phosphatase Enzyme that mediates the hydrolytic cleavage of phosphate esters.

Phototroph Organism that obtains its energy from light.

Phytotoxic Capable of inhibiting the growth of plants or algae.

Plasmid Small circular DNA molecule that is extrachromosomal and replicates autonomously.

Pleomorphic Capable of changing shape.

Polylinker Segment of DNA that contains closely spaced recognition sites for several restriction endonucleases.

Polymerase Enzyme that adds nucleotides to the 3'-hydroxyl terminus or removes nucleotides from the 5'-phosphate terminus of nicked DNA.

Primary degradation The initial alteration of a compound.

Primer Short section of DNA required to be attached to mRNA to initiate the activity of reverse transcriptase.

Prokaryote Haploid organism characterized by lacking a nuclear membrane.

Protoplasm The cell constituents within the cell wall.

Protoplast A viable cell that lacks a cell wall.

Psychrophile Organism that grows best at temperatures below 20°C.

Recycling fermentor An apparatus for the continuous culture of microorganisms in a steady state whereby the cells are returned to the culture vessel while medium and waste materials are removed.

Regulatory sequence A DNA segment involved in regulating a gene.

Regulatory site Area on an enzyme reversibly occupied by a noncompetitive inhibitor.

Relaxed plasmid Plasmid that is present in a cell as multiple copies.

Replicon DNA molecule that contains an origin of replication.

Repressor protein Protein that binds to the operator region of a gene and blocks its transcription.

Respiration pathway Metabolic pathway in which oxygen or other inorganic compound or ion serves as the terminal electron acceptor.

Resting cell A viable cell which is not actively growing or dividing.

Restriction endonuclease Enzyme that recognizes and cleaves specific sequences of nucleotides within double-stranded DNA.

Reverse transcriptase (RNA-dependent DNA polymerase) Enzyme that catalyzes the formation of double-stranded DNA from the information on mRNA.

Reverse inhibitor Chemical that binds to an enzyme but which may be removed with resulting activation of the enzyme; see competitive inhibitor, noncompetitive inhibitor.

Ribonucleic acid (RNA) A polynucleotide composed of ribonucleotide units.

Ribosomal RNA (rRNA) RNA molecule attached to the ribosome that serves as a framework for the binding of the polypeptide subunits of a protein.

Ribosome Site of protein biosynthesis.

RNA-dependent DNA polymerase See reverse transcriptase.

RNA polymerase Enzyme that catalyzes the formation of RNA from the information on DNA or RNA.

Semicontinuous reactors Reactors that are operated by adding or removing reactant or products over discrete time intervals, but where the reactions are allowed to proceed continuously. See fill and draw reactor.

Sequential induction Control of a long metabolic pathway such that sections of the pathway are under separate regulatory control and each section is induced by the product of a prior section.

Sludge volume index (SVI) A measure of settleability of suspended biomass that is based on the volume of solids settled from a mixed liquor over a given time interval. Standard methods exist for this test.

Stationary phase Growth phase during which no net change in cell numbers is seen; number of cells generated and dying is equivalent.

Steady state The condition where properties of a system of any given point in the system are the same over time.

Steric hindrance The inability of atoms or groups on a molecule to rotate freely because of mutual repelling due to van der Waals forces.

Sticky end Linear double-stranded DNA in which one strand extends beyond the other.

Stringent plasmid Plasmid which is present in a cell in one or, at the most, three copies.

Structural gene A gene that codes for a protein.

Suspended biomass The state of biomass growth where sufficient mechanical energy is introduced by the cells or from external sources to favor free suspension of cells or flocs of biomass in the mixed liquor or broth and to avoid the formation of a fixed biofilm on reactor surfaces.

Synthetase Enzyme that mediates condensation of two separate molecules coupled with cleavage of ATP.

T_4 DNA ligase Enzyme that links together complementary fragments of double-stranded DNA.

T_4 RNA ligase Enzyme that links together complementary fragments of single-stranded DNA or RNA.

Taxonomy The science of arranging organisms into logical groups and describing in detail the basic taxonomic unit, the species.

Terminal deoxynucleotidyl transferase Enzyme that adds deoxynucleotides to the 3'-hydroxyl end of DNA.

Thermophile Organism that grows at temperatures above 50°C.

Thiokinase Enzyme that catalyzes the ATP-dependent formation of thiol esters.

Thylakoid Internal membrane structure in cyanobacteria that contains the photosynthetic apparatus.

Thymine Pyrimidine base unit of a deoxyribonucleoside.

Transcription Process of converting information coded by DNA into RNA.

Transduction Bacteriophage-mediated transfer of genetic material into a cell.

Transferase Enzyme that catalyzes the transfer of an intact group of atoms from a donor to an acceptor molecule.

Transfer RNA (tRNA) RNA molecule that serves to bring a specific amino acid into proximity with the developing polypeptide.

Transformation Process of transfer of exogenous DNA into a cell.

Translation Process of protein biosynthesis according to the code carried by the mRNA.

Transposon A segment of DNA that can be moved from one area on a chromosome to another.

Tricarboxylic acid cycle Respiration pathway utilized by aerobic organisms.

Unsteady state The condition where properties of a system at any given point in the system are changing over time.

Uracil Pyrimidine base unit of a ribonucleoside.

Vector A replicon to which a fragment of DNA may be attached so that the fragment may be replicated.

Vesicle Cavity filled with liquid or gas.

Viable Capable of growing.

Yield coefficient (Y) The ratio of the change in biomass concentration, X, and the change in substrate concentration, C_a, over an interval of time. This is a measure of biomass production for unit substrate removal.

References

1. Ahmed, M., and D. D. Focht (1973). Degradation of polychlorinated biphenyls by two species of *Achromobacter*. *Can. J. Microbiol.* 19:47—52.
2. Alexander, M. (1965). Biodegradation: Problems of molecular recalcitrance and microbial fallibility. *Adv. Appl. Microbiol.* 7:35—80.
3. Alexander, M. (1965). Persistence and biological reactions of pesticides in soils. *Soil Sci. Soc. Am. Proc.* 29:1—7.
4. Anderson, D. A., and R. J. Sobieski (1980). *Introduction to Microbiology*. C. V. Mosby Co., St. Louis.
5. Anderson, J. J., and S. Dagley (1980). Catabolism of aromatic acids in *Trichosporon cutaneum*. *J. Bacteriol.* 141:534—543.
6. Anderson, J. P. E., and E. P. Lichtenstein (1972). Effects of various soil fungi and insecticides on the capacity of *Mucor alternans* to degrade DDT. *Can. J. Microbiol.* 18:553—560.
7. Anderson, J. P. E., E. P. Lichtenstein, and W. F. Whittingham (1970). Effect of *Mucor alternans* on the persistence of DDT and dieldrin in culture and in soil. *J. Econ. Entomol.* 63:1595—1599.
8. Ando, K., A. Kato, and S. Suzuki (1970). Isolation of 2,4-dichlorophenol from a soil fungus and its biological significance. *Biochem. Biophys. Res. Commun.* 39:1104—1107.
9. Aoki, K., K. Ohtsuka, R. Shinke, and H. Nishira (1984). Rapid biodegradation of aniline by *Frateuria* species ANA-18 and its aniline metabolism. *Agric. Biol. Chem.* 48:865—872.
10. Appleton, H. T., S. Banerjee, and H. C. Sikka (1980). Fate of 3,3'-dichlorobenzidine in the aquatic environment. pp. 251—

272 In Appleton, H. T. (ed.). *Dyn., Exposure Hazard Asses. Toxic Chem.* (Pap. Symp.), Ann Arbor Science Publ. Inc., Ann Arbor, MI.

11. Attaway, H. H., N. D. Camper, and M. J. B. Paynter (1969). Transformation of the herbicide methyl-N-(3,4-dichlorophenyl)-carbamate (swep) in soil. *Bull. Environ. Contam. Toxicol.* 4: 240—245.

12. Attaway, H. H., N. D. Camper, and M. J. B. Paynter (1982). Anaerobic microbial degradation of diuron by pond sediment. *Pestic. Biochem. Physiol.* 17:96—101.

13. Audus, L. J. (1964). Herbicide behavior in the soil. II. Interaction with soil microorganisms. pp. 163—206 in Andus, L. J. (ed.). *The Physiology and Biochemistry of Herbicides.* Academic Press, Inc., New York.

14. Axcell, B. C., and P. J. Geary (1973). The metabolism of benzene by bacteria. Purification and some properties of the enzyme *cis*-1,2-dihydroxycyclohexa-3,5-diene (nicotinamide adenine dinucleotide) oxidoreductase (*cis*-benzene glycol dehydrogenase). *Biochem. J.* 136:927—934.

15. Axcell, B. C., and P. J. Geary (1975). Purification and some properties of a soluble benzene-oxidizing system from a strain of *Pseudomonas. Biochem. J.* 146:173—183.

16. Bachofer, R., F. Lingens, and W. Schafer (1975). Conversion of aniline into pyrocatechol by a *Nocardia* sp.; incorporation of oxygen-18. *FEBS Lett.* 50:288—290.

17. Bailey, R. E., S. J. Gonslor, and W. L. Rhinehart (1983). Biodegradation of the monochlorobiphenyls and biphenyl in river water. *Environ. Sci. Technol.* 17:617—621.

18. Ballschmiter, K., C. Unglert, and H. J. Neu (1977). Abbau von chlorierten aromaten: Mikrobiologischer abbau der poly-chlorierten biphenyle (PCB). III. Chlorierte benzoesäuren als metabolite der PCB. *Chemosphere* 6:51—56.

19. Ballschmiter, K., C. Unglert, and P. Heinzmann (1977). Formation of chlorophenols by microbial transformation of chlorobenzenes. *Angew. Chem. Int. Ed. Engl.* 16:645.

20. Barker, P. S., and F. O. Morrison (1965). The metabolism of TDE by *Proteus vulgaris. Can. J. Zool.* 43:652—654.

21. Barker, P. S., F. O. Morrison, and R. S. Whitaker (1965). Conversion of DDT to DDD by *Proteus vulgaris*, a bacterium isolated from the intestinal flora of a mouse. *Nature* 205:621—622.

22. Bartha, R. (1971). Fate of herbicide-derived chloroanilines in soil. *J. Agric. Food Chem.* 19:385—387.

23. Bartha, R. (1968). Biochemical transformations of anilide herbicides in soil. *J. Agric. Food Chem.* 16:602—604.

24. Bartha, R., and D. Pramer (1967). Pesticide transformation
     to aniline and azo compounds in soil. *Science* 156:1617–1618.

25. Bartha, R., H. A. B. Linke, and D. Pramer (1968). Pesticide
     transformations: Production of chloroazobenzenes from chloro-
     anilines. *Science* 161:582–583.

26. Baxter, R. A., P. E. Gilbert, R. A. Lidgett, J. H. Mainprize,
     and H. A. Vodden (1975). The degradation of polychlorinated
     biphenyls by microorganisms. *Sci. Total Environ.* 4:53–61.

26a. Baxter, R. M., and B. A. Sutherland (1984). Biochemical
     and photochemical processes in the degradation of chlorinated
     biphenyls. *Environ. Sci. Technol.* 18:608–610.

27. Bayly, R. C., and S. Dagley (1969). Oxoenoic acids as meta-
     bolites in the bacterial degradation of catechols. *Biochem. J.*
     111:303–307.

28. Beall, M. L., Jr. (1976). Persistence of aerially applied hexa-
     chlorobenzene on grass and soil. *J. Environ. Qual.* 5:367–369.

29. Beck, J., and K. E. Hansen (1974). The degradation of
     quintozene, pentachlorobenzene, hexachlorobenzene and penta-
     chloroaniline in soil. *Pestic. Sci.* 5:41–48.

30. Belasco, I. J., and H. L. Pease (1969). Investigation of diu-
     ron- and linuron-treated soils for 3,3',4,4'-tetrachloroazoben-
     zene. *J. Agric. Food Chem.* 17:1414–1417.

31. Bell, G. R. (1960). Studies on a soil *Achromobacter* which
     degrades 2,4-dichlorophenoxyacetic acid. *Can. J. Microbiol.*
     6:325–337.

32. Bell, G. R. (1957). Some morphological and biochemical char-
     acteristics of a soil bacterium which decomposes 2,4-dichloro-
     phenoxyacetic acid. *Can. J. Microbiol.* 3:821–840.

33. Beltrame, P., P. L. Beltrame, P. Carniti, and D. Pitea (1982).
     Kinetics of biodegradation of mixtures containing 2,4-dichloro-
     phenol in a continuous stirred reactor. *Water Res.* 16:429–433.

34. Benezet, H. J., and C. O. Knowles (1981). Degradation of
     chlordimeform by algae. *Chemosphere* 10:909–917.

35. Berg, A., K. Carlstrom, J. A. Gustafsson, and M. Ingelman-
     Sundberg (1975). Demonstration of a cytochrome P-450-depen-
     dent steriod 15-beta-hydroxylase in *Bacillus megaterium*. *Bio-
     chem. Biophys. Res. Commun.* 66:1414–1423.

36. Bevenue, A., and H. Beckman (1967). Pentachlorophenol: A
     discussion of its properties and its occurrence as a residue
     in human and animal tissues. *Res. Rev.* 19:83–134.

37. Blackburn, J. W., W. L. Troxler, and G. S. Sayler (1984).
     Prediction of the fates of organic chemicals in a biological
     treatment process—an overview. *Environ. Prog.* 3:163–176.

38. Blackburn, J. W., W. L. Troxler, K. N. Truong, R. P. Zink,
     S. C. Meckstroth, J. R. Florance, A. Groen, G. S. Sayler,
     R. W. Beck, R. A. Minear, A. Breen, and O. Yagi. (1985).

Prediction of the fates of organic chemicals in activated sludge wastewater treatment processes. EPA/600/52-85/102, NTIS PB 85-247674, U.S. Environmental Protection Agency, Cincinnati, OH.

39. Bohinski, R. C. (1973). *Modern Concepts in Biochemistry*. Allyn and Bacon, Inc. Boston, MA.

40. Bollag, J. M., M. A. Loos, and M. Alexander (1967). Enzymatic degradation of phenoxyalkanoate herbicides. *Bacteriol. Proc.* A42:8.

41. Bollag, J. M. (1974). Microbial transformation of pesticides. *Adv. Appl. Microbiol.* 18:75–130.

42. Bollag, J. M. (1972). Biochemical transformation of pesticides by soil fungi. *CRC Crit. Rev. Microbiol.* 2:35–58.

43. Bollag, J. M., and S. Russel (1976). Aerobic versus anaerobic metabolism of halogenated anilines by a *Paracoccus* sp. *Microb. Ecol.* 3:65–73.

44. Bollag, J. M., C. S. Helling, and M. Alexander (1968). 2,4-D metabolism. Enzymatic hydroxylation of chlorinated phenols. *J. Agric. Food Chem.* 16:826–828.

45. Bollag, J. M., C. S. Helling, and M. Alexander (1967). Metabolism of 4-chloro-2-methylphenoxyacetic acid by soil bacteria. *Appl. Microbiol.* 15:1393–1398.

46. Bollag, J. M., G. G. Briggs, J. E. Dawson, and M. Alexander (1968). Enzymatic degradation of chlorocatechols. *J. Agric. Food Chem.* 16:829–833.

47. Bordeleau, L. M., and R. Bartha (1972). Biochemical transformations of herbicide-derived anilines: Purification and characterization of causative enzymes. *Can. J. Microbiol.* 18:1865–1871.

48. Bordeleau, L. M., and R. Bartha (1970). Azobenzene residues from aniline-based herbicides; evidence for labile intermediates. *Bull. Environ. Contam. Toxicol.* 5:34–37.

49. Bordeleau, L. M., and R. Bartha (1972). Biochemical transformation of herbicide-derived anilines: Requirements of molecular configuration. *Can. J. Microbiol.* 18:1873–1882.

50. Bordeleau, L. M., and R. Bartha (1971). Ecology of a herbicide transformation: Synergism of two fungi. *Soil Biol. Biochem.* 3:281–284.

51. Bordeleau, L. M., H. A. B. Linke, and R. Bartha (1972). Herbicide-derived chloroazobenzene residues: Pathway of formation. *J. Agric. Food Chem.* 20:573–578.

51a. Bounds, H. C., and A. R. Colmer (1965). Detoxification of some herbicides by *Streptomyces*. *Weeds* 13:249–252.

52. Bouwer, E. J., and P. L. McCarty (1982). Removal of trace chlorinated organic compounds by activated carbon and fixed-film bacteria. *Environ. Sci. Technol.* 16:836–843.

53. Boyd, S. A., and D. R. Shelton (1984). Anaerobic biodegradation of chlorophenols in fresh and acclimated sludge. *Appl. Environ. Microbiol.* 47:272–277.

54. Boyd, S. A., D. R. Shelton, D. Berry, and J. M. Tiedje (1983). Anaerobic biodegradation of phenolic compounds in digested sludge. *Appl. Environ. Microbiol.* 46:50–54.

55. Braunberg, R. C., and V. Beck (1968). Interaction of DDT and the gastrointestinal microflora of the rat. *J. Agric. Food Chem.* 16:451–453.

56. Briggs G. G., and N. Walker (1973). Microbial metabolism of 4-chloroaniline. *Soil Biol. Biochem.* 5:695–697.

57. Brown, D., and P. Laboureur (1983). The aerobic biodegradability of primary aromatic amines. *Chemosphere* 12:405–414.

58. Burge, W. D. (1973). Transformation of propanil-derived 3,4-dichloroaniline in soil to 3,3',4,4'-tetrachloroazobenzene as related to soil peroxidase activity. *Soil Sci. Soc. Am. Proc.* 37:392–395.

59. Burge, W. D. (1971). Anaerobic decomposition of DDT in soil. Acceleration by volatile components of alfalfa. *J. Agric. Food Chem.* 19:375–378.

60. Burger, K., I. C. Macrae, and M. Alexander (1962). Decomposition of phenoxyalkyl carboxylic acids. *Soil Sci. Soc. Am. Proc.* 26:243–246.

61. Burns and Roe Industrial Services Organization (1982). Fate of priority pollutants in publicly owned treatment works. Final Report. Vol. 1. U. S. Environment Protection Agency, EPA-440/1-82/303. Washington, D.C.

62. Buser, H. R., and H. P. Bosshardt (1975). Studies on the possible formation of polychloroazobenzenes in quintozene treated soil. *Pestic. Sci.* 6:35–41.

63. Buswell, J. A. (1975). Metabolism of phenol and cresols by *Bacillus stearothermophilus. J. Bacteriol.* 124:1077–1083.

64. Buswell, J. A., and J. S. Clark (1976). Oxidation of aromatic acids by a facultative thermophilic *Bacillus* sp. *J. Gen. Microbiol.* 96:209–213.

65. Byast, T. H., and R. J. Hance (1975). Degradation of 2,4,5-T by South Vietnamese soils incubated in the laboratory. *Bull. Environ. Contam. Toxicol.* 14:71–76.

66. Cain, R. B. (1961). The metabolism of protocatechuic acid by a vibrio. *Biochem. J.* 79:298–312.

66a. Cain, R. B. (1980). The uptake and catabolism of lignin-related aromatic compounds and their regulation in microorganisms. pp. 21–60 in Kirk, T. K., T. Higuchi, and H. Chang (eds.). *Lignin Biodegradation: Microbiology, Chemistry, and Potential Applications.* CRC Press, Inc., Boca Raton, FL.

67. Cain, R. B., R. F. Bilton, and J. A. Darrah (1968). The metabolism of aromatic acids by microorganisms. Metabolic pathways in the fungi. *Biochem. J.* 108:797—828.

68. Camoni, I., A. di Muccio, D. Pontecorvo, M. Rubbiani, V. Silano, L. Vergori, C. von Hunolstein, G. Antonini, N. Orsi, and P. Valenti (1983). Lack of in vitro oxidation of 2,3,7,8-tetrachlorodibenzo-*p*-dioxin (TCDD) in the presence of laccase from *Polyporus versicolor* fungus. *Chemosphere* 12:945—949.

68a. Carey, A. E., and G. R. Harvey (1978). Metabolism of polychlorinated biphenyls by marine bacteria. *Bull. Environ. Contam. Toxicol.* 20:527—534.

69. Castro, T. F., and T. Yoshida (1974). Effect of organic matter on the biodegradation of some organochlorine insecticides in submerged soils. *Soil Sci. Plant Nutr.* 20:363—370.

70. Catelani, D., and A. Colombi (1974). Metabolism of biphenyl. Structure and physicochemical properties of 2-hydroxy-6-oxo-6-phenylhexa-2,4-dienoic acid, the *meta*-cleavage product from 2,3-dihydroxybiphenyl by *Pseudomonas putida*. *Biochem. J.* 143:431—434.

71. Catelani, D., A. Colombi, C. Sorlini, and V. Treccani (1973). Metabolism of biphenyl. 2-Hydroxy-6-oxo-phenylhexa-2,4-dienoate: The *meta* cleavage product from 2,3-dihydroxybiphenyl by *Pseudomonas putida*. *Biochem. J.* 134:1063—1066.

71a. Catelani, D., C. Sorlini, and V. Treccani (1971). The metabolism of biphenyl by *Pseudomonas putida*. *Experientia* 27:1173—1174.

72. Catterall, F. A., K. Murray, and P. A. Williams (1971). The configuration of the 1,2-dihydronaphthalene formed in the bacterial metabolism of naphthalene. *Biochem. Biophys. Acta* 237:361—364.

73. Cerniglia, C. E. (1982). Initial reactions in the oxidation of anthracene by *Cunninghamella elegans*. *J. Gen. Microbiol.* 128:2055—2061.

74. Cerniglia, C. E., and D. T. Gibson (1980). Fungal oxidation of (+/-)-9,10-dihydrobenzo(a)pyrene: Formation of diastereomeric benzo(a)pyrene 9,10-diol 7,8-epoxides. *Proc. Natl. Acad. Sci. USA* 77:4554—4558.

75. Cerniglia, C. E., and D. T. Gibson (1978). Metabolism of naphthalene by cell free extracts of *Cunninghamella elegans*. *Arch. Biochem. Biophys.* 186:121—127.

76. Cerniglia, C. E., and D. T. Gibson (1977). Metabolism of naphthalene by *Cunninghamella elegans*. *Appl. Environ. Microbiol.* 34:363—370.

76a. Cerniglia, C. E., D. T. Gibson, and C. van Baalen (1979). Algal oxidation of aromatic hydrocarbons: Formation of 1-naphthol from naphthalene by *Agmenellum quadruplicatum*, strain PR-6. *Biochem. Biophys. Res. Commun.* 88:50—58.

77. Cerniglia, C. E., J. C. Morgan, and D. T. Gibson (1979). Bacterial and fungal oxidation of dibenzofuran. *Biochem. J.* 180:175—185.

78. Cerniglia, C. E., R. L. Hegert, P. J. Szaniszlo, and D. T. Gibson (1978). Fungal transformation of naphthalene. *Arch. Microbiol.* 117:135—143.

78a. Cerniglia, C. E., C. van Baalen, and D. T. Gibson (1980). Metabolism of naphthalene by the cyanobacterium *Oscillatoria* sp., strain JCM. *J. Gen. Microbiol.* 116:485—494.

79. Chacko, C. I., and J. L. Lockwood (1967). Accumulation of DDT and dieldrin by microorganisms. *Can. J. Microbiol.* 13: 1123—1126.

80. Chacko, C. I., J. L. Lockwood, and M. Zabik (1966). Chlorinated hydrocarbon pesticides: Degradation by microbes. *Science* 154:893—895.

81. Chakrabarty, A. H. (1980). Plasmids and dissimilation of synthetic environmental pollutants. pp. 21—30 In *Plasmids and Transposons*. Academic Press, New York.

82. Channa Reddy, C. M. Sugumaran, and C. S. Vaidyanathan (1976). Metabolism of benzoate by a soil pseudomonad. *Ind. J. Biochem. Biophys.* 13:165—169.

83. Chapman, P. J. (1976). Microbial degradation of halogenated compounds. *Biochem. Soc. Trans.* 4:463—466.

84. Chapman, P. J. (1972). An outline of reaction sequences used for the bacterial degradation of phenolic compounds. pp. 17—55 In *Degradation of Synthetic Organic Molecules in the Biosphere.* National Academy of Sciences, Washington, D.C.

84a. Chapman, P. J., and S. Dagley (1962). Oxidation of homogentisic acid by cell-free extracts of a vibrio. *J. Gen. Microbiol.* 28:251—256.

85. Chatterjee, D. K., J. J. Kilbane, and A. M. Chakrabarty (1982). Biodegradation of 2,4,5-trichlorophenoxyacetic acid in soil by a pure culture of *Pseudomonas cepacia*. *Appl. Environ. Microbiol.* 44:514—516.

86. Chatterjee, D. K., S. T. Kellogg, S. Hamada, and A. M. Chakrabarty (1981). Plasmid specifying total degradation of 3-chlorobenzoate by a modified *ortho* pathway. *J. Bacteriol.* 146: 639—646.

87. Chisaka, H., and P. C. Kearney (1970). Metabolism of propanil in soils. *J. Agric. Food Chem.* 18:854—858.

88. Chowdhury, A., D. Vockel, P. N. Moza, W. Klein, and F. Korte (1981). Balance of conversion and degradation of 2,6-dichlorobenzonitrile-^{14}C in humus soil. *Chemosphere* 10:1101—1108.

89. Chu, J., and E. J. Kirsch (1973). Utilization of halophenols by a pentachlorophenol metabolizing bacterium. *Dev. Ind. Microbiol.* 14:264—273.

90. Chu, J., and E. J. Kirsch (1972). Metabolism of pentachlorophenol by an axenic bacterial culture. *Appl. Microbiol.* 23: 1033—1035.

91. Clark, R. R., E. S. K. Chian, and R. A. Griffin (1979). Degradation of polychlorinated biphenyls by mixed microbial cultures. *Appl. Environ. Microbiol.* 37:680—685.

92. Claus, D., and N. Walker (1964). The decomposition of toluene by soil bacteria. *J. Gen. Microbiol.* 36:107—122.

92a. Colwell, R. R., and G. S. Sayler (1978). Bacterial degradation of industrial chemicals in aquatic environments. pp. 111—134 In R. Mitchell (ed.). *Water Pollution Microbiology.* Wiley Interscience, New York.

93. Cook, K. A., and R. B. Cain (1974). Regulation of aromatic metabolism in the fungi: Metabolic control of the 3-oxoadipate pathway in the yeast *Rhodotorula mucilaginosa. J. Gen. Microbiol.* 85:37—50.

94. Corbett, M. D., and B. R. Corbett (1981). Metabolism of 4-chloronitrobenzene by the yeast *Rhodosporidium* sp. *Appl. Environ. Microbiol.* 41:942—949.

95. Crawford, R. L. (1975). Novel pathway for degradation of protocatechuic acid in *Bacillus* species. *J. Bacteriol.* 121:531—536.

96. Crawford, R. L., P. E. Olson, and T. D. Frick (1979). Catabolism of 5-chlorosalicylate by a *Bacillus* isolated from the Mississippi River. *Appl. Environ. Microbiol.* 38:379—384.

97. Crosby, D. G. (1981). Environmental chemistry of pentachlorophenol. *Pure Appl. Chem.* 53:1051—1080.

98. Cserjesi, A. J. (1967). The adaptation of fungi to pentachlorophenol and its biodegradation. *Can. J. Microbiol.* 13:1243—1249.

99. Cserjesi, A. J. (1972). Detoxification of chlorinated phenols. *Int. Biodeterior. Bull.* 8:135—138.

100. Curtis, R. F., C. Dennis, J. M. Gee, M. G. Gee, N. M. Griffiths, D. G. Land, J. L. Peel, and D. Robinson (1974). Chloroanisoles as a cause of musty taint in chickens and their microbiological formation from chlrophenols in broiler house litters. *J. Agric. Food Sci.* 25:811—828.

101. Curtis, R. F., D. G. Land, N. M. Griffiths, M. Gee, D. Robinson, J. L. Peel, C. Dennis, and J. M. Gee (1972). 2,3,4,6-tetrachloroanisole association with musty taint in chickens and microbiological formation. *Nature* 235:223—224.

102. Dagley, S., and D. A. Stopher (1959). A new mode of fission of the benzene nucleus by bacteria. *Biochem. J.* 73:16P.

103. Dagley, S., and D. T. Gibson (1965). The bacterial degradation of catechol. *Biochem. J.* 95:466–474.

104. Dagley, S., W. C. Evans, and D. W. Ribbons (1960). New pathways in the oxidative metabolism of aromatic compounds by microorganisms. *Nature* 188:560–566.

105. Dalton, R. L., A. W. Evans, and R. C. Rhodes (1966). Disappearance of diuron from cotton field soils. *Weeds* 14:31–33.

106. Davies, J. I., and W. C. Evans (1964). Oxidative metabolism of naphthalene by soil pseudomonads. *Biochem. J.* 91:251–261.

107. Dawes, I. W., and I. W. Sutherland (1976). *Microbial Physiology*. John Wiley & Sons, New York.

108. De Vos, R. H., M. C. Ten Noever De Brauw, and P. D. A. Olthof (1974). Residues of pentachloronitrobenzene and related compounds in greenhouse soils. *Bull. Environ. Contam. Toxicol.* 11:567–571.

109. DeLaune, R. D., R. P. Gambrell, and K. S. Reddy (1983). Fate of pentachlorophenol in estuarine sediment. *Environ. Pollut.* (Ser. B)6:297–308.

110. Deo, P. G., and M. Alexander (1976). Ring hydroxylation of p-chlorophenylacetate by an *Arthrobacter* strain. *Appl. Environ. Microbiol.* 32:195–196.

111. DeRose, H. R., and A. S. Newman (1947). The comparison of the persistence of certain plant growth-regulators when applied to soil. *Soil Sci. Soc. Am. Proc.* 12:222–226.

112. Deuel, L. E., Jr., K. W. Brown, F. C. Turner, D. G. Westfall, and J. D. Price (1977). Persistence of propanil, DCA, and TCAB in soil and water under flooded rice culture. *J. Environ. Qual.* 6:127–132.

113. Dewey, O. R., R. V. Lyndsay, and G. S. Hartley (1962). Biological destruction of 2,3,6-trichlorobenzoic acid in soil. *Nature* 195:1232.

114. DiGeronimo, M. J., M. Nikaido, and M. Alexander (1979). Utilization of chlorobenzoates by microbial populations in sewage. *Appl. Environ. Microbiol.* 37:619–625.

115. Dimond, J. B., G. Y. Belyea, R. E. Kadunce, A. S. Getchell, and J. A. Blease (1970). DDT residues in robins and earthworms associated with contaminated forest soils. *Can. Entomol.* 102:1122–1130.

116. Dodge, R. H., C. E. Cerniglia, and D. T. Gibson (1979). Fungal metabolsim of biphenyl. *Biochem. J.* 178:223–230.

117. Dorn, E., and H. J. Knackmuss (1978). Chemical structure and biodegradability of halogenated aromatic compounds. Substituent effects on 1,2-dioxygenation of catechol. *Biochem. J.* 174:85–94.

118. Dorn, E., and H. J. Knackmuss (1978). Chemical structure and biodegradability of halogenated aromatic compounds. Two

catechol 1,2-dioxygenases from a 3-chlorobenzoate-grown pseudomonad. *Biochem. J.* 174:73–84.

119. Dorn, E., M. Hellwig, W. Reineke, and H. J. Knackmuss (1974). Isolation and characterization of a 3-chlorobenzoate degrading pseudomonad. *Arch. Microbiol.* 99:61–70.

120. Dubey, H. D., and J. F. Freeman (1964). Influence of soil properties and microbial activity on the phytotoxicity of linuron and diphenamid. *Soil Sci.* 97:334–340.

121. Duffy, J. R., and N. Wong (1967). Residues of organochlorine insecticides and their metabolites in soil in the Atlantic provinces of Canada. *J. Agric. Food Chem.* 15:457–464.

122. Duncan, C. G., and F. J. Deverall (1964). Degradation of wood preservatives by fungi. *Appl. Microbiol.* 12:57–62.

123. Durham, D. R., and L. N. Ornston (1980). Homologous structural genes and similar induction patterns in *Azotobacter* spp. and *Pseudomonas* spp. *J. Bacteriol.* 143:834–840.

124. Dutton, P. L., and W. C. Evans (1968). The photometabolism of benzoic acid by *Rhodopseudomonas palustris*: A new pathway of aromatic ring metabolism. *Biochem. J.* 105:5P–6P.

125. Dutton, P. L., and W. C. Evans (1969). The metabolism of aromatic compounds by *Rhodopseudomonas palustris*. A new, reductive, method of aromatic ring metabolism. *Biochem. J.* 113:525–536.

126. Duxbury, J. M., J. M. Tiedje, M. Alexander, and J. E. Dawson (1970). 2,4-D metabolism: Enzymatic conversion of chloromaleylacetic acid to succinic acid. *J. Agric. Food Chem.* 18:199–201.

127. E. C. Jordan Co. (1982). Fate of priority pollutants in publicly owned treatment works. 30-day study. U. S. Environmental Protection Agency. EPA-440/1-82/302. Washington, D.C.

128. Edgehill, R. U., and R. K. Finn (1983). Microbial treatment of soil to remove pentachlorophenol. *Appl. Environ. Microbiol.* 45:1122–1125.

129. Edgehill, R. U., and R. K. Finn (1983). Activated sludge treatment of synthetic wastewater containing pentachlorophenol. *Biotechnol. Bioeng.* 25:2165–2176.

130. Edgehill, R. U., and R. K. Finn (1982). Isolation, characterization and growth kinetics of bacteria metabolizing pentachlorophenol. *Eur. J. Appl. Microbiol. Biotechnol.* 16:179–184.

131. Ellis, P. A., and N. D. Camper (1982). Aerobic degradation of diuron by aquatic microorganisms. *J. Environ. Sci. Health* B17:277–289.

132. Ellis, P. A., N. D. Camper, and J. M. Shively (1980). Degradation of selected herbicides by aquatic microorganisms. Technical Report No. 84, Water Resources Research Institute, Clemson University, 36 p.

133. Engelhardt, G., P. R. Wallnofer, and R. Plapp (1972). Identification of N,O-dimethylhydroxylamine as a microbial degradation product of the herbicide, linuron. *Appl. Microbiol.* 23:664—666.

134. Engelhardt, G., P. R. Wallnofer, G. Fuchsbichler, and W. Baumeister. 1977. Bacterial transformations of 4-chloroaniline. *Chemosphere* 6:85—92.

135. Engst, R., and M. Kujawa (1967). Enzymatischer abbau des DDT durch schimmelpilze. 2. Mitt. reaktionsverlauf des enzymatischen DDT-abbaues. *Die Nahrung* 11:751—760.

136. Engst, R., and M. Kujawa (1968). Enzymatischer abbau des DDT durch schimmelpilze. 3. Mitt. darstellung des 2,2-bis (p-chlorphenyl)-acetaldehyds (DDHO) und seine bedeutung im abbaucyclus. *Die Nahrung* 12:783—785.

136a. Engst, R., M. Kujawa, and G. Muller (1967). Enzymatischer abbau des DDT durch schimmelpilze. 1. Mitt. isolierung und identifizierung eines DDT abbauenden schimmelpilzes. *Die Nahrung* 11:401—403.

137. Engst, R., R. M. Macholz, and M. Kujawa (1977). Recent state of lindane metabolism. *Resid. Rev.* 68:59—90.

138. Esaac, E. G., and F. Matsumura (1980). Metabolism of insecticides by reductive systems. *Pharm. Ther.* 9:1—26.

139. Ettinger, M. B., and C. C. Ruchhoft (1950). Persistence of chlorophenols in polluted river water and sewage dilutions. *Sewage Ind. Wastes* 22:1214—1217.

140. Etzel, J. E., and E. J. Kirsch (1974). Biological treatment of contrived and industrial wastewater containing pentachlorophenol. *Dev. Ind. Microbiol.* 16:287—295.

141. Evans, W. C. (1947). Oxidation of phenol and benzoic acid by some soil bacteria. *Biochem. J.* 41:373—382.

142. Evans, W. C., B. S. W. Smith, H. N. Fernley, and J. I. Davies (1971). Bacterial metabolism of 2,4-dichlorophenoxyacetate. *Biochem. J.* 122:543—551.

143. Evans, W. C., B. S. W. Smith, P. Moss, and H. N. Fernley (1971). Bacterial metabolism of 4-chlorophenoxyacetate. *Biochem. J.* 122:509—517.

144. Evans, W. C., B. S. W. Smith, R. P. Linstead, and J. A. Elvidge (1951). Chemistry of the oxidative metabolism of certain aromatic compounds by microorganisms. *Nature* 168:772—775.

145. Evans, W. C., H. N. Fernley, and E. Griffiths (1965). Oxidative metabolism of phenanthrene and anthracene by soil pseudomonads. *Biochem. J.* 95:819—831.

146. Evans, W. C., and B. S. W. Smith (1954). The photochemical inactivation and microbial metabolism of the chlorophenoxyacetic acid herbicides. *Biochem. J.* 57.

147. Farmer, V. C., M. E. K. Henderson, and J. D. Russell (1959). Reduction of certain aromatic acids to aldehydes and alcohols by *Polystictus versicolor*. *Biochem. Biophys. Acta* 35:202–211.

148. Faulkner, J. K., and D. Woodcock (1964). Metabolism of 2,4-dichlorophenoxyacetic acid ('2,4-D') by *Aspergillus niger* van Tiegh. *Nature* 203:865.

149. Faulkner, J. K., and D. Woodcock (1965). Fungal detoxication. Part VII. Metabolism of 2,4-dichlorophenoxyacetic and 4-chloro-2-methylphenoxyacetic acids by *Aspergillus niger*. *J. Chem. Soc.* 1965:1187–1191.

150. Faulkner, J. K., and D. Woodcock (1961). Fungal detoxication. Part V. Metabolism of o- and p-chlorophenoxyacetic acids by *Aspergillus niger*. *J. Chem. Soc.* 5397–5400.

151. Fernley, H. N., E. Griffiths, and W. C. Evans (1964). Oxidative metabolism of phenanthrene and anthracene by soil bacteria: The initial ring fission step. *Biochem. J.* 91:15P–16P.

152. Fernley, H. N., and W. C. Evans (1959). Metabolism of 2,4-dichlorophenoxyacetic acid by a soil *Pseudomonas*: Isolation of γ-chloromuconic acid as an intermediate. *Biochem. J.* 73:22P.

153. Fisher, J. D. (1974). Metabolism of the herbicide pronamide in soil. *J. Agric. Food Chem.* 22:606–608.

154. Fisher, P. R., J. Appleton, and J. M. Pemberton (1978). Isolation and characterization of the pesticide-degrading plasmid pJP1 from *Alcaligenes paradoxus*. *J. Bacteriol.* 135:798–804.

155. Fletcher, C. I., and D. D. Kaufman (1979). Hydroxylation of monochloroaniline pesticide residues by *Fusarium oxysporum*. *J. Agric. Food Chem.* 27:1127–1130.

156. Focht, D. D. (1972). Microbial degradation of DDT metabolites to carbon dioxide, water, and chloride. *Bull. Environ. Contam. Toxicol.* 7:52–56.

156a. Focht, D. D., and M. Alexander (1970). DDT metabolites and analogs: Ring fission by *Hydrogenomonas*. *Science* 170:91–92.

156b. Focht, D. D., and M. Alexander (1971). Aerobic cometabolism of DDT analogues by *Hydrogenomonas* sp. *J. Agric. Food Chem.* 19:20–22.

157. Francis, A. J., R. J. Spanggord, G. I. Ouchi, R. Bramhall, and N. Bohonos (1976). Metabolism of DDT analogues by a *Pseudomonas* sp. *Appl. Environ. Microbiol.* 32:213–216.

158. Francis, A. J., R. J. Spanggord, G. I. Ouchi, and N. Bohonos (1978). Cometabolism of DDT analogs by a *Pseudomonas* sp. *Appl. Environ. Microbiol.* 35:364–367.

159. Franzke, C. L., M. Kujawa, and R. Engst (1970). Enzyma-
     tischer abbau des DDT durch schimmelpilze. 4. Mitt. eingluss
     des DDT auf das wachstum von *Fusarium oxysporum* sowie
     auf die pilzesterase. *Die Nahrung* 14:339−346.
160. French, A. L., and R. A. Hoopingarner (1970). Dechlorina-
     tion of DDT by membranes isolated from *Escherichia coli*. *J.
     Econ. Entomol.* 63:756−759.
161. Fries, G. R., G. S. Marrow, and C. H. Gordon (1969). Me-
     tabolism of *o,p'*-DDT by rumen microorganisms. *J. Agr. Food
     Chem.* 17:860−862.
162. Furukawa, K. (1982). Microbial degradation of polychlorinated
     biphenyls (PCBs). pp. 33−57 In A. M. Chakrabarty (ed.).
     Biodegradation and Detoxification of Environmental Pollutants.
     CRC Press, Inc., Boca Raton, FL.
163. Furukawa, K., and F. Matsumura (1976). Microbial metabolism
     of polychlorinated biphenyls. Studies on the relative degrad-
     ability of polychlorinated biphenyls components by *Alcaligenes*
     sp. *J. Agric. Food Chem.* 24:251−256.
164. Furukawa, K., F. Matsumura, and K. Tonomura. 1978. *Al-
     caligenes* and *Acinetobacter* strains capable of degrading
     polychlorinated biphenyls. *Agric. Biol. Chem.* 42:543−548.
165. Furukawa, K., K. Tonomura, and A. Kamibayashi. 1978.
     Effect of chlorine substitution on the biodegradability of poly-
     chlorinated biphenyls. *Appl. Environ. Microbiol.* 35:223−227.
166. Furukawa, K., K. Tonomura, and A. Kamibayashi (1979).
     Metabolism of 2,4,4'-trichlorobiphenyl by *Acinetobacter* sp.
     P6. *Agric. Biol. Chem.* 43:1577−1583.
167. Furukawa, K., N. Tonomura, and A. Kamibayashi (1983).
     Metabolic breakdown of Kaneclors (polychlorobiphenyls) and
     their products by *Acinetobacter* sp. *Appl. Environ. Microbiol.*
     46:140−145.
168. Furukawa, K., N. Tomizuka, and A. Kamibayashi (1979).
     Effect of chlorine substitution on the bacterial metabolism of
     various polychlorinated biphenyls. *Appl. Environ. Microbiol.*
     38:301−310.
169. Gamar, Y., and J. K. Gaunt (1971). Bacterial metabolism of
     4-chloro-2-methylphenoxyacetate. Formation of glyoxylate by
     side-chain cleavage. *Biochem. J.* 122:527−531.
170. Gaunt, J. K., and W. C. Evans (1971). Metabolism of 4-chloro-
     2-methylphenoxyacetate by a soil pseudomonad. Preliminary
     evidence for the metabolic pathway. *Biochem. J.* 122:519−526.
171. Gaunt, J. K., and W. C. Evans (1971). Metabolism of 4-chloro-
     2-methylphenoxyacetate by a soil pseudomonad. *Biochem. J.*
     122:533−542.
171a. Gaunt, J. K., and W. C. Evans (1961). Metabolism of 4-chloro-
     2-methylphenoxyacetic acid by a soil microorganism. *Biochem.
     J.* 79:25P−26P.

172. Gee, J. M., and J. L. Peel (1974). Metabolism of 2,3,4,6-tetrachlorophenol by microorganisms from broiler house litter. *J. Gen. Microbiol.* 85:237—243.

173. Geissbuhler, H. C., C. Haselbach, H. Aebi, and L. Ebner (1963). The fate of N'-(4-chlorophenoxy)-phenyl-N,N-dimethyl-urea C-1983 in soils and plants. III. Breakdown in soils and plants. *Weed Res.* 3:277—297.

174. Geissbuhler, H., H. Martin, and G. Voss (1975). The substituted ureas. pp. 209—291 In Kearney, P. C. and D. D. Kaufman (eds). *Herbicides: Chemistry, Degradation, and Mode of Action*, 2nd Ed., Vol. 1, Marcel Dekker, Inc., New York.

175. Gibson, D. T. (1976). Initial reactions in the bacterial degradation of aromatic hydrocarbons. *Zbl. Bakt. Hyg., I. ABT. Orig. B* 162:157—168.

176. Gibson, D. T. (1972). Initial reactions in the degradation of aromatic hydrocarbons. pp. 116—136 In *Degradation of Synthetic Organic Molecules in the Biosphere.* National Academy of Sciences, Washington, D.C.

177. Gibson, D. T. (1978). Microbial transformations of aromatic pollutants. pp. 187—204 In Hutzinger, O., I. H. van Lelyveld and B. C. J. Zoeteman (eds.). *Aquatic Pollutants: Transformation and Biological Effects.* Pergamon Press, New York.

177a. Gibson, D. T., and V. Subramanian (1984). Microbial degradation of aromatic hydrocarbons. pp. 181—252 In Gibson, D. T. (ed.). *Microbial Degradation of Organic Compounds.* Marcel Dekker, Inc., New York.

178. Gibson, D. T., G. E. Cardini, F. C. Maseles, and R. E. Kallio (1970). Incorporation of oxygen-18 into benzene by *Pseudomonas putida. Biochemistry* 9:1631—1635.

179. Gibson, D. T., Hensley, M., H. Hoshioka, and T. J. Mabry (1970). Formation of (+)-cis2,3-dihydroxy-1-methylcyclohexa-4,6-diene from toluene by *Pseudomonas putida. Biochemistry* 9:1626—1630.

180. Gibson, D. T., J. M. Wood, P. J. Chapman, and S. Dagley (1967). Bacterial degradation of aromatic compounds. *Biotechnol. Bioeng.* 9:33—44.

181. Gibson, D. T., J. R. Koch, and R. E. Kallio (1968). Oxidative degradation of aromatic hydrocarbons by microorganisms. I. Enzymatic formation of catechol from benzene. *Biochemistry* 7:2653—2662.

182. Gibson, D. T., J. R. Koch, C. L. Schuld, and R. E. Kallio (1968). Oxidative degradation of aromatic hydrocarbons by microorganisms. II. Metabolism of halogenated aromatic hydrocarbons. *Biochemistry* 7:3795—3802.

183. Gibson, D. T., R. L. Roberts, M. C. Wells, and V. M.
      Kobal (1973). Oxidation of biphenyl by a *Beijerinckia* species.
      Biochem. Biophys. Res. Commun. 50:211–219.

184. Glass, B. L. (1972). Relation between the degradation of
      DDT and the iron redox system in soils. *J. Agric. Food Chem.*
      20:324–327.

184a. Gledhill, W. E. (1975). Biodegradation of 3,4,4'-trichlorocar-
      banilide, TCC, in sewage and activated sludge. *Water Res.*
      9:649–654.

184b. Goldman, P., G. W. A. Milne, and D. B. Keister (1968).
      Carbon-halogen bond cleavage. III. Studies on bacterial hali-
      dohydrolases. *J. Biol. Chem.* 243:428–434.

185. Golovleva, L. A., R. N. Pertsova, A. M. Boronin, V. G.
      Grishchenkov, B. P. Baskunov, and A. V. Polyakova (1982).
      Degradation of polychloroaromatic insecticides by *Pseudomonas
      aeruginosa* containing biodegradation plasmids. *Microbiology*
      51:772–777.

186. Grant, D. J. W., and J. V. Wilson (1973). Degradation and
      hydrolysis of amides by *Corynebacterium pseudodiphtheriticum*
      NCIB 10803. *Microbios* 8:15–22.

186a. Grayson, M., and D. Eckroth (eds.) (1980). *Kirk-Othmer
      Encyclopedia of Chemical Technology.* Vol. 12, Wiley & Sons,
      New York.

187. Grishchenkov, V. G., I. E. Fedechkina, B. P. Gaskunov,
      L. A. Anisimova, A. M. Boronin, and L. A. Golovleva (1984).
      Degradation of 3-chlorobenzoic acid by *Pseudomonas putida*
      strain. *Microbiology* 52:602–606.

188. Gross, S. R., R. D. Gafford, and E. L. Tatum (1956). The
      metabolism of protocatechuic acid by *Neurospora. J. Biol.
      Chem.* 219:781–796.

189. Groves, K., and K. S. Chough (1970). Fate of the fungicide
      2,6-dichloro-4-nitroaniline (DCNA) in plants and soils. *J.
      Agric. Food Chem.* 18:1127–1128.

189a. Guenzi, W. D., and W. E. Beard (1967). Anaerobic biodegra-
      dation of DDT to DDD in soil. *Science* 156:1116–1117.

190. Guenzi, W. D., and W. E. Beard (1968). Anaerobic conver-
      sion of DDT to DDD and aerobic stability of DDT in soil.
      Soil Sci. Soc. Am. Proc. 32:522–524.

191. Guyer, M., and G. Hegeman (1969). Evidence for a reductive
      pathway for the anaerobic metabolism of benzoate. *J. Bacter-
      iol.* 99:906–907.

192. Haider, K. (1979). Degradation and metabolization of lindane
      and other hexachlorocyclohexane isomers by anaerobic and
      aerobic soil microorganisms. *Z. Naturforsch.* 34C:1066–1069.

193. Haller, H. D. (1978). Degradation of mono-substituted ben-
      zoates and phenols by wastewater. *J. Water Pollut. Control
      Fed.* 50:2771–2777.

194. Haller, H. D., and R. K. Finn (1979). Biodegradation of 3-chlorobenzoate and formation of black color in the presence and absence of benzoate. *Eur. J. Appl. Microbiol. Biotechnol.* 8:191–205.

195. Hartmann, J., W. Reineke, and H. J. Knackmuss (1979). Metabolism of 3-chloro-, 4-chloro-, and 3,5-dichlorobenzoate by a pseudomonad. *Appl. Environ. Microbiol.* 37:421–428.

196. Hegeman, G. D. (1972). The evolution of metabolic pathways in bacteria. pp. 56-72 In *Degradation of Synthetic Organic Molecules in the Biosphere.* National Academy of Sciences, Washington, D.C.

197. Helling, C. S., A. R. Isensee, E. A. Woolson, P. D. J. Ensor, G. E. Jones, J. R. Plimmer, and P. C. Kearney (1973). Chlorodioxins in pesticides, soils, and plants. *J. Environ. Qual.* 2:171–178.

198. Helling, C. S., J. M. Bollag, and J. E. Dawson (1968). Cleavage of ether-oxygen bond in phenoxyacetic acid by an *Arthrobacter* sp. *J. Agric. Food Chem.* 16:538–539.

199. Helm, V., and H. Reber (1979). Investigation on the regulation of aniline utilization in *Pseudomonas multivorans* strain AN 1. *Eur. J. Appl. Microbiol. Biotechnol.* 7:191–199.

200. Herbst, E., I. Scheunert, W. Klein, and F. Korte (1977). Fate of PCBs-^{14}C in sewage treatment - laboratory experiments with activated sludge. *Chemosphere* 6:725–730.

201. Hicks G. F., Jr., and T. R. Corner (1973). Location and consequences of 1,1,1-trichloro-2,2-bis(p-chlorophenyl)ethane uptake by *Bacillus megaterium. Appl. Microbiol.* 25:381–387.

202. Hill, D. W., and P. L. McCarty (1967). Anaerobic degradation of selected chlorinated hydrocarbon pesticides. *J. Water Pollut. Control Fed.* 39:1259–1277.

203. Hill, G. D., J. W. McGahen, H. M. Baker, D. W. Finnerty, and C. W. Bingeman (1955). The fate of substituted urea herbicides in agricultural soils. *Agron. J.* 47:93–104.

204. Hock, W. K., and H. D. Sisler (1969). Metabolism of chloroneb by *Rhizoctonia solani* and other fungi, *J. Agric. Food Chem.* 17:123–128.

205. Hogn, T., and L. Jaenicke (1972). Benzene metabolism of *Moraxella* species. *Eur. J. Biochem.* 30:369–375.

206. Horowitz, A., J. M. Suflita, and J. M. Tiedje (1983). Reductive dehalogenation of halobenzoates by anaerobic lake sediment microorganisms. *Appl. Environ. Microbiol.* 45:1459–1465.

207. Horvath, R. S. (1970). Microbial cometabolism of 2,4,5-trichlorophenoxyacetic acid. *Bull. Environ. Contam. Toxicol.* 5:537–541.

208. Horvath, R. S. (1971). Cometabolism of the herbicide 2,3,6-trichlorobenzoate. *J. Agric. Food Chem.* 19:291–293.

209. Horvath, R. S. (1970). Cometabolism of methyl- and chloro-substituted catechols by an *Achromobacter* sp. possessing a new *meta*-cleaving oxygenase. *Biochem. J.* 119:871–876.

210. Horvath, R. S., and M. Alexander (1970). Cometabolism: A technique for the accumulation of biochemical products. *Can J. Microbiol.* 15:1131–1132.

211. Horvath, R. S., and M. Alexander (1970). Cometabolism of *m*-chlorobenzoate by an *Arthrobacter*. *Appl. Microbiol.* 20: 254–258.

212. Horvath, R. S., J. E. Dotzlaf, and R. Kreger (1975). Cometabolism of *m*-chlorobenzoate by natural microbial populations grown under cosubstrate enrichment conditions. *Bull. Environ. Contam. Toxicol.* 13:357–361.

213. Hughes, A. F., and C. T. Corke (1974). Formation of tetrachloroazobenzene in some Canadian soils treated with propanil and 3,4-dichloroaniline. *Can. J. Microbiol.* 20:35–39.

214. Hütter, R., and M. Philippi (1982). Studies on microbial metabolism of TCDD under laboratory conditions. pp. 87–93 In Hutzinger, O. R., W. Frei, E. Merian, and F. Pocchiari (eds.). *Chlorinated Dioxins and Related Compounds. Impact on the Environment*. Pergamon Press, New York.

214a. Hutzinger, O., and W. Verkamp (1981). Xenobiotic chemicals with pollution potential. pp. 3–46 In T. Leisinger, A. M. Cook, R. Hütter and J. Nuesch (eds). *Microbial Degradation of Xenobiotic and Recalcitrant Compounds*. Academic Press, London.

215. Hutzinger, O., S. Safe, and V. Zitko (1974). Commercial PCB preparations: Properties and composition. pp. 7–39 In *The Chemistry of PCB's*, CRC Press, Cleveland.

216. Ichihara, A., K. Adachi, K. Hosokawa, and Y. Takeda (1962). The enzymatic hydroxylation of aromatic carboxylic acids; substrate specificities of anthranilate and benzoate oxidases. *J. Biol. Chem.* 237:2296–2302.

217. Ide, A., Y. Niki, F. Sakamoto, I. Watanabe, and H. Watanabe (1972). Decomposition of pentachlorophenol in paddy soil. *Agric. Biol. Chem.* 36:1937–1944.

218. Ingols, R. S., P. E. Gaffney, and P. C. Stevenson (1966). Biological activity of halophenols. *J. Water Pollut. Control Fed.* 38:629–635.

219. Isensee, A. R., D. D. Kaufman, and G. E. Jones (1982). Fate of 3,4-dichloroaniline in a rice (*Oryza sativa*)-paddy microecosystem. *Weed Sci.* 30:608–613.

220. Iwan, J., G.-A. Hoyer, D. Rosenberg, and D. Goller (1976). Transformations of 4-chloro-o-toluidine in soil: Generation of coupling products by one-electron oxidation. *Pestic. Sci.* 7: 621–631.

220a. Iwata, Y., W. E. Westlake, and F. A. Gunther (1972). Varying persistence of polychlorinated biphenyls in six California soils under laboratory conditions. *Bull. Environ. Contam. Toxicol.* 9:204–211.

221. Jacobson, S. N., and M. Alexander (1981). Enhancement of the microbial dehalogenation of a model chlorinated compound. *Appl. Environ. Microbiol.* 42:1062–1066.

222. Janke, D. O. V. Maltseva, L. A. Golovleva, and W. Fritsche (1984). On the relation between cometabolic monochloroaniline turnover and intermediary metabolism in *Rhodococcus* AN 117. *Z. Allg. Mikrobiol.* 24:305–316.

223. Janke, D., P. Baskunov, M. Y. Nefedova, A. M. Zyankun, and L. A. Golovleva (1984). Incorporation of $^{18}O_2$ during cometabolic degradation of 3-chloroaniline by *Rhodococcus* sp. AN 117. *Z. Allg. Mikrobiol.* 24:253–259.

224. Jeffrey, A. M., H. J. C. Yeh, D. M. Jerina, T. R. Patel, J. F. Davey, and D. T. Gibson (1975). Initial reactions in the oxidation of naphthalene by *Pseudomonas putida*. *Biochemistry* 14:575–584.

225. Jenson, H. L., and H. I. Petersen (1952). Decomposition of hormone herbicides by bacteria. *Acta Agric. Scand.* 2:215–231.

226. Jerina, D. M., H. Selander, H. Yagi, M. C. Wells, J. F. Davey, V. Mahadevan, and D. T. Gibson (1976). Dihydrodiols from anthracene and phenanthrene. *J. Am. Chem. Soc.* 98:5988–5996.

227. Jerina, D. M., J. W. Daly, A. Jeffrey, and D. T. Gibson (1971). *Cis*-1,2-dihydroxy-1,2-dihydronaphthalene: A bacterial metabolite from naphthalene. *Arch. Biochem. Biophys.* 142:394–396.

228. Johnsen, R. E. (1976). DDT metabolism in microbial systems. *Residue Rev.* 61:1–28.

229. Johnson, B. T., and J. O. Kennedy (1973). Biomagnification of *p,p'*-DDT and methoxychlor by bacteria. *Appl. Microbiol.* 26:66–71.

230. Johnson, B. T., R. N. Goodman, and H. S. Goldberg (1967). Conversion of DDT to DDD by pathogenic and saprophytic bacteria associated with plants. *Science* 157:560–561.

231. Johnston, H. W., G. G. Briggs, and M. Alexander (1972). Metabolism of 3-chlorobenzoic acid by a pseudomonad. *Soil Biol. Biochem.* 4:187–190.

232. Juengst, F. W., Jr., and M. Alexander (1976). Conversion of 1,1,1-trichloro-2,2-bis(*p*-chlorophenyl)ethane (DDT) to water-soluble products by microorganisms. *J. Agric. Food Chem.* 24:111–115.

233. Kaiser, K. L. E., and P. T. S. Wong (1974). Bacterial degradation of polychlorinated biphenyls. I. Identification of some metabolic products from Aroclor 1242. *Bull. Environ. Contam. Toxicol.* 11:291−296.

234. Kallman, B. J., and A. K. Andrews (1963). Reductive dechlorination of DDT to DDD by yeast. *Science* 141:1050−1051.

235. Kaminski, U., D. Janke, H. Prauser, and W. Fritsche (1983). Degradation of aniline and monochloroanilines by *Rhodococcus* sp. AN 117 and a pseudomonad: A comparative study. *Z. Allg. Mikrobiol.* 23:235−246.

236. Kaneko, M. K. Morimoto, and S. Nambu (1976). The response of activated sludge to a polychlorinated biphenyl (KC-500). *Water Res.* 10:157−163.

237. Karns, J. S., J. J. Kilbane, S. Duttagupta, and A. M. Chakrabarty. 1983. Metabolism of halophenols by 2,4,5-trichlorophenoxyacetic acid-degrading *Pseudomonas cepacia*. *Appl. Environ. Microbiol.* 46:1176−1181.

238. Katagiri, M., and O. Hayaishi (1957). Enzymatic degradation of β-ketoadipic acid. *J. Biol. Chem.* 226:439−448.

239. Kaufman, D. D. (1967). Degradation of carbamate herbicides in soil. *J. Agric. Food Chem.* 14:582−591.

240. Kaufman, D. D., and J. Blake (1973). Microbial degradation of several acetamide, acylanilide, carbamate, toluidine and urea pesticides. *Soil Biol. Biochem.* 5:297−308.

241. Kaufman, D. D., and P. C. Kearney (1965). Microbial degradation of isopropyl-N-3-chlorophenylcarbamate and 2-chloroethyl-N-3-chlorophenylcarbamate. *Appl. Microbiol.* 13:443−446.

242. Kaufman, D. D., J. R. Plimmer, and U. I. Klingebiel (1973). Microbial oxidation of 4-chloroaniline. *J. Agric. Food Chem.* 21:127−132.

243. Kaufman, D. D., J. R. Plimmer, J. Iwan, and U. I. Kingebiel (1972). 3,3',4,4'-tetrachloroazoxybenzene from 3,4-dichloroaniline in microbial culture. *J. Agric. Food Chem.* 20:916−919.

244. Kearney, P. C., and D. D. Kaufman (1965). Enzyme from soil bacterium hydrolyzes phenylcarbamate herbicides. *Science* 147:740−741.

245. Kearney, P. C., and J. R. Plimmer (1972). Metabolism of 3,4-dichloroaniline in soils. *J. Agric. Food Chem.* 20:584−585.

246. Kearney, P. C., E. A. Woolson, and C. P. Ellington, Jr. (1972). Persistence and metabolism of chlorodioxins in soils. *Environ. Sci. Technol.* 6:1017−1019.

247. Keil, J. E., S. H. Sandifer, C. D. Graber, and L. E. Priester (1972). DDT and polychlorinated biphenyl (Aroclor 1242). Effects of uptake on *E. coli* growth. *Water Res.* 6:837−841.

248. Khan, S. U., P. B. Marriage, and W. J. Saidak (1976). Persistence and movement of diuron and 3,4-dichloroaniline in an orchard soil. *Weed Sci.* 24:583—586.

249. Kilbane, J. J., D. K. Chatterjee, J. S. Karns, S. T. Kellogg, and A. M. Chakrabarty (1982). Biodegradation of 2,4,5-trichlorophenoxyacetic acid by a pure culture of *Pseudomonas cepacia. Appl. Environ. Microbiol.* 44:72—78.

250. Kincannon, D. F., E. L. Stover, and Y. P. Chung (1981). Biological treatment of organic compounds found in industrial aqueous effluents. Am. Chem. Soc. Nat. Meet., Atlanta Ga.

251. Kincannon, D. F., and E. L. Stover (1983). Determination of activated sludge biokinetic constants for chemical and plastic industrial wastewaters. NTIS Report PB83-245233, (EPA Report EPA-600/2-83-073A).

252. Kirkpatrick, D., S. R. Biggs, B. Conway, C. M. Finn, D. R. Hawkins, T. Honda, M. Ishida, and G. P. Powell (1981). Metabolism of N-(2,3-dichlorophenyl)-3,4,5,6-tetrachlorophthalamic acid (techlofthalam) in paddy soil and rice. *J. Agric. Food Chem.* 29:1149—1153.

253. Kirsch, E. J., and J. E. Etzel (1973). Microbial decomposition of pentachlorophenol. *J. Water Pollut. Control Fed.* 45:359—364.

254. Kitagawa, M. (1956). Studies on the oxidation mechanism of methyl group. *J. Biochem.* 43:553—563.

255. Kiyohara, H., and K. Nagao (1978). The catabolism of phenanthrene and naphthalene by bacteria. Pseudomonads. *J. Gen. Microbiol.* 105:69—75.

256. Kiyohara, H., K. Nagao, and R. Nomi (1976). Degradation of phenanthrene through o-phthalate by an *Aeromonas* sp. *Agric. Biol. Chem.* 40:1075—1082.

257. Klages, U., and F. Lingens (1980). Degradation of 4-chlorobenzoic acid by a *Pseudomonas* sp. *Zbl. Bakt. Hyg., I. Abt. Orig. C* 1:215—223.

258. Klages, U., A. Markus, and F. Lingens (1981). Degradation of 4-chlorophenylacetic acid by a *Pseudomonas* species. *J. Bacteriol.* 146:64—68.

259. Klecka, G. M., and D. T. Gibson (1979). Metabolism of dibenzo-(1,4)dioxan by a *Pseudomonas* species. *Biochem. J.* 180:639—645.

260. Klecka, G. M., and D. T. Gibson (1980). Metabolism of dibenzo-p-dioxin and chlorinated dibenzo-p-dioxins by a *Beijerinckia* species. *Appl. Environ. Microbiol.* 39:228—296.

261. Knackmuss, H. J. (1981). Degradation of halogenated and sulfonated hydrocarbons. pp. 189—212 In Leisinger, T., R. Hütter, A. M. Cook, and J. Nuesch (eds.). *Microbial Degra-*

dation of Xenobiotics and Recalcitrant Compounds, Academic Press, New York.

262. Knackmuss, H. J., and M. Hellwig (1978). Utilization and cooxidation of chlorinated phenols by *Pseudomonas* sp. B13. *Arch. Microbiol.* 117:1—7.

263. Ko, W. H., and J. L. Lockwood (1968). Accumulation and concentration of chlorinated hydrocarbon pesticides by microorganisms in soil. *Can. J. Microbiol.* 14:1075—1078.

264. Kocher, J., F. Lingens, and W. Koch (1976). Untersuchungen zum abbau des herbizids chlorphenprop-methyl im boden und durch mikroorganismen. *Weed Res.* 16:93—100.

265. Kong, H. L., and G. S. Sayler (1983). Degradation and total mineralization of monohalogenated biphenyls in natural sediment and mixed bacterial culture. *Appl. Environ. Microbiol.* 46:666—672.

265a. Kujawa, M., and R. Engst (1970). Enzymatischer abbau des DDT durch schimmelplize. 5. Mitt. versuche zur fraktionierung des kultur-filtrats. *Die Nahrung* 14:347—355.

266. Kuwatsuka, S., and M. Igarashi (1975). Degradation of PCP in soils. II. The relationship between the degradation of PCP and the properties of soils, and the identification of the degradation products of PCP. *Soil Sci. Plant Nutr.* 21:405—414.

266a. Lack, L. (1959). The enzymatic oxidation of gentisic acid. *Biochim. Biophys. Acta* 34:117—123.

267. Lal, R., and D. M. Saxena (1982). Accumulation, metabolism, and effects of organochlorine insecticides on microorganisms. *Microbiol. Rev.* 46:95—127.

268. Langlois, B. E., J. A. Collins, and K. G. Sides (1970). Some factors affecting degradation of organochlorine pesticides by bacteria. *J. Dairy Sci.* 53:1671—1675.

269. Lanzilotta, R. P., and D. Pramer (1970). Herbicide transformation. I. Studies with whole cells of *Fusarium solani*. *Appl. Microbiol.* 19:301—306.

270. Lanzilotta, R. P., and D. Pramer (1970). Herbicide transformation. II. Studies with an acylamidase of *Fusarium solani*. *Appl. Microbiol.* 19:307—313.

271. Lanzilotta, R. P., R. Bartha, and D. Pramer (1967). Microbial transformations of the herbicide 3',4'-dichloropropionalide. *Bacteriol. Proc.* A45:8.

272. Larsson, P. (1981). Transport of ^{14}C-labelled PCB compounds from sediment to water and from water to air in laboratory model systems. *Water Res.* 17:1317—1326.

273. Leadbetter, E. R., and J. W. Foster (1959). Incorporation of molecular oxygen in bacterial cells utilizing hydrocarbons for growth. *Nature* 184:1428—1429.

274. Leatham, G. F., R. L. Crawford, and T. K. Kirk (1983). Degradation of phenolic compounds and ring cleavage of catechol by *Phanerochaete chrysosporium*. *Appl. Environ. Microbiol.* 46:191—197.

275. Leather, G. R., and C. L. Foy (1977). Metabolism of bifenox in soil and plants. *Pestic. Biochem. Physiol.* 7:437—442.

276. Lehninger, A. L. (1982). *Principles of Biochemistry*. Worth Publishers, Inc. New York.

277. Leutritz, J., Jr. (1965). Biodegradability of pentachlorophenol. *Forest Prod. J.* 15:269—272.

278. Lichtenstein, E. P., T. W. Fuhremann, and K. R. Schulz (1971). Persistence and vertical distribution of DDT, lindane and aldrin residues. *J. Agric. Food Chem.* 19:718—721.

279. Lillis, V., K. S. Dodgson, G. F. White, and W. J. Payne (1983). Initiation of activation of a preemergent herbicide by a novel alkylsufatase of *Pseudomonas putida* FLA. *Appl. Environ. Microbiol.* 46:988—994.

280. Liu, D. (1982). Assessment of continuous biodegradation of commercial PCB formulations. *Bull. Environ. Contam. Toxicol.* 29:200—207.

281. Liu, D. (1980). Enhancement of PCBs biodegradation by sodium ligninsulfonate. *Water Res.* 14:1467—1475.

282. Liu, D., K. Thomson, and W. M. J. Strachan (1981). Biodegradation of pentachlorophenol in a simulated aquatic environment. *Bull. Environ. Contam. Toxicol.* 26:85—90.

283. Loos, M. A. (1975). Phenoxyalkanoic acids. pp. 1—128 in Kearney, P. C. and D. D. Kaufman (eds), *Herbicides: Chemistry, Degradation, and Mode of Action*, 2nd ed. Vol. 1, Marcel Dekker Inc., New York.

284. Loos, M. A., J. M. Bollag, and M. Alexander (1967). Phenoxyacetate herbicide detoxication by bacterial enzymes. *J. Agric. Food Chem.* 15:858—860.

285. Loos, M. A., R. N. Roberts, and M. Alexander (1967). Phenols as intermediates in the decomposition of phenoxyacetates by an *Arthrobacter* species. *Can. J. Microbiol.* 13:679—690.

286. Loos, M. A., R. N. Roberts, and M. Alexander (1967). Formation of 2,4-dichlorophenol and 2,4-dichloroanisole from 2,4-dichlorophenoxyacetate by *Arthrobacter* sp. *Can. J. Microbiol.* 13:691—699.

287. Lunt, D., and W. C. Evans (1970). The microbial metabolism of biphenyl. *Biochem. J.* 118:54P—55P.

288. Lyr, H. (1963). Enzymatische detoxifikation chlorieter phenole. *Phytopathol. Z.* 47:73—83.

289. MacDonald, D. L., R. Y. Stanier, and J. L. Ingraham (1954). The enzymatic formation of β-carboxymuconic acid. *J. Biol. Chem.* 210:809—820.

290. Macrae, I. C., and M. Alexander (1964). Use of gas chromatography for the demonstration of a pathway of phenoxy herbicide degradation. *Agron. J.* 56:91–92.

291. Macrae, I. C., and M. Alexander (1963). Metabolism of phenoxyalkyl carboxylic acids by a *Flavobacterium* species. *J. Bacteriol.* 86:1231–1235.

292. Macrae, I. C., M. Alexander, and A. D. Rovira (1963). The decomposition of 4-(2,4-dichlorophenoxy)butyric acid by *Flavobacterium* sp. *J. Gen. Microbiol.* 32:69–76.

293. Malaney, G. W. (1960). Oxidation abilities of aniline-acclimated activated sludge. *J. Water Pollut. Control Fed.* 32:1300–1311.

294. Maniatis, T., E. F. Fritsch, and J. Sambrook (1982). *Molecular Cloning. A Laboratory Manual.* Cold Spring Harbor Laboratory, New York.

295. Marinucci, A. C., and R. Bartha (1979). Biodegradation of 1,2,3- and 1,2,4-trichlorobenzene in soil and in liquid enrichment culture. *Appl. Environ. Microbiol.* 38:811–817.

296. Markus, A., U. Klages, and F. Lingens (1982). Chemische sythese von 3-chlor-4-hydroxy-, and 4-chlor-2-hydroxyphenylessigsaure. *Hoppe-Seyler's Z. Physiol. Chem.* 363:431–437.

297. Martens, R. (1978). Degradation of the herbicide [^{14}C]-diclofopmethyl in soil under different conditions. *Pestic. Sci.* 9:127–134.

298. Masse, R., F. Messier, L. Peloquin, C. Ayotte, and M. Sylvestre (1984). Microbial biodegradation of 4-chlorobiphenyl, a model compound of chlorinated biphenyls. *Appl. Environ. Microbiol.* 47:947–951.

299. Matsumura, F., and C. R. Krishna Murti (1982). *Biodegradation of Pesticides.* Plenum Press, New York.

300. Matsumura, F., and G. M. Boush (1968). Degradation of insecticides by a soil fungus, *Trichoderma viride*. *J. Econ. Entomol.* 61:610–612.

301. Matsumura, R., and H. J. Benezet (1973). Studies on the bioaccumulation and microbial degradation of 2,3,7,8-tetrachlorodibenzo-p-dioxin. *Environ. Health Perspect.* 5:253–258.

302. Matsumura, F., K. C. Patil, and G. M. Boush (1971). DDT metabolized by microorganisms from Lake Michigan. *Nature* 230:325–326.

303. Matter-Muller. C., W. Gujer, W. Giger, and W. Stumm (1980). Nonbiological elimination mechanisms in a biological sewage treatment plant. *Prog. Water Tech.* 12:299–314.

304. McCall, P. J., S. A. Vrona, and S. S. Kelley (1981). Fate of uniformly carbon-14 ring labeled 2,4,5-trichlorophenoxyacetic acid and 2,4-dichlorophenoxyacetic acid. *J. Agric. Food Chem.* 29:100–107.

305. McClure, G. W. (1974). Degradation of anilide herbicides by propham-adapted microorganisms. *Weed Sci.* 22:323—329.

306. McCormick, L. L., and A. E. Hiltbold (1966). Microbiological decomposition of atrazine and diuron in soil. *Weeds* 14:77—82.

307. Meagher, R. B., and L. N. Ornston (1973). Relationships among enzymes of the β-ketoadipate pathway. I. Properties of *cis,cis*-muconate-lactonizing enzyme and muconolactone isomerase from *Pseudomonas putida*. *Biochemistry* 12:3523—3530.

308. Mendel, J. L., and M. S. Walton (1966). Conversion of *p,p'*-DDT to *p,p'*-DDD by intestinal flora of the rat. *Science* 151:1527—1528.

309. Mendel, J. L., A. K. Klein, J. T. Chen, and M. S. Walton (1967). Metabolism of DDT and some other chlorinated organic compounds by *Aerobacter aerogenes*. *J. Assoc. Offic. Anal. Chemists* 50:897—903.

310. Minard, R. D., Fussel, and J. M. Bollag (1977). Chemical transformation of 4-chloroaniline to a triazine in a bacterial culture medium. *J. Agric. Food Chem.* 25:841—844.

311. Miskus, R. P., Blair, and J. E. Casida (1965). Conversion of DDT to DDD by bovine rumen fluid lake water, and reduced porphyrins. *J. Agric. Food Chem.* 13:481—483.

312. Miyazaki, S., G. M. Boush, and F. Matsumura (1970). Microbial degradation of chlorobenzilate (ethyl 4,4'-dichlorobenzilate) and chloropropylate (isopropyl 4,4'-dichlorobenzilate). *J. Agric. Food Chem.* 18:87—91.

313. Miyazaki, S., G. M. Boush, and F. Matsumura (1969). Metabolism of ^{14}C-chlorobenzilate and ^{14}C-chloropropylate by *Rhodotorula gracilis*. *Appl. Microbiol.* 18:972—976.

314. Miyazaki, S., H. C. Sikka, and R. S. Lynch (1975). Metabolism of dichlobenil by microorganisms in the aquatic environment. *J. Agric. Food Chem.* 23:365—368.

314a. Montgomery, M., T. C. Yu, and V. H. Freed (1972). Kinetics of dichlobenil degradation in soil. *Weed Res.* 12:31—36.

315. Moos, L. P., E. J. Kirsch, R. F. Wukasch, and C. P. L. Grady Jr. (1983). Pentachlorophenol biodegradation. I. Aerobic. *Water Res.* 11:1575—1584.

316. Morrison, R. T., and R. N. Boyd (1973). *Organic Chemistry*. Allyn and Bacon, Inc., Boston, 1258 pp.

317. Moza, R. I., Weisgerber, and W. Klein (1976). Fate of 2,2'-dichlorobiphenyl-^{14}C in carrots, sugar beets, and soil under outdoor conditions. *J. Agric. Food Chem.* 24:881—885.

318. Muller, W. P., and F. Korte (1975). Microbial degradation of benzo[a]pyrene, monolinuron, and dieldrin in waste compositions. *Chemosphere* 4:195—198.

319. Murthy, N. B. K., and D. D. Kaufman (1978). Degradation of pentachloronitrobenzene (PCNB) in anaerobic soils. *J. Agric. Food Chem.* 26:1151—1156.

320. Murthy, N. B. K., D. D. Kaufman, and G. G. Fries (1979). Degradation of pentachlorophenol (PCP) in aerobic and anaerobic soil. *J. Environ. Sci. Health* B14:1-14.

321. Nakagawa, H., H. Inoue, and Y. Takeda (1963). Characteristics of catechol oxygenase from *Brevibacterium fusarium*. *J. Biochem.* 54:65—74.

322. Nakanishi, T., and H. Oku (1969). Metabolism and accumulation of pentachloronitrobenzene by phytopathogenic fungi in relation to selective toxicity. *Phytopathology* 59:1761—1762.

323. Nakazawa, T., and T. Yokota (1973). Benzoate metabolism in *Pseudomonas putida* (*arvilla*) mt-2: Demonstration of two benzoate pathways. *J. Bacteriol.* 115:262—267.

324. Nash, R. G., and E. A. Woolson (1967). Persistence of chlorinated hydrocarbon insecticides in soils. *Science* 157:924—927.

325. Neilson, A. H., A. S. Allard, P. A. Hynning, M. Remberger, and L. Lander (1983). Bacterial methylation of chlorinated phenols and guaiacols: Formation of veratroles from guaiacols and high-molecular weight chlorinated lignin. *Appl. Environ. Microbiol.* 45:774—783.

326. Neujahr, H. Y. (1983). Effect of anions, chaotropes, and phenol on the attachment of flavin adenine dinucleotide to phenol hydroxylase. *Biochemistry* 22:580—584.

327. Neujahr, H. Y., and J. M. Varga (1970). Degradation of phenols by intact cells and cell-free preparations of *Trichosporon cutaneum*. *Eur. J. Biochem.* 13:37—44.

328. Niki, Y., and S. Kuwatsuka (1976). Degradation products of chlomethoxynil (X-52) in soil. *Soil Sci. Plant Nutr.* 22:233—245.

329. Niki, Y., and S. Kuwatsuka (1976). Degradation of diphenyl ether herbicides in soils. *Soil Sci. Plant Nutr.* 22:223—232.

330. Nishizuka, Y., A. Ichiyama, S. Nakamura, and O. Hayaishi (1962). A new metabolic pathway of catechol. *J. Biol. Chem.* 237:PC268—PC270.

331. Nozaka, J., and M. Kusunose (1968). Metabolism of hydrocarbons in microorganisms. Part I. Oxidation of p-xylene and toluene by cell-free enzyme preparations of *Pseudomonas aeruginosa*. *Agric. Biol. Chem.* 32:1033—1039.

332. Nozaka, J., and M. Kusunose (1969). Metabolism of hydrocarbons in microorganisms. Part II. Degradation of toluene by cell-free extracts of *Pseudomonas mildenbergii*. *Agric. Biol. Chem.* 33:962—964.

333. Ohmori, T., T. Ikai, Y. Minoda, and K. Yamada (1973). Utilization of polyphenyl and polyphenyl-related compounds by microorganisms. Part I. *Agric. Biol. Chem.* 37:1599—1605.

334. Old, R. W., and S. B. Primrose (1981). *Principles of Gene Manipulation and Introduction to Genetic Engineering*. University of California Press, Berkeley.

334a. Ohmori, T., T. Ikai, Y. Minoda, and K. Yamada (1973).
Utilization of polyphenyl and polyphenyl-related compounds
by microorganisms. Part I. *Agric. Biol. Chem.* 37:1599–1605.

335. Ondrako, J. M., and L. N. Ornston (1980). Biological distribution and physiological role of the β-ketoadipate transport system. *J. Gen. Microbiol.* 120:199–209.

336. Ornston, L. N. (1966). The conversion of catechol and protocatechuate to β-ketoadipate by *Pseudomonas putida*. II.
Enzymes of the protocatechuate pathway. *J. Biol. Chem.* 241:
3787–3794.

337. Ornston, L. N. (1966). The conversion of catechol and protocatechuate to β-ketoadipate by *Pseudomonas putida*. III.
Enzymes of the catechol pathway. *J. Biol. Chem.* 241:3795–
3799.

338. Ornston, L. N. (1966). The conversion of catechol and protocatechuate to β-ketoadipate by *Pseudomonas putida*. IV.
Regulation. *J. Biol. Chem.* 241:3800–3810.

339. Ornston, L. N., and D. Parke (1976). Properties of an inducible uptake system for β-ketoadipate in *Pseudomonas putida*.
J. Bacteriol. 125:475–488.

340. Ornston, L. N., and R. Y. Stanier (1966). The conversion
of catechol and protocatechuate to β-ketoadipate by *Pseudomonas putida*. I. Biochemistry. *J. Biol. Chem.* 241:3776–3786.

341. Ornston, L. N. (1964). Mechanism of β-ketoadipate formation
by bacteria. *Nature* 204:1279–1283.

342. Ottey, L., and E. L. Tatum (1957). The cleavage of β-ketoadipic acid by *Neurospora crassa*. *J. Biol. Chem.* 229:77–83.

343. Ou, L. T., and H. C. Sikka (1977). Extensive degradation
of silvex by synergistic action of aquatic microorganisms.
J. Agric. Food Chem. 25:1336–1339.

344. Owen, R. B., Jr., J. B. Dimond, and A. S. Getchell (1977).
DDT: Persistence in northern spodosols. *J. Environ. Qual.*
6:359–360.

345. Parke, D., R. B. Meagher, and L. N. Ornston (1973). Relationships among enzymes of the β-ketoadipate pathway. III.
Properties of crystalline γ-carboxymuconolactone decarboxylase from *Pseudomonas putida*. *Biochemistry*. 12:3537–3542.

346. Parr, J. E., and S. Smith (1974). Degradation of DDT in
an Everglades muck as affected by lime, ferrous iron, and
anaerobiosis. *Soil Sci.* 118:45–52.

347. Parr, J. F., G. H. Willis, and S. Smith (1970). Soil anaerobiosis: II. Effect of selected environments and energy sources
on the degradation of DDT. *Soil Sci.* 110:306–312.

348. Patel, R. N., R. B. Meagher, and L. N. Ornston (1973).
Relationships among enzymes of the β-ketoadipate pathway.

II. Properties of crystalline β-carboxy-*cis,cis*-muconate-lactonizing enzyme from *Pseudomonas putida*. *Biochemistry* 12: 3531—3537.

349. Patil, K. C., F. Matsumura, and G. M. Boush (1970). Degradation of endrin, aldrin, and DDT by soil microorganisms. *Appl. Microbiol.* 19:879—881.

350. Patil, K. C., F. Matsumura, and G. M. Boush (1972). Metabolic transformation of DDT, dieldrin, aldrin, and endrin by marine microorganisms. *Environ. Sci. Technol.* 6:629—632.

350a. Pfaender, F. K., and M. Alexander (1972). Extensive microbial degradation of DDT in vitro and DDT metabolism by natural communities. *J. Agric. Food Chem.* 20:842—846.

351. Philippi, M., J. Schmid, H. K. Wipf, and R. Hütter (1982). A microbial metabolite of TCDD. *Experientia* 38:659—661.

352. Philippi, M., V. Krasnobajew, J. Zeyer, and R. Hütter (1981). Fate of TCDD in microbial cultures and in soil under laboratory conditions. pp. 221—233 in Leisinger, T., R Hütter, A. M. Cook and J. Nuesch, (eds.). *Microbial Degradation of Xenobiotics and Recalcitrant Compounds*. Academic Press, New York.

353. Pierce, R. H., Jr., and D. M. Victor (1977). The fate of pentachlorophenol in an aquatic ecosystem. Paper Presented At the Symp. on Pentachlorophenol, June 27—29, 1977, Pensacola, FL. USEPA Gulf Breeze Res. Lab and Univ. of West Florida.

354. Pignatello, J. J., M. M. Martinson, J. G. Steiert, R. E. Carlson, and R. L. Crawford (1983). Biodegradation and photolysis of pentachlorophenol in artificial freshwater streams. *Appl. Environ. Microbiol.* 46:1028—1031.

355. Plimmer, J. R., P. C. Kearney, and D. W. von Endt (1968). Mechanism of conversion of DDT to DDD by *Aerobacter aerogenes*. *J. Agric. Food Chem.* 16:594—597.

356. Pocchiari, F. (1978). 2,3,7,8-Tetrachlorodibenzo-*para*-dioxin decontamination. *Ecol. Bull. (Stockholm)* 27:67—70.

357. Proctor, M. H., and S. Scher (1960). Decomposition of benzoate by a photosynthetic bacterium. *Biochem. J.* 76:33P.

358. Quensen III, J. F., and F. Matsumura (1983). Oxidative degradation of 2,3,7,8-tetrachlorodibenzo-*p*-dioxin by microorganisms. *Environ. Toxicol. Chem.* 2:261—268.

359. Raman, T. S., and E. R. B. Shanmugasundaram (1962). Metabolism of some aromatic acids by *Aspergillus niger*. *J. Bacteriol.* 84:1339—1340.

360. Rappé, C., H. R. Buser, and H. P. Bosshardt (1979). Dioxins, dibenzofurans and other polyhalogenated aromatics — production, use, formation and destruction. *Ann. NY Acad. Sci.* 320:1—18.

361. Reber, H., V. Helm, and N. G. K. Karanth (1979). Comparative studies on the metabolism of aniline and chloroanilines by *Pseudomonas multivorans* strain An 1. *Eur. J. Appl. Microbiol. Biotechnol.* 7:181—189.

362. Reber, J. J., and G. Thierbach (1980). Physiological studies on the oxidation of 3-chlorobenzoate by *Acinetobacter calcoaceticus* strain Bs 5. *Eur. J. Appl. Microbiol. Biotechnol.* 10:223—233.

363. Reineke, W., and H. J. Knackmuss (1978). Chemical structure and biodegradability of halogenated aromatic compounds. Substituent effects on 1,2-dioxygenation of benzoic acid. *Biochem. Biophys. Acta* 542:412—423.

364. Reineke, W., and H. J. Knackmuss (1980). Hybrid pathway for chlorobenzoate metabolism in *Pseudomonas* sp. B13 derivatives. *J. Bacteriol.* 142:467—473.

365. Reineke, W., and H. J. Knackmuss (1978). Chemical structure and biodegradability of halogenated aromatic compounds. Substituent effects on dehydrogenation of 3,5-cyclohexadiene-1,2-diol-1-carboxylic acid. *Biochem. Biophys. Acta* 542:424—429.

366. Reineke, W., and H. J. Knackmuss (1984). Microbial metabolism of haloaromatics: Isolation and properties of a chlorobenzene-degrading bacterium. *Appl. Environ. Microbiol.* 47:395—402.

367. Reineke, W., W. Otting, and H. J. Knackmuss (1978). Cis-dihydrodiols microbially produced from halo- and methylbenzoic acids. *Tetrahedron* 34:1707—1714.

368. Reiner, A. M. (1972). Metabolism of aromatic compounds in bacteria. Purification and properties of the catechol-forming enzyme, 3,5-cyclohexadiene-1,2-diol-1-carboxylic acid (NAD$^+$) oxidoreductase (decarboxylating). *J. Biol. Chem.* 247:4960—4965.

369. Reiner, E. A., J. Chu, and E. J. Kirsch (1978). Microbial metabolism of pentachlorophenol. pp. 67—81 In Rao, K. R. (ed.). *Pentachlorophenol.* Plenum Press, New York.

370. Rhodes, R. C., H. L. Pease, and R. K. Brantley (1971). Fate of ^{14}C-labeled chloroneb in plants and soils. *J. Agric. Food Chem.* 19:745—749.

371. Rice, C. P., H. C. Sikka, and R. S. Lynch (1974). Persistence of dichlobenil in a farm pond. *J. Agric. Food Chem.* 22:533—534.

372. Roberts, T. R., and G. Stoydin (1976). Degradation of the insecticide SD 8280, 2-chloro-1-(2,4-dichlorophenyl)vinyl dimethyl phosphate, in soils. *Pestic. Sci.* 7:145—149.

373. Rogoff, M. H., and I. Wender (1957). The microbiology of coal. I. Bacterial oxidation of phenanthrene. *J. Bacteriol.* 73:264—268.

374. Rogoff, M. H., and I. Wender (1957). 3-Hydroxy-2-naphthoic acid as an intermediate in bacterial dissimilation of anthracene. *J. Bacteriol.* 74:108–109.

375. Rogoff, M. H., and J. J. Reid (1956). Bacterial decomposition of 2,4-dichlorophenoxyacetic acid. *J. Bacteriol.* 71:303–307.

376. Rosenberg, A., and M. Alexander (1980). 2,4,5-trichlorophenoxyacetic acid (2,4,5-T) decomposition in tropical soil and its cometabolism by bacteria *in vitro*. *J. Agric. Food Chem.* 28:705–709.

377. Rosenberg, A., and M. Alexander (1980). Microbial metabolism of 2,4,5-trichlorophenoxyacetic acid in soil, soil suspensions, and axenic culture. *J. Agric. Food Chem.* 28:297–302.

378. Ross, J. A., and B. G. Tweedy (1973). Malonic acid conjugation by soil microorganisms of a pesticide-derived aniline moiety. *Bull. Environ. Contam. Toxicol.* 10:234–236.

379. Rott, B., S. Nitz, and F. Korte (1979). Microbial decomposition of sodium pentachlorophenolate. *J. Agric. Food Chem.* 27:306–310.

380. Ruisinger, S., U. Klages, and F. Lingens (1976). Abbau der 4-chlorbenzoesaure durch eine *Arthrobacter*-species. *Arch. Microbiol.* 110:253–256.

381. Russel, S., and J. M. Bollag (1977). Formylation and acetylation of 4-chloroaniline by a *Streptomyces* sp. *Acta Microbiol. Pol.* 26:59–64.

382. Sala-Trepat, J. M., and W. C. Evans (1971). The *meta* cleavage of catechol by *Azotobacter* species. 4-Oxalocrotonate pathway. *Eur. J. Biochem.* 20:400–413.

382a. Sala-Trepat, J. M., K. Murray, and P. A. Williams (1971). The physiological significance of the two divergent metabolic steps in the *meta* cleavage of catechols by *Pseudomonas putida* NCIB 10015. *Biochem. J.* 124:20P–21P.

383. Sala-Trepat, J. M., K. Murray, and P. A. Williams (1972). The metabolic divergence in the *meta* cleavage of catechols by *Pseudomonas putida* NCIB 10015. *Eur. J. Biochem.* 28:347–356.

384. Saxena, A., and R. Bartha (1983). Microbial mineralization of humic acid-3,4-dichloroaniline complexes. *Soil Biol. Biochem.* 15:59–62.

385. Sayler, G. S., A. Breen, J. W. Blackburn, and O. Yagi (1984). Predictive assessment of priority pollutant bio-oxidation kinetics in activated sludge. *Environ. Prog.* 3:153–163.

386. Schmidt, E., and H. J. Knackmuss (1980). Chemical structure and biodegradability of halogenated aromatic compounds. Conversion of chlorinated muconic acids into maleoylacetic acid. *Biochem. J.* 192:339–347.

386a. Schmidt, E., G. Remberg, and H. J. Knackmuss (1980). Chemical structure and biodegradability of halogenated aromatic compounds. Halogenated muconic acids as intermediates. *Biochem. J.* 192:331–337.

387. Schwetz, B. A., J. M. Norris, G. L. Sparschu, V. K. Rowe, P. J. Gehring, J. L. Emerson, and C. G. Gerbig (1973). Toxicology of chlorinated dibenzo-p-dioxins. *Environ. Health Perspect.* 5:87–99.

388. Schwien, U., and E. Schmidt (1982). Improved degradation of monochlorophenols by a constructed strain. *Appl. Environ. Microbiol.* 44:33–39.

389. Seuferer, S. L., H. D. Braymer, and J. J. Dunn (1979). Metabolism of diflubenzuron by soil microorganisms and mutagenicity of the metabolites. *Pestic. Biochem. Physiol.* 10:174–180.

390. Shailubhai, K., S. R. Sahasrabudhe, K. A. Vora, and V. V. Modi (1984). Degradation of chlorobenzoates by *Aspergillus niger*. *Experientia* 40:406–407.

391. Shamat, N. A., and W. J. Maier (1980). Kinetics of biodegradation of chlorinated organics. *J. Water Pollut. Control Fed.* 52:2158–2166.

392. Sharabi, H. E. D., and L. M. Bordeleau (1969). Biochemical decomposition of the herbicide N-(3,4-dichlorophenyl)-2-methylpentanamide and related compounds. *Appl. Microbiol.* 18:369–375.

393. Sharpee, K. W., J. M. Duxbury, and M. Alexander (1973). 2,4-Dichlorophenoxyacetate metabolism by *Arthrobacter* sp.: Accumulation of a chlorobutenolide. *Appl. Microbiol.* 26:445–447.

394. Shiaris, M. P., and G. S. Sayler (1982). Biotransformation of PCB by natural assemblages of freshwater microorganisms. *Environ. Sci. Technol.* 16:367–369.

395. Shoda, M., and S. Udaka (1980). Preferential utilization of phenol rather than glucose by *Trichosporon cutaneum* possessing a partially constitutive catechol 1,2-oxygenase. *Appl. Environ. Microbiol.* 39:1129–1133.

396. Sistrom, W. R., and R. Y. Stanier (1954). The mechanism of formation of β-ketoadipic acid by bacteria. *J. Biol. Chem.* 210:821–836.

397. Smith, A. E. (1977). Degradation of the herbicide dichlorfop-methyl in prairie soils. *J. Agric. Food Chem.* 25:893–898.

398. Smith, A. E. (1974). Breakdown of the herbicide dicamba and its degradation products 3,6-dichlorosalicylic acid in prairie soils. *J. Agric. Food Chem.* 22:601–605.

399. Smith, A. E. (1973). Transformation of dicamba in Regina heavy clay. *J. Agric. Food Chem.* 21:708–710.

399a.   Smith, A. E. (1973). Degradation of dicamba in prairie soils. *Weeds Res.* 13:373–378.

400.   Smith, A. E. (1976). The hydrolysis of herbicidal phenoxyalkanoic esters to phenoxyalkanoic acids in Saskatchewan soils. *Weeds Res.* 16:19–22.

401.   Smith, A. E., and D. R. Cullimore (1975). Microbiological degradation of the herbicide dicamba in moist soils at different temperatures. *Weed Res.* 15:59–62.

402.   Smith, A. E., and D. V. Phillips (1976). Degradation of 4-(2,4-dichlorophenoxy)butyric acid (2,4-DB) by *Phytophthora megasperma.* *J. Agric. Food Chem.* 24:294–296.

403.   Smith, A. E., and G. G. Briggs (1978). The fate of the herbicide chlortoluron and its possible degradation products in soils. *Weed Res.* 18:1–7.

404.   Smith, R. V., and J. P. Rosazza (1974). Microbial models of mammalian metabolism. Aromatic hydroxylation. *Arch. Biochem. Biophys.* 161:551–558.

405.   Soderquist, C. J., and D. G. Crosby (1975). Dissipation of 4-chloro-2-methylphenoxyacetic acid (MCPA) in a rice field. *Pestic. Sci.* 6:17–33.

405a.   Spicher, G. (1954). Beitrage zur kenntnis der wirksamkeit des 2,4-D-zersetzers *Flavobacterium peregrinum* st. et sp. *Zentbl. Bakt. Parasit. Abt. II.* 108:225–231.

406.   Spokes, J. R. , and N. Walker (1974). Chlorophenol and chlorobenzoic acid cometabolism by different genera of soil bacteria. *Arch. Microbiol.* 96:125–134.

407.   Stanier, R. Y. (1947). Simultaneous adaptation: A new technique for the study of metabolic pathways. *J. Bacteriol.* 54:339–348.

408.   Stanier, R. Y., and L. N. Ornston (1973). The β-ketoadipate pathway. *Adv. Microb. Physiol.* 9:89–151.

409.   Stanier, R. Y., B. P. Sleeper, M. Tsuchida, and D. L. MacDonald (1950). The bacterial oxidation of aromatic compounds. III. The enzymatic oxidation of catechol and protocatechuic acid to β-ketoadipic acid. *J. Bacteriol.* 59:137–151.

410.   Stanier, R. Y., N. J. Palleroni, and M. Doudoroff (1966). The aerobic pseudomonads: A taxonomic study. *J. Gen. Microbiol.* 43:159–271.

411.   Stanlake, G. J., and R. K. Finn (1982). Isolation and characterization of a pentachlorophenol-degrading bacterium. *Appl. Environ. Microbiol.* 44:1421–1427.

411a.   Stapp, C., and G. Spicher (1954). Untersuchungen uber die wirkung von 2,4-D im boden. *Zentbl. Bakt. Parasit. Abt. II.* 108:113–126.

412. Steenson, T. I., and N. Walker (1957). The pathway of breakdown of 2,4-dichloro- and 4-chloro-2-methylphenoxyacetic acid by bacteria. *J. Gen. Microbiol.* 16:146—155.

413. Steenson, T. I., and N. Walker (1958). Adaptive patterns in the bacterial oxidation of 2,4-dichloro- and 4-chloro-2-methyl-phenoxyacetic acid. *J. Gen. Microbiol.* 18:692—697.

414. Steenson, T. I., and N. Walker (1956). Observations on the bacterial oxidation of chlorophenoxyacetic acids. *Plant Soil* 8:17—32.

415. Stenersen, J. H. V. (1965). DDT-metabolism in resistant and susceptible stable-flies and in bacteria. *Nature* 207:660—661.

416. Stover, E. L., and D. F. Kincannon (1981). Biological treatability of specific organic compounds found in chemical industry wastewaters. 36th Ind. Waste Conf., Purdue Univ., W. La-fayette, Ind.

417. Suflita, J. M., A. Horowitz, D. R. Shelton, and J. M. Tiedje (1982). Dehalogenation: A novel pathway for the anaerobic biodegradation of haloaromatic compounds. *Science* 218:1115—1117.

418. Suflita, J. M., J. A. Robinson, and J. M. Tiedje (1983). Kinetics of microbial dehalogenation of haloaromatic substrates in methanogenic environments. *Appl. Environ. Microbiol.* 45:1466—1473.

419. Surovtseva, E. G., G. K. Vasil'eva, A. I. Vol'nova, and B. P. Baskunov (1981). Degradation of monochloroanilines via the *meta* pathway by *Alcaligenes faecalis*. *Proc. Acad. Sci. USSR* 254:487—490.

420. Süss, A., G. Fuchsbichler, and C. Eben (1978). Abbau von anilin, 4-chloranilin und 3,4-dichloranilin in verschiedenen boden. *Z. Pflanzenernaehr. Bodenkd.* 141:57—66.

421. Sutherland, J. B., D. L. Crawford, and A. L. Pometto III (1981). Catabolism of substituted benzoic acids by *Streptomyces* species. *Appl. Environ. Microbiol.* 41:442—48.

422. Suzuki, T. (1978). Enzymatic methylation of pentachlorophenol and its related compounds by cell-free extracts of *Mycobacterium* sp. isolated from soil. *J. Pestic. Sci.* 3:441—443.

423. Suzuki, T. (1977). Metabolism of pentachlorophenol by a soil microbe. *J. Environ. Sci. Health* B12:113—127.

424. Suzuki, T. (1983). Metabolism of pentachlorophenol (PCP) by soil microorganisms. *Bull. Natl. Inst. Agric. Sci. (Japan)* C(38):69—120.

425. Suzuki, T. (1983). Methylation and hydroxylation of pentachlorophenol by *Mycobacterium* sp. isolated from soil. *J. Pestic. Sci.* 8:419—428.

426. Sylvestre, M., and J. Fauteux (1982). A new facultative anaerobe capable of growth on chlorobiphenyls. *J. Gen. Appl. Microbiol.* 28:61—72.

427. Sylvestre, M., R. Masse, F. Messier, J. Fauteux, J. G.
     Bisaillon, and R. Beaudet (1982). Bacterial nitration of 4-chloro-
     biphenyl. *Appl. Environ. Microbiol.* 44:871–877.
428. Tahara, S., Z. Hafsah, A. Ono, E. Asaishi, and J. Mizutani
     (1981). Metabolites of 2,4-dichloro-1-nitrobenzene by *Mucor
     javanicus*. *Agric. Biol. Chem.* 45:2253–2258.
429. Tarrant, R. F., D. G. Moore, W. B. Bollen, and B. R. Loper
     (1972). DDT residues in forest floor and soil after aerial
     spraying, Oregon—1965–68. *Pestic. Monit. J.* 6:65–72.
430. Taylor, B. F., and M. J. Heeb (1972). The anaerobic degra-
     dation of aromatic compounds by a denitrifying bacterium.
     Arch. Mikrobiol. 83:165–171.
431. Taylor, B. F., W. L. Campbell, and I. Chinoy (1970). Anaer-
     obic degradation of the benzene nucleus by a facultatively
     anaerobic microorganism. *J. Bacteriol.* 102:430–437.
432. Taylor, H. F., and R. L. Wain (1962). Side-chain degradation
     of certain omega-phenoxyalkanecarboxylic acids by *Nocardia
     coeliaca* and other microorganisms isolated from soils. *Proc. R.
     Soc. Lond. B*156:172–186.
433. Tiedje, J. M., and M. Alexander (1969). Enzymatic cleavage
     of the ether bond of 2,4-dichlorophenoxyacetate. *J. Agric.
     Food Chem.* 17:1080–1084.
434. Tiedje, J. M., J. M. Duxbury, M. Alexander, and J. E.
     Dawson (1969). 2,4-D metabolism: Pathway of degradation of
     chlorocatechols by *Arthrobacter* sp. *J. Agric. Food Chem.*
     17:1021–1026.
435. Tillmanns, G. M., P. R. Wallnöfer, G. Engelhardt, K. Olie,
     and O. Hutzinger (1978). Oxidative dealkylation of five phenyl-
     urea herbicides by the fungus *Cunninghamella echinulata*
     Thaxter. *Chemosphere* 7:59–64.
436. Torstensson, N. T., J. Stark, and B. Goransson (1975). The
     effect of repeated application of 2,4-D and MCPA on their
     breakdown in soil. *Weed Res.* 15:159–164.
437. Truong, K. N., and J. W. Blackburn (1984). The stripping
     of organic chemicals in biological treatment processes. *Environ.
     Prog.* 3:143–152.
438. Tucker, E. S., V. W. Saeger, and O. Hicks (1975). Activated
     sludge primary biodegradation of polychlorinated biphenyls.
     Bull. Environ. Contam. Toxicol. 14:705–713.
439. Tulp, M. Th. M., R. Schmitz, and O. Hutzinger (1978). The
     bacterial metabolism of 4,4'-dichlorobiphenyl, and its suppres-
     sion by alternative carbon sources. *Chemosphere* 7:103–108.
440. Tulp, M. Th. M., G. M. Tillmanns, and O. Hutzinger (1977).
     Environmental chemistry of PCB-replacement compounds. V.
     The metabolism of chloroisopropylbiphenyls in fish, frogs, fungi
     and bacteria. *Chemosphere* 6:223–230.

441. Unligil, H. H. (1968). Depletion of pentachlorophenol by fungi. *Forest Prod. J.* 18:45−50.

442. Van Alfen, N. K., and T. Kosuge (1974). Microbial metabolism of the fungicide 2,6-dichloro-4-nitroaniline. *J. Agric. Food Chem.* 22:221−224.

442a. Verloop, A., and W. B. Nimmo (1970). Metabolism of dichlobenil in sandy soil. *Weed Res.* 10:65−70.

443. Vlitos, A. J. (1953). Biological activation of sodium 2-(2,4-dichlorophenoxy)ethyl sulfate. *Contrib. Boyce Thompson Inst.* 17:127−149.

444. Voerman, S., and A. F. H. Besemer (1975). Persistence of dieldrin, lindane, and DDT in a light sandy soil and their uptake by grass. *Bull. Environ. Contam. Toxicol.* 13:501−505.

445. Wain, R. L., and H. F. Taylor (1965). Phenols as plant growth regulators. *Nature* 207:167−169.

446. Wakeham, S. G., A. C. Davis, and J. L. Karas (1983). Mesocosm experiments to determine the fate and persistence of volatile organic compounds in coastal seawater. *Environ. Sci. Technol.* 17:611−617.

447. Walker, N. (1954). Preliminary observations on the decomposition of chlorophenols in soil. *Plant Soil* 5:194−204.

448. Walker, N. (1973). Metabolism of chlorophenols by *Rhodotorula glutinis*. *Soil Biol. Biochem.* 5:525−530.

449. Walker, N., and D. Harris (1969). Aniline utilization by a soil pseudomonad. *J. Appl. Bacteriol.* 32:457−462.

450. Walker, N., and D. Harris (1970). Metabolism of 3-chlorobenzoic acid by *Azotobacter* species. *Soil Biol. Biochem.* 2:27−32.

451. Walker, R. L., and A. S. Newman (1956). Microbial decomposition of 2,4-dichlorophenoxyacetic acid. *Appl. Microbiol.* 4:201−206.

451a. Walker, N., and G. H. Wiltshire (1955). The decomposition of 1-chloro- and 1-bromonaphthalene by soil bacteria. *J. Gen. Microbiol.* 12:478−483.

452. Wallnöfer, P. (1969). The decomposition of urea herbicides by *Bacillus sphaericus* isolated from soil. *Weed Res.* 9:333−339.

453. Wallnöfer, P. R., and J. Bader (1970). Degradation of urea herbicides by cell-free extracts of *Bacillus sphaericus*. *Appl. Microbiol.* 19:714−717.

454. Wallnöfer, P. R., G. Engelhardt, S. Safe, and O. Hutzinger (1973). Microbial hydroxylation of 4-chlorobiphenyl and 4,4'-dichlorobiphenyl. *Chemosphere* 2:69−72.

455. Wallnöfer, P. R., G. Tillmanns, and G. Engelhardt (1977). Degradation of acylanilide pesticides by *Aspergillus niger*. *Pest. Biochem. Physiol.* 7:481−485.

456. Wallnöfer, P. R., S. Safe, and O. Hutzinger (1973). Micro-
      bial hydroxylation of the herbicide N-(3,4-dichlorophenyl)-
      methacrylamide (Dicryl). *J. Agric. Food Chem.* 21:502–504.

457. Wallnöfer, P. R., S. Safe, and O. Hutzinger (1972). Die
      hydroxylation des herbizids karsil [N-(3,4-dichlorophenyl)-2-
      methylpetanamid] durch *Rhizopus japonicus*. *Chemosphere*
      1:155–158.

458. Ward, C. T., and F. Matsumura (1978). Fate of 2,3,7,8-
      tetrachlorodibenzo-*p*-dioxin (TCDD) in a model aquatic envi-
      ronment. *Arch. Environ. Contam. Toxicol.* 7:349–357.

459. Watanabe, I. (1973). Isolation of pentachlorophenol decom-
      posing bacteria from soil. *Soil Sci. Plant Nutr.* 19:109–116.

460. Watson, J. R. (1977). Seasonal variation in the biodegrada-
      tion of 2,4-D in river water. *Weed Res.* 11:153–157.

461. Webley, D. M., R. B. Duff, and V. C. Farmer (1958). The
      influence of chemical structure on beta-oxidation by soil no-
      cardias. *J. Gen. Microbiol.* 18:733–746.

462. Webley, D. M., R. B. Duff, and V. C. Farmer (1957). For-
      mation of a beta-hydroxy acid as an intermediate in the micro-
      biological conversion of monochlorophenoxybutyric acids to
      the corresponding substituted acetic acids. *Nature* 179:1130–
      1131.

463. Wedemeyer, G. (1967). Dechlorination of 1,1,1-trichloro-2,2-
      bis(*p*-chlorophenyl)ethane by *Aerobacter aerogenes*. *Appl.
      Microbiol.* 15:569–574.

464. Wedemeyer, G. (1966). Dechlorination of DDT by *Aerobacter
      aerogenes*. *Science* 152:647.

465. Wedemeyer, G. (1967). Biodegradation of dichlorodiphenyl-
      trichloroethane: Intermediates in dichlorodiphenylacetic acid
      metabolism by *Aerobacter aerogenes*. *Appl. Microbiol.* 15:
      1494–1495.

466. Weinbach, E. C. (1957). Biochemical basis for the toxicity
      of pentachlorophenol. *Proc. Natl. Acad. Sci. USA* 43:393–397.

466a. Westmacott, D., and S. J. L. Wright (1975). Studies on the
      breakdown of *p*-chlorophenyl methylcarbamate. II. In cultures
      of a soil *Arthrobacter* sp. *Pestic. Sci.* 6:61–68.

467. Wheelis, M. L., N. J. Palleroni, and R. Y. Stanier (1967).
      The metabolism of aromatic acids by *Pseudomonas testosteroni*
      and *P. acidovorans*. *Arch. Mikrobiol.* 59:302–314.

468. Wiese, M. V., and J. M. Vargas, Jr. (1073). Interconversion
      of chloroneb and 2,4-dichloro-4-methoxyphenol by soil micro-
      organisms. *Pestic. Biochem. Physiol.* 3:214–222.

468a. Williams, P. A., K. Murray, and J. M. Sala-Trepat (1971).
      The coexistence of two metabolic pathways in the *meta* clea-
      vage of catechol by *Pseudomonas putida* NCIB 10105. *Biochem.
      J.* 124:19P–20P.

469. Wilson, R. G., Jr., and H. H. Cheng (1978). Fate of 2,4-D in a Naff silt loam soil. *J. Environ. Qual.* 7:281—286.

470. Wolf, D. C., and J. P. Martin (1976). Decomposition of fungal mycelia and humic-type polymers containing carbon-14 from ring and sidechain labeled 2,4-D and chlorpropham. *Soil Sci. Soc. Am. J.* 40:700—704.

471. Wolfe, N. L., R. G. Zepp, and D. F. Paris (1978). Carbaryl, propham and chlorpropham: A comparison of the rates of hydrolysis and photolysis with the rate of biolysis. *Water Res.* 12:565—571.

471a. Worsey, M. J., and P. A. Williams (1975). Metabolism of toluene and xylenes by *Pseudomonas putida* (*arvilla*) MT-2: Evidence for a new function of the TOL plasmid. *J. Bacteriol.* 124:7—13.

472. Wright, S. J. L., and A. Forey (1972). Metabolism of the herbicide barban by a soil penicillium. *Soil Biol. Biochem.* 4: 207—213.

473. Wright, S. J. L., A. F. Stainthorpe, and J. D. Downs (1977). Interactions of the herbicide propanil and a metabolite, 3,4-dichloroaniline, with blue-green algae. *Acta Phytopathol. Acad. Sci. Hung.* 12:51—60.

474. Yagi, O., and R. Sudo (1980). Degradation of polychlorinated biphenyls by microorganisms. *J. Water Pollut. Control Fed.* 52:1035—1043.

475. Yamaguchi, M., T. Yamaguchi, and H. Fujisawa (1975). Studies on mechanism of double hydroxylation. I. Evidence for participation of NADH-cytochrome C reductase in the reaction of benzoate 1,2-dioxygenase (benzoate hydroxylase). *Biochem. Biophys. Res. Commun.* 67:264—271.

476. Yamazaki, I. (1966). Function of peroxidase as an oxygen-activating enzyme. pp. 433—442 In Block, K. and O. Hayaishi (eds.). *Biological and Chemical Aspects of Oxygenases.* Maruzen Co. Ltd., Tokyo.

477. Yeh, W. K., and L. N. Ornston (1980). Origins of metabolic diversity: Substitution of homologous sequences into genes for enzymes with different catalytic activities. *Proc. Natl. Acad. Sci. USA* 77:5365—5369.

478. Yeh, W. K., G. Davis, P. Fletcher, and L. N. Ornston (1978). Homologous amino acid sequences in enzymes mediating sequential metabolic reactions. *J. Biol. Chem.* 253:4920—4923.

479. Yeh, W. K., P. Fletcher, and L. N. Ornston (1980). Evolutionary divergence of co-selected β-ketoadipate enol-lactone hydrolases in *Acinetobacter calcoaceticus*. *J. Biol. Chem.* 255:6342—6346.

480. Yeh, W. K., P. Fletcher, and L. N. Ornston (1980). Homo-
      logies in the NH$_2$-terminal amino acid sequences of γ-carboxy-
      muconolactone decarboxylases and muconolactone isomerases.
      J. Biol. Chem. 255:6347–6354.
481. Yih, R. Y., and C. Swithenbank (1971). Identification of me-
      tabolites of N-(1,1-dimethylpropynyl)-3,5-dichlorobenzamide in
      soil and alfalfa. J. Agric. Food Chem. 19:314–319.
482. Yoshida, T., and T. F. Castro (1975). Degradation of 2,4-D,
      2,4,5-T, and picloram in two Philippine soils. Soil Sci. Plant
      Nutr. 21:397–404.
483. You, I. S., and R. Bartha (1982). Stimulation of 3,4-dichloro-
      aniline mineralization by aniline. Appl. Environ. Microbiol. 44:
      678–681.
484. You, I. S., and R. Bartha (1982). Metabolism of 3,4-dichloro-
      aniline by Pseudomonas putida. J. Agric. Food Chem. 30: 274–
      277.
485. Yule, W. N. (1973). Intensive studies of DDT residues in
      forest soil. Bull. Environ. Contam. Toxicol. 9:57–64.
486. Zaitsev, G. M., and U. N. Karasevich (1981). Utilization of
      4-chlorobenzoic acid by Arthrobacter globiformis. Microbiology
      50:23–27.
487. Zeyer, J., and P. C. Kearney (1982). Microbial degradation
      of parachloroaniline as sole carbon and nitrogen source. Pest.
      Biochem. Physiol. 17:215–223.
488. Zeyer, J., and P. C. Kearney (1982). Microbial metabolism of
      propanil and 3,4-dichloroaniline. Pest. Biochem. Physiol. 17:
      224–231.
489. Zoro, J. A., J. M. Hunter, G. Eglinton, and G. C. Ware
      (1974). Degradation of p,p'-DDT in reducing environments.
      Nature 247:235–237.

Bibliography

Ahlborg, U. G. (1978). Dechlorination of pentachlorophenol *in vivo* and *in vitro*. pp. 115–130 in Rao, K. R. (ed.). *Pentachlorophenol*. Plenum Press, New York.

Ahmed, M., and D. D. Focht (1973). Oxidation of polychlorinated biphenyls by *Achromobacter* PCB. *Bull. Environ. Contam. Toxicol.* 10:70–72.

Akhtar, M. N., D. R. Boyd, N. J. Thompson, D. T. Gibson, V. Mahadevan, and D. M. Jerina (1975). Absolute stereochemistry of the dihydroanthracene-*cis*- and -*trans*-1,2-diols produced from anthracene by mammals and bacteria. *J. Chem. Soc. Perkin Trans.* I. 1975:2506–2511.

Alexander, M. (1981). Biodegradation of chemicals of environmental concern. *Science* 211:132–138.

Alexander, M. (1981). Microbial degradation of pesticides. Final Report. Office of Naval Research Contract N0001478C–0044, Task No. NR 205–032. 15 pp.

Alexander, M. (1973). Nonbiodegradable and other recalcitrant molecules. *Biotechnol. Bioeng.* 15:611–647.

Alexander, M. (1975). Environmental and microbiological problems arising from recalcitrant molecules. *Microb. Ecol.* 2:17–27.

Alexander, M., and M. I. H. Aleem (1961). Effect of chemical structure on microbial decomposition of aromatic herbicides. *J. Agric. Food Chem.* 9:44–47.

Alexander, M., and B. K. Lustigman (1966). Effect of chemical structure on microbial degradation of substituted benzenes. *J. Agric. Food Chem.* 14:410–413.

Anderson, M. O., and H. Okrend (1968). Degradation of 2,4-D by *Aerobacter aerogenes*. *Bacteriol. Proc.* A25:5.

Andrews, J. F. (1968). A mathematical model for the continuous culture of microorganisms utilizing inhibitory substrates. *Biotechnol. Bioeng.* 10:707–723.

Aranha, H. G., and L. R. Brown (1981). Effect of nitrogen source on end products of naphthalene degradation. *Appl. Environ. Microbiol.* 42:74–78.

Arsenault, R. D. (1976). Pentachlorophenol and contained chlorinated dibenzodioxins in the environment. *J. Am. Wood-Preserv. Assoc.* 72:122–148.

Atlas, R. M. (1981). Microbial degradation of petroleum hydrocarbons: An environmental perspective. *Microbiol. Rev.* 45:180–129.

Audus, L. M., and K. V. Symonds (1955). Further studies on the breakdown of 2,4-dichlorophenoxyacetic acid by a soil bacterium. *Ann. Appl. Biol.* 42:174–182.

Auret, B. J., D. R. Boyd, P. M. Robinson, and C. G. Watson (1971). The NIH shift during the hydroxylation of aromatic substrates by fungi. *Chem. Commun.* 24:1585–1587.

Bachofer, R. (1976). Mikrobieller abbau von saureanilid-fungiziden. Microbial breakdown of acid anilide fungicides. *Zbl. Bakt. Hyg., I. Abt. Orig. B* 162:153–156.

Baird, R., L. Carmona, and R. L. Jenkins (1977). Behavior of benzidine and other aromatic amines in aerobic wastewater treatment. *J. Water Pollut. Control Fed.* 49:1609–1615.

Baker, R. J. (1969). Characteristics of chlorine compounds. *J. Water Pollut. Control Fed.* 41:482–485.

Balba, M. T. M., E. Senior, and D. B. Nedwell (1981). Anaerobic metabolism of aromatic compounds by microbial associations isolated from salt marsh sediment. *Biochem. Soc. Trans.* 9:230–231.

Ballschmitter, K., M. Zell, and H. J. Neu (1978). Persistence of PCB's in the ecosphere: Will some PCB-components "never" degrade? *Chemosphere* 7:173–176.

Banerjee, S., S. H. Yalkowsky, and S. C. Valvani (1980). Water solubility and octanol/water partition coefficients of organics. Limitations of the solubility-partition coefficient correlation. *Environ. Sci. Technol.* 14:1227–1229.

Barnhart, C. L. H., and J. R. Vestal (1983). Effects of environmental toxicants on metabolic activity of natural microbial communities. *Appl. Environ. Microbiol.* 46:970–977.

Barnsley, E. A. (1975). The induction of the enzymes of naphthalene metabolism in pseudomonads by salicylate and 2-aminobenzoate. *J. Gen. Microbiol.* 88:193–196.

Barnsley, E. A. (1976). Naphthalene metabolism by pseudomonads: The oxidation of 1,2-dihydroxynaphthalene to 2-hydroxychromene-2-carboxylic acid and the formation of 2'-hydroxybenzalpyruvate. *Biochem. Biophys. Res. Commun.* 72:1116–1121.

Bartels, I., H. J. Knackmuss, and W. Reineke (1984). Suicide inactivation of catechol 2,3-dioxygenase from *Pseudomonas putida* MT-2 by 3-halocatechols. *Appl. Environ. Microbiol.* 47:500–505.

Bartha, R., and L. M. Bordeleau (1969). Transformaion of herbicide-derived chloroanilines by cell-free peroxidases in soil. *Bacteriol. Proc.* A26:4.

Bartha, R., H. Linke, and D. Pramer (1968). Transformation of aniline herbicides and chloroanilines in soil. *Bacteriol. Proc.* A26:5.

Bartholomew, G. W., and F. K. Pfaender (1983). Influence of spatial and temporal variations on organic pollutant biodegradation rates in an estuarine environment. *Appl. Environ. Microbiol.* 45:103–109.

Baughman, G. L., and D. F. Paris (1981). Microbial bioconcentration of organic pollutants from aquatic systems—A critical review. *CRC Crit. Rev. Microbiol.* 8:205–228.

Baxter, R. M., and D. A. Sutherland (1984). Biochemical and photochemical processes in the degradation of chlorinated biphenyls. *Environ. Sci. Technol.* 18:608–610.

Bayley, S. A., D. W. Norris, and P. Broda (1979). The relationship of degradative and resistance plasmids of *Pseudomonas* belonging to the same incompatibility group. *Nature* 280:338–339.

Bayly, R. C., and M. G. Barbour (1984). The degradation of aromatic compounds by the *meta* and gentisate pathways. pp. 253–294 In Gibson, D. T. (ed.). *Microbial Degradation of Organic Compounds.* Marcel Dekker, Inc., New York.

Beltrame, P., P. L. Beltrame, P. Carniti, and D. Pitea (1980). Kinetics of phenol degradation by activated sludge in a continuous-stirred reactor. *J. Water Pollut. Control Fed.* 52:126–133.

Beynon, K. I., D. H. Hutson, and A. N. Wright (1973). The metabolism and degradation of vinyl phosphate insecticides. *Res. Rev.* 47:55–142.

Bilbo, A. J., and G. M. Wyman (1953). Steric hindrance to coplanarity in o-fluorobenzidines. *J. Am. Chem. Soc.* 75:5312–5314.

Bilton, R. F., and R. B. Cain (1965). The metabolism of aromatic compounds by yeasts and moulds. *J. Gen. Microbiol.* 41:xv.

Bilton, R. F., and R. B. Cain (1968). The metabolism of aromatic acids by microorganisms. A reassessment of the role of O-benzoquinone as a product of protocatechuate metabolism by fungi. *Biochem. J.* 108:829–832.

Blades-Fillmore, L. A., W. H. Clement, and S. D. Faust (1982). The effect of sediment on the biodegradation of 2,4,6-trichlorophenol in Delaware River Water. *J. Environ. Sci. Health* A17:797–818.

Bocks, S. M. (1967). Fungal metabolism—III. The hydroxylation of anisole, phenoxyacetic acid, phenylacetic acid and benzoic acid by *Aspergillus niger*. *Phytochemistry* 6:785–789.

Boethling, R. S., and M. Alexander (1979). Effect of concentration of organic chemicals on their biodegradation by natural microbial communities. *Appl. Environ. Microbiol.* 37:1211–1216.

Boethling, R. S., and M. Alexander (1979). Microbial degradation of organic compounds at trace levels. *Environ. Sci. Technol.* 13:989–991.

Bollag, J. M., E. J. Czaplicki, and R. D. Minard (1975). Bacterial metabolism of 1-naphthol. *J. Agric. Food Chem.* 23:85–90.

Bordeleau, L. M., and R. Bartha (1972). Biochemical transformations of herbicide-derived anilines in culture medium and in soil. *Can. J. Microbiol.* 18:1857–1864.

Bordeleau, L. M., H. A. B. Linke, and R. Bartha (1969). Pathway of chloroazobenzene formation chloroaniline-based herbicides in soil. *Bacteriol. Proc.* A21:4.

Borighem, G., and J. Vereecken (1978). Study of the biodegradation of phenol in river water. *Ecol. Modelling* 4:51–59.

Bouwer, E. J., P. L. McCarty, and J. C. Lance (1981). Trace organic behavior in soil columns during rapid infiltration of secondary wastewater. *Water Res.* 15:151–159.

Boyle, T. P., E. F. Robinson-Wilson, J. D. Petty, and W. Weber (1980). Degradation of pentachlorophenol in simulated lentic environment. *Bull. Environ. Contam. Toxicol.* 24:177–184.

Brink, R. H., Jr. (1976). Studies with chlorophenols, acrolein, dithiocarbamates and dibromonitrilopropionamide in bench-scale biodegradation units. pp. 785–791 in Sharpley, J. M. and A. M. Kaplan (eds.). *Proceedings of the Third International Biodegradation Symposium*, Applied Science Publ., London.

Britton, L. N. (1984). Microbial degradation of aliphatic hydrocarbons. pp. 89–129 in Gibson, D. T. (ed.). *Microbial Degradation of Organic Compounds*. Marcel Dekker, Inc., New York.

Broda, P., R. Downing, P. Lehrbach, I. McGregor, and P. Meulien (1981). Metabolic plasmid organization and distribution. pp. 511–517 In Levy, S. B., R. C. Clowes, and E. L. Koenig (eds.). *Molecular Biology, Pathogenicity, and Ecology of Bacterial Plasmids*. Plenum Press, New York.

Broecker, B., and R. Zahn (1977). The performance of activated sludge plants compared with the results of various bacterial toxicity tests—A study with 3,5-dichlorophenol. *Water Res.* 11: 165–172.

Brown, D. S., and E. W. Flagg (1981). Empirical prediction of organic pollutant sorption in natural sediments. *J. Environ. Qual.* 10:382–386.

Burchfield, H. P., and E. E. Storrs (1976). Mechanism of action of fungicides and their reactivities with cellular and environmental substrates. pp. 1043–1055 in Sharpley, J. M., and A. M. Kaplan (eds.). *Proceedings of the Third International Biodegradation Symposium*. Applied Science Publ., London.

Buswell, J. A., and D. G. Twomey (1974). Aromatic acid oxidation by a thermophilic bacterium. *Proc. Soc. Gen. Microbiol.* 1:48.

Butler, G. L. (1977). Algae and pesticides. *Res. Rev.* 66:19—62.

Cain, R. B. (1962). New aromatic ring-splitting enzyme, protocatechuic acid-4,5-oxygenase. *Nature* 193:842—844.

Cain, R. B., D. W. Ribbons, and W. C. Evans (1961). The metabolism of protocatechuic acid by certain microorganisms. A reassessment of the evidence for the participation of 2:6-dioxa-3:7-dioxobicyclo[3:3:0]-octane as an intermediate. *Biochem. J.* 79:312—316.

Camoni, I., A. Di Muccio, D. Pontecorvo, F. Taggi, and L. Vergori (1982). Laboratory investigation for the microbiological degradation of 2,3,7,8-tetrachlorodibenzo-p-dioxin in soil by addition of organic compost. pp. 95—103 In Hutzinger, O., R. W. Frei, E. Merian and F. Pocchiari (eds.). *Chlorinated Dioxins and Related Compounds. Impact on the Environment.* Pergamon Press. New York.

Canovas, J. L., J. Aagaard, and R. Y. Stanier (1966). Studies on the reaction mechanism of dioxygenases. pp. 113—123 In Bloch, K., and O. Hayaishi (eds.). *Biological and Chemical Aspects of Oxygenases.* Maruzen Co. Ltd., Tokyo.

Canovas, J. L., L. N. Ornston, and R. Y. Stanier (1967). Evolutionary significance of metabolic control systems. *Science* 156: 1695—1699.

Carey, A. E., and G. R. Harvey (1978). Metabolism of polychlorinated biphenyls by marine bacteria. *Bull. Environ. Contam. Toxicol.* 20:527—534.

Castro, C. E. (1977). Biodehalogenation. *Environ. Health Perspect.* 21:279—283.

Catterall, F. A., J. M. Sala-Trepat, and P. A. Williams (1971). The coexistence of two pathways for the metabolism of 2-hydroxy-muconic semialdehyde in a naphthalene-grown pseudomonad. *Biochem. Biophys. Res. Commun.* 43:463—469.

Cerniglia, C. E., and D. T. Gibson (1979). Oxidation of benzo(a)pyrene by the filamentous fungus *Cunninghamella elegans*. *J. Biol. Chem.* 254:12174—12180.

Cerniglia, C. E., and D. T. Gibson (1980). Fungal oxidation of benzo(a)pyrene and (+/-)-*trans*-7,8-dihydroxy-7,8-dihydrobenzo(a)pyrene. Evidence for the formation of a benzo(a)pyrene 7,8-diol-9,10-epoxide. *J. Biol. Chem.* 255:5159—5163.

Cerniglia, C. E., D. T. Gibson, and C. van Baalen (1980). Oxidation of naphthalene by cyanobacteria and microalgae. *J. Gen. Microbiol.* 116:495—500.

Cerniglia, C. E., W. Mahaffey, and D. T. Gibson (1980). Fungal oxidation of benzo(a)pyrene: Formation of (-)-*trans*-7,8-dihydroxy-7,8-dihydrobenzo(a)pyrene by *Cunninghamella elegans*. *Biochem. Biophys. Res. Commun.* 94:226—232.

Cerniglia, C. E., C. van Baalen, and D. T. Gibson (1980). Oxidation of biphenyl by the cyanobacterium *Oscillatoria* sp., strain JCM. *Arch. Microbiol.* 125:203–207.

Cerniglia, C. E., J. P. Freeman, and C. van Baalen (1981). Biotransformation and toxicity of aniline and aniline derivatives in cyanobacteria. *Arch. Microbiol.* 130:272–275.

Chakrabarty, A. M. (1976). Plasmids in *Pseudomonas*. *Ann. Rev. Genet.* 10:7–30.

Chakrabarty, A. M. (1972). Genetic basis of the biodegradation of salicylate in *Pseudomonas*. *J. Bacteriol.* 112:815–823.

Chakrabarty, A. M. (1978). Transposition of plasmid DNA segments specifying hydrocarbon degradation and their expression in various microorganisms. *Proc. Natl. Acad. Sci. USA* 75:3109–3112.

Chakrabarty, A. M. (1982). Genetic mechanisms in the dissimilation of chlorinated compounds. pp. 127–139 in A. M. Chakrabarty (ed.). *Biodegradation and Detoxification of Environmental Pollutants*. CRC Press, Inc., Boca Raton, Florida.

Chambers, C. W., and P. W. Kabler (1964). Biodegradability of phenols as related to chemical structure. *Dev. Ind. Microbiol.* 5:85–93.

Chatterjee, D. K., and A. M. Chakrabarty (1981). Plasmids in the biodegradation of PCB's and chlorobenzoates. pp. 213–219 in Leisinger, T., R. Hütter, A. M. Cook and J. Nuesch, (eds.). *Microbial Degradation of Xenobiotics and Recalcitrant Compounds*. Academic Press, New York.

Chatterjee, D. K., and A. M. Chakrabarty (1983). Genetic homology between independently isolated chlorobenzoate-degradative plasmids. *J. Bacteriol.* 153:532–534.

Chatterjee, D, K., S. T. Kellog, D. R. Watkins, and A. M. Chakrabarty (1981). Plasmids in the biodegradation of chlorinated aromatic compounds. pp. 519–528 In Levy, S. B., R. C. Clowes and E. L. Koenig (eds.). *Molecular Biology, Pathogenicity, and Ecology of Bacterial Plasmids*. Plenum Press, New York.

Chu, I., D. C. Villeneuve, V. Secours, and A. Viau (1977). Metabolism of chloronaphthalenes. *J. Agric. Food Chem.* 25:881–883.

Clark, D. E., J. E. Young, R. L. Younger, L. M. Hunt, and J. K. McLaran (1964). The fate of 2,4-dichlorophenoxyacetic acid in sheep. *J. Agric. Food Chem.* 12:43–45.

Clarke, P. H. (1980). Experiments in microbial evolution: New enzymes, new metabolic activities. *Proc. R. Soc. Lond.* B207:385–404.

Clarke, P. H. (1984). The evolution of degradative pathways, pp. 11–27 In Gibson, D. T. (ed.). *Microbial Degradation of Organic Compounds*. Marcel Dekker, Inc., New York.

Clifford, D. R., and D. Woodcock (1964). Metabolism of phenoxy-acetic acid by *Aspergillus niger* van Tiegh. *Nature* 203:763.

Collinsworth, W. L., P. J. Chapman, and S. Dagley (1973). Stereospecific enzymes in the degradation of aromatic compounds by *Pseudomonas putida*. *J. Bacteriol.* 113:922–931.

Colwell, R. R. (1983). Biotechnology in the marine sciences. *Science* 222:19–24.

Cook, A. M., and R. Hutter (1981). Degradation of S-triazines: A critical view of biodegradation. pp. 237–249 In Leisinger, T., R. Hütter, A. M. Cook, and J. Nuesch (eds). *Microbial Degradation of Xenobiotics and Recalcitrant Compounds*. Academic Press, New York.

Cook, A. M., H. Grossenbacher, and R. Hütter (1983). Isolation and cultivation of microbes with biodegradative potential. *Experientia* 39:1191–1198.

Cooper, R. A., and M. A. Skinner (1980). Catabolism of 3- and 4-hydroxyphenylacetate by the 3,4-dihydroxyphenylacetate pathway in *Escherichia coli*. *J. Bacteriol.* 143:302–306.

Coveney, M. F., and R. G. Wetzel (1984). Improved double-vial radiorespirometric technique for mineralization of ^{14}C-labeled substrates. *Appl. Environ. Microbiol.* 47:1154–1157.

Cripps, R. E., and T. R. Roberts (1978). Microbial degradation of herbicides, pp. 669–730 In Hill, I. R. and S. J. L. Wright (eds.). *Pesticide Microbiology* Academic Press, New York.

Cripps, R. E., and R. J. Watkinson (1978). Polycyclic aromatic hydrocarbons: Metabolism and environmental aspects. pp. 113–134 In Watkinson, R. J. (ed.). *Developments in Biodegradation of Hydrocarbons* -1. Applied Science Publishers Ltd., London.

Crosby, D. G. (1972). Environmental photooxidation of pesticides. pp. 206–278 In *Degradation of Synthetic Organic Molecules in the Biosphere*. National Academy of Sciences, Washington, D.C.

Crosby, D. G., and A. S. Wong (1977). Environmental degradation of 2,3,7,8-tetrachlorodibenzo-p-dioxin (TCDD). *Science* 195:1337–1338.

Cserjesi, A. J., and E. L. Johnson (1972). Methylation of pentachlorophenol by *Trichoderma virgatum*. *Can. J. Microbiol.* 18:45–49.

Dagley, S. (1971). Catabolism of aromatic compounds by microorganisms. *Adv. Microb. Physiol.* 6:1–46.

Dagley, S. (1972). Microbial degradation of stable chemical structures: General features of metabolic pathways. pp. 1–16 In *Degradation of Synthetic Organic Molecules in the Biosphere*. National Academy of Sciences, Washington, D.C.

Dagley, S. (1975). A biochemical approach to some problems of environmental pollution. *Essays Biochem.* 11:81–138.

Dagley, S. (1975). Microbial degradation of organic compounds in the biosphere. *Am. Sci.* 63:681–689.

Dagley, S. (1977). Microbial degradation of organic compounds in the biosphere. *Survey Prog. Chem.* 8:121–170.

Dagley, S. (1978). Determinants of biodegradability. *Q. Rev. Biophys.* 11:577–602.

Dagley, S. (1978). Microbial catabolism, the carbon cycle and environmental pollution. *Naturwissenschaften* 65:85–95.

Dagley, S. (1981). New perspectives in aromatic catabolism. pp. 181–188 In Leisinger, T., R. Hütter, A. M. Cook, and J. Nuesch (eds), *Microbial Degradation of Xenobiotics and Recalcitrant Compounds.* Academic Press, New York.

Dagley, S., and M. D. Patel (1957). Oxidation of p-cresol and related compounds by a *Pseudomonas. Biochem. J.* 66:227–233.

Dagley, S., P. J. Chapman, D. T. Gibson, and J. M. Wood (1964). Degradation of the benzene nucleus by bacteria. *Nature* 202: 775–778.

Dagley, S., J. Thomas, and D. T. Gibson (1964). Oxidation of cresols by soil pseudomonads. *Bacteriol. Proc.* P96:104.

Dagley, S., P. J. Geary, and J. M. Wood (1968). The metabolism of protocatechuate by *Pseudomonas testosteroni. Biochem. J.* 109:559–568.

Daly, J. W., D. M. Jerina, and B. Witkop (1972). Arene oxides and the NIH shift: The metabolism, toxicity and carcinogenicity of aromatic compounds. *Experientia* 28: 1219–1264.

Davies, J. I., and W. C. Evans (1962). Ring fission of the naphthalene nucleus by certain soil pseudomonads. *Biochem. J.* 85: 21P–22P.

Davis, E. M., H. E. Murray, J. G. Liehr, and E. L. Powers (1981). Basic microbial degradation rates and chemical byproducts of selected organic compounds. *Water Res.* 15:1125–1127.

Dean-Raymond, D., and R. Bartha (1975). Biodegradation of some polynuclear aromatic petroleum components by marine bacteria. *Dev. Ind. Microbiol.* 16:97–110.

Dearden, M. B., C. R. E. Jefcoate, and J. R. L. Smith (1968). Hydroxylation of aromatic compounds induced by the activation of oxygen. pp. 260–278 In Mayo, F. R. (ed.). *Oxidation of Organic Compounds.* Am. Chem. Soc., Washington.

De Kreuk, J. F., and A. O. Hanstveit (1981). Determination of the biodegradability of the organic fraction of chemical wastes. *Chemosphere* 10:561–571.

Dennis, W. J. Jr., Y. H. Chang, and W. J. Cooper (1979). Catalytic dechlorination of organochlorine compounds. V. Polychlorinated biphenyls - Aroclor 1254. *Bull. Environ. Contam. Toxicol.* 22:750–753.

Der Yang, R., and A. E. Humphrey (1975). Dynamic and steady state studies of phenol biodegradation in pure and mixed cultures. *Biotechnol. Bioeng.* 17:1211–1235.

DiGeronimo, M. J., M. Nikaido, and M. Alexander (1978). Most-probable-number technique for the enumeration of aromatic degraders in natural environments. *Microb. Ecol.* 4:263—266.

Don, R. H., and J. M. Pemberton (1981). Properties of six pesticide degradation plasmids isolated from *Alcaligenes paradoxus* and *Alcaligenes eutrophus*. *J. Bacteriol.* 145:681—686.

Donnelly, M. I., and S. Dagley (1980). Production of methanol from aromatic acids by *Pseudomonas putida*. *J. Bacteriol.* 142:916—924.

Donnelly, M. I., P. J. Chapman, and S. Dagley (1981). Bacterial degradation of 3,4,5-trimethoxyphenylacetic and 3-ketoglutaric acids. *J. Bacteriol.* 147:477—481.

Drinkwine, A. D., and J. R. Fleeker (1981). Metabolism of 2,5-dichloro-4-hydroxyphenoxyacetic acid in plants. *J. Agric. Food Chem.* 29:763—766.

Dunn, N. W., and I. C. Gunsalus (1973). Transmissible plasmid coding early enzymes of naphthalene oxidation in *Pseudomonas putida*. *J. Bacteriol.* 114:974—979.

Durham, D. R., L. A. Stirling, L. N. Ornston, and J. J. Perry (1980). Intergeneric evolutionary homology revealed by the study of protocatechuate 3,4-dioxygenase from *Azotobacter vinelandii*. *Biochemistry* 19:149—155.

Edwards, C. A. (1966). Insecticide residues in soils. *Res. Rev.* 13:83—132.

El-Dib, M. A., and O. A. Aly (1976). Persistence of some phenylamide pesticides in the aquatic environment—III. Biological degradation. *Water Res.* 10: 1055—1059.

Engelhardt, G., P. R. Wallnofer, and H. G. Rast (1981). Bacterial degradation of veratrylglycerol-β-arylethers as model compounds for soil-bound pesticide residues. pp. 293—296 In Leisinger, T., R. Hütter, A. M. Cook, and J. Nuesch (eds.). *Microbial Degradation of Xenobiotics and Recalcitrant Compounds*. Academic Press, New York.

Engst, R., M. Kujawa, and G. Muller (1967). Enzymatischer abbau des DDT durch schimmelpilze. I. Mitt. isolierung und identifizierung eines DDT abbaudenden schimmelpilzes. *Nahrung* 11: 401—403.

Engst, R., R. M. Macholz, and M. Kujawa (1979). Recent state of lindane metabolism. Part II. *Res. Rev.* 72:71—95.

Ensley, B. D., Jr. (1984). Microbial metabolism of condensed thiophenes. pp. 309—317 In Gibson, D. T. (ed.). *Microbial Degradation of Organic Compounds*. Marcel Dekker, Inc. New York.

Estabrook, R. S., J. B. Schenkman, W. Cammer, and H. Remmer (1966). Cytochrome P-450 and mixed function oxidations. pp. 153—178 In Bloch, K., and O. Hayashi (eds.). *Biological and Chemical Aspects of Oxygenases*. Maruzen Co. Ltd., Tokyo.

Evans, W. (1977). Biochemistry of the bacterial catabolism of aromatic compounds in anaerobic envornments. *Nature* 270:17–22.

Evans, W. C. (1947). Oxidation of phenol and benzoic acid by some soil bacteria. *Biochem. J.* 41:373–382.

Evans. W. C. (1963). The microbiological degradation of aromatic compounds. *J. Gen. Microbiol.* 32:177–184.

Evans, W. C., and P. Moss (1957). The metabolism of the herbicide, p-chlorophenoxyacetic acid by a soil microorganism—The formation of a β-chloromuconic acid on ring fission. *Biochem. J.* 65:8P.

Falb, R. D. (1976). Future prospects for immobilized enzymes in biodegradation. pp. 995–999 in Sharpley, J. M., and A. M. Kaplan (eds.). *Proceedings of the Third International Biodegradation Symposium.* Applied Science Publ., London.

Falco, J. W., K. T. Sampson, and R. F. Carsel (1977). Physical modeling of pesticide degradation. *Dev. Ind. Microbiol.* 18:193–202.

Fannin, T. E., M. D. Marcus, D. A. Anderson, and H. L. Bergman (1981). Use of a fractional factorial design to evaluate interactions of environmental factors affecting biodegradation rates. *Appl. Environ. Microbiol.* 42:936–943.

Farrel, R. (1979). Degradative plasmids: Molecular nature and mode of evolution. pp. 97–109 In Timmis, K. N., and A. Puhler (eds.). *Plasmids of Medical, Environmental and Commercial Importance.* Elsevier/North Holland Biomedical Press.

Feist, C. F., and G. D. Hegeman (1969). Phenol and benzoate metabolism by *Pseudomonas putida*: Regulation of tangential pathways. *J. Bacteriol.* 100:869–877.

Fenical, W. (1975). Halogenation in the Rhodophyta: A review. *J. Phycol.* 11:245–259.

Ferebee, R. N., and R. K. Guthrie (1973). The effects of selected herbicides on bacterial populations in an aquatic environment. *Water. Res. Bull.* 9:1125–1134.

Fernley, H. N., and W. C. Evans (1958). Oxidative metabolism of polycyclic hydrocarbons by soil pseudomonads. *Nature* 1832:373–375.

Ferris, J. P., M. J. Fasco, F. L. Stylianopoulou, D. M. Jerina, J. W. Daly, and A. M. Jeffrey (1973). Monooxygenase activity in *Cunninghamella bainieri*: Evidence for a fungal system similar to liver microsomes. *Arch. Biochem. Biophys.* 156:97–103.

Ferris, J. P., L. H. Macdonald, M. A. Patrie, and M. A. Martin (1976). Aryl hydrocarbon activity in the fungus *Cunninghamella bainieri*: Evidence for the presence of cytochrome P-450. *Arch. Biochem. Biophys.* 175:443–452.

Ferry, J. P., and R. S. Wolfe (1976). Anaerobic degradation of benzoate to methane by a microbial consortium. *Arch. Microbiol.* 107:33–40.

Fewson, C. A. (1981). Biodegradation of aromatics with industrial relevance. pp. 141−179 In Leisinger, T., R. Hütter, A. M. Cook and J. Nuesch (eds.). *Microbial Degradation of Xenobiotics and Recalcitrant Compounds*. Academic Press, New York.

Finn, R. K. (1983). Use of specialized microbial strains in the treatment of industrial waste and in soil decontamination. *Experientia* 39:1231−1236.

Fogel, S., R. L. Lancione, and A. E. Sewall (1982). Enhanced biodegradation of methoxychlor in soil under sequential environmental conditions. *Appl. Environ. Microbiol.* 44:113−120.

Fowden, L. (1968). The occurrence and metabolism of carbon-halogen compounds. *Proc. R. Soc. (Lond.) B.* 171:5−18.

Franklin, F. C. H., Bagdasarian, and K. N. Timmis (1981). Manipulation of degradative genes of soil bacteria. pp. 109−130 In Leisinger, T., R. Hütter, A. M. Cook and J. Nuesch (eds.). *Microbial Degradation of Xenobiotics and Recalcitrant Compounds* Academic Press, New York.

Franklin, F. C. H., M. Bagdasarian, M. M. Bagdasarian, and K. N. Timmis (1981). Molecular and functional analysis of the TOL plasmid pWWO from *Pseudomonas putida* and cloning of genes for the entire regulated aromatic ring *meta* cleavage pathway. *Proc. Natl. Acad. Sci. USA* 78:7458−7462.

Freed, V. H., C. T. Chiou, and R. Haque (1977). Chemodynamics: Transport and behavior of chemicals in the environment−A problem in environmental health. *Environ. Health Perspec.* 20: 55−70.

Friello, D. A., J. R. Mylroie, and A. M. Chakrabarty (1976). Use of genetically engineered multi-plasmid microorganisms for rapid degradation of fuel hydrocarbons. pp. 205−214 In Sharpley, J. M. and A. M. Kaplan (eds.). *Proceedings of the Third International Biodegradation Symposium*. Applied Science Publ., London.

Fujiwara, M., L. A. Golovleva, Y. Saeki, M. Nozaki, and O. Hayaishi (1975). Extradiol cleavage of 3-substituted catechols by an intradiol dioxygenase, pyrocatechase, from a pseudomonad. *J. Biol. Chem.* 250:4848−4855.

Furukawa, K., and A. M. Chakrabarty (1982). Involvement of plasmids in total degradation of chlorinated biphenyls. *Appl. Environ. Microbiol.* 44:619−626.

Gaal, A., and H. Y. Neujahr (1981). Induction of phenol-metabolizing enzymes in *Trichosporon cutaneum*. *Arch. Microbiol.* 130: 54−58.

Gale, G. R. (1952). The oxidation of benzoic acid by mycobacteria. II. The metabolism of postulated intermediates in the benzoate oxidation chain by four avirulent and two virulent organisms. *J. Bacteriol.* 64:131−135.

Gambrell, R. P., C. N. Reddy, V. Collard, G. Green, and W. H. Patrick, Jr. (1984). The recovery of DDT, kepone, and permethrin added to soil and sediment suspensions incubated under controlled redox potential and pH conditions. *J. Water Pollut. Control Fed.* 56:174–182.

Gerike, P. (1984). The biodegradability testing of poorly water soluble compounds. *Chemosphere* 13:169–190.

Gerike, P., W. K. Fischer, and W. Holtmann (1980). Biodegradability determinations in trickling filter units compared with the OECD confirmatory test. *Water Res.* 14:753–758.

Gerike, P., W. Holtmann, and W. Jasiak (1984). A test for detecting recalcitrant metabolites. *Chemosphere* 13:121–141.

Ghisalba, O. (1983). Chemical wastes and their biodegradation—An overview. *Experientia* 39:1247–1257.

Gibson, D. T. (1968). Microbial degradation of aromatic compounds. *Science* 161:1093–1097.

Gibson, D. T. (1971). The microbial oxidation of aromatic hydrocarbons. *CRC Crit. Rev. Microbiol.* 1:199–223.

Gibson, D. T. (1976). Microbial degradation of polycyclic aromatic hydrocarbons. pp. 57–66 In Sharpley, J. M. and A. M. Kaplan (eds.). *Proceedings of the Thrid International Biodegradation Symposium.* Applied Science Publishers, London.

Gibson, D. T. (1980). Microbial metabolism. pp. 161–192 In Hutzinger, O. (ed). *Handbook of Environmental Chemistry.* Vol. 2A. Springer-Verlag. New York.

Gilbert, P. A. (1979). Biodegradability and the estimation of environmental concentration. *Ecotoxicol. Environ. Safety* 3:111–115.

Goldman, P. (1972). Enzymology of carbon-halogen bonds. pp. 147–165 In *Degradation of Synthetic Organic Molecules in the Biosphere.* National Academy of Sciences, Washington, D.C.

Golovleva, L. A., and G. K. Skryabin (1981). Microbial degradation of DDT. pp. 287–291 In Leisinger, T., R. Hütter, A. M. Cook and J. Nuesch (eds.) *Microbial Degradation of Xenobiotics and Recalcitrant Compounds,* Academic Press, New York.

Grayson, M. and D. Eckroth (eds.) (1980). *Kirk-Othmer Encyclopedia of Chemical Technology.* Vol. 12. John Wiley & Sons, New York.

Griffiths, E., and W. C. Evans (1965). A cell-free perhydroxylase system from soil pseudomonads, with activity on aromatic hydrocarbons. *Biochem. J.* 95:51P–52P.

Griffiths, E., D. Rodrigues, J. I. Davies, and W. C. Evans (1964). Ability of *Vibrio* 0/1 to synthesize either catechol 1,2-oxygenase or catechol 2,3-oxygenase, depending on the primary inducer. *Biochem. J.* 91:16P.

Grimes, D. J., and S. M. Morrison (1975). Bacterial bioconcentration of chlorinated hydrocarbon insecticides from aqueous systems. *Microb. Ecol.* 2:43–59.

Grover, P. L., A Hewer, and P. Sims (1972). Formation of K-region epoxides as microsomal metabolites of pyrene and benzo[a]pyrene. *Biochem. Pharmacol.* 21:2713–2726.

Gunsalus, I. C. (1972). Early reactions in the degradation of camphor: P-450$_{CAM}$ hydroxylase. pp. 137–146 In *Degradation of Synthetic Organic Molecules in the Biosphere*. National Academy of Sciences, Washington, D.C.

Gunsalus, I. C., and K. M. Yen (1981). Metabolic plasmid organization and distribution. pp. 449–409 In Levy, S. B., R. C. Clowes and E. L. Koenig (eds.). *Molecular Biology, Pathogenicity, and Ecology of Bacterial Plasmids*. Plenum Press, New York.

Guroff, G., J. W. Daly, D. M. Jerina, J. Renson, B. Witkop, and S. Udenfriend (1967). Hydroxylation-induced migration: The NIH shift. *Science* 157:1524–1530.

Haas, D. (1983). Genetic aspects of biodegradation by pseudomonads. *Experientia* 39:1199–1213.

Haber, C. L., L. N. Allen, S. Zhao, and R. S. Hanson (1983). Methylotrophic bacteria: Biochemical diversity and genetics. *Science* 221:1147–1153.

Haider, K., G. Jagnow, R. Kohnen, and S. U. Lim (1974). Degradation of chlorinated benzene, phenol and cyclohexane derivatives by soil bacteria that utilize benzene and phenol under aerobic conditions. *Arch. Microbiol.* 96:183–200.

Haque, A., I. Scheunert, and F. Korte (1978). Isolation and identification of a metabolite of pentachlorophenol-^{14}C in rice plants. *Chemosphere* 1:65–69.

Harder, W. (1981). Enrichment and characterization of degrading organisms. pp. 77–96 In Leisinger, T., R. Hütter, A. M. Cook and J. Nuesch (eds.). *Microbial Degradation of Xenobiotics and Recalcitrant Compounds*. Academic Press, New York.

Harder, W., and L. Dijkhuizen (1982). Strategies of mixed substrate utilization in microorganisms. *Phil. Trans. R. Soc. Lond. B.* 297:459–480.

Hargrave, B. T., and G. A. Phillips (1974). Adsorption of ^{14}C-DDT to particle surfaces. pp. II–13–1B In De Freitas, A. S. W., D J. Kushner and S. U. Quadri (eds.). *Proceedings of the International Conference on Transport of Persistent Chemicals in Aquatic Ecosystems*. Nat. Res. Council Can.

Hayaishi, O. (1966). E. Enzymic studies on the mechanism of double hydroxylation. *Pharmacol. Rev.* 18:71–75.

Healy, J. B., Jr., and L. Y. Young (1979). Anaerobic biodegradation of eleven aromatic compounds to methane. *Appl. Environ. Microbiol.* 38:84–89.

Hegman, G. (1967). The metabolism of p-hydroxybenzoate by *Rhodopseudomonas palustris* and its regulation. *Arch. Mikrobiol.* 59:143–148.

Henderson, M. E. K. (1963). Fungal metabolism of certain aromatic compounds related to lignin. *Pure Appl. Chem.* 7:589—602.

Hickman, G. T., and J. T. Novak (1984). Acclimation of activated sludge to pentachlorophenol. *J. Water Pollut. Control Fed.* 56: 364—369.

Higgins, J. I., R. C. Hammonds, and D. Scott (1984). Transformation of C1 compounds by microorgansims. pp. 43—87 In Gibson, D. T. (ed.) *Microbial Degradation of Organic Compounds.* Marcel Dekker, Inc., New York.

Hill, I. R. (1978). Microbial transformation of pesticides. pp. 137—202 In Hill, I. R. and S. J. L. Wright, (eds). *Pesticide Microbiology.* Academic Press, New York.

Hill, I. R., and S. J. L. Wright (1978). The behavior and fate of pesticides in microbial environments. pp. 79—136 In Hill, I. R. and S. J. L. Wright (eds). *Pesticide Microbiology.* Academic Press, New York.

Hockenbury, M. R., and C. P. L. Grady, Jr. (1977). Inhibition of nitrification - effects of selected organic compounds. *J. Water Pollut. Control Fed.* 49:768—777.

Holden, A. V. (1972). The effects of pesticides on life in fresh waters. *Proc. R. Soc. Lond.* B180:383—394.

Hopper, D. J. (1978). Microbial degradation of aromatic hydrocarbons. pp. 85—112 In Watkins, R. J. (ed). *Developments in Biodegradation of Hydrocarbons - 1.* Applied Science Publishers Ltd., London.

Hopper, D. J., and P. J. Chapman (1970). Gentisic acid and its 3- and 4-methyl-substituted homologues as intermediates in the bacterial degradation of *m*-cresol, 3,5-xylenol and 2,5-xylenol. *Biochem. J.* 122:19—28.

Horvath, R. S. (1972). Microbial cometabolism and the degradation of organic compounds in nature. *Bacteriol. Rev.* 36:146—155.

Hsia, M. T. S., and B. L. Kreamer (1981). Metabolism studies of 3,3',4,4'-tetrachloroazobenzene. I. *In vitro* metabolic pathways with rat liver microsomes. *Chem. Biol. Interact.* 34:19—29.

Hsu, T. S., and R. Bartha (1976). Hydrolyzable and nonhydrolyzable 3,4-dichloroaniline-humus complexes and their respective rates of biodegradation. *J. Agric. Food Chem.* 24:118—122.

Huang, J. C. (1974). Water-sediment distribution of chlorinated hydrocarbon pesticides in various environmental conditions. pp. II—23—20 In De Freitas, A. S. W., D. J. Kushner and S. U. Quadri (eds). *Proceedings of the International Conference on Transport of Persistent Chemicals in Aquatic Ecosystems.* Nat. Res. Council Can.

Hughes, D. E. (1965). The metabolism of halogen-substituted benzoic acids by *Pseudomonas fluorescens. Biochem. J.* 96:181—188.

Hulbert, M. H., and S. Krawiec (1977). Cometabolism: A critique. *J. Theor. Biol.* 69:287–291.

Hutton, D. G., and S. Temple (1979). Priority pollutant removal: Comparison of DuPont PACT process and activated sludge. *Proc. Ind. Waste Symp.*, 52nd Water Pollut. Control Fed. Meeting, Houston.

Hutzinger, O., and A. A. M. Roof (1980). Hydrocarbons and halogenated hydrocarbons in the aquatic environment: Some thoughts on the philosophy and practice of environmental analytical chemistry. pp. 9–28 In Afghan, B. K. and D. MacKay (eds). *Hydrocarbons and Halogenated Hydrocarbons in the Aquatic Environment*. Plenum Publishing Corp., New York.

Ingebritsen, T. S., and P. Cohen (1983). Protein phosphatases: Properties and role in cellular regulation. *Science* 221:331–338.

Irvine, R. L., and A. W. Busch (1969). Factors responsible for non-biodegradability of industrial wastes. *J. Water Pollut. Control Fed.* 41:R482–R491.

Isensee, A. R., and G. E. Jones (1975). Distribution of 2,3,7,8-tetrachlorodibenzo-p-dioxin (TCDD) in aquatic model ecosystem. *Environ. Sci. Technol.* 9:668–672.

Itoh, M., S. Takahaski, M. Iritani, and Y. Kaneko (1980). Degradation of three isomers of cresol and monohydroxybenzoate by *Eumycetes*. *Agric. Biol. Chem.* 44:1037–1042.

Iwata, Y. W. E. Westlake, and F. A. Gunther (1973). Varying persistence of polychlorinated biphenyls in six California soils under laboratory conditions. *Bull. Environ. Contam. Toxicol.* 9:204–211.

Jacobson, S. N., N. L. O'Hara, and M. Alexander (1980). Evidence for cometabolism in sewage. *Appl. Environ. Microbiol.* 40:917–921.

Jamison, V. W., R. L. Raymond, and J. O. Hudson (1971). Hydrocarbon co-oxidation by *Nocardia corallina* strain V-49. *Dev. Ind. Microbiol.* 12:99–105.

Jang, L. K., P. W. Chang, J. E. Findley, and T. F. Yen (1983). Selection of bacteria with favorable transport properties through porous rock for the application of microbial-enhanced oil recovery. *Appl. Environ. Microbiol.* 46:1066–1072.

Janke, D., and W. Fritsche (1979). Dechlorierung von 4-chlorphenol nach extradioler ringspaltung durch *Pseudomonas putida*. *Z. Allg. Microbiol.* 19:193–141.

Janke, D., R. Pohl, and W. Fritsche (1981). Regulation of phenol degradation in *Pseudomonas putida*. *Z. Allg. Microbiol.* 21:295–303.

Jannasch, H. W. (1967). Growth of marine bateria at limiting concentrations of organic carbon in seawater. *Limnol. Oceanogr.* 12:264–271.

Jeenes, D. J., and P. A. Williams (1982). Excision and integration of degradative pathways genes from TOL plasmid pWWO. *J. Bacteriol.* 150:188—194.

Jeenes, D. J., W. Reineke, H. J. Knackmuss, and P. A. Williams (1982). TOL plasmid pWWO in constructed halobenzoate-degrading *Pseudomonas* strains: Enzyme regulation and DNA structure. *J. Bacteriol.* 150:180—187.

Jenson, R. A. (1976). Enzyme recruitment in evolution of new function. *Ann. Rev. Microbiol.* 30:409—425.

Joel, A. R., and C. P. L. Grady, Jr. (1977). Inhibition of nitrification - effects of aniline after biodegradation. *J. Water Pollut. Control Fed.* 49:778—788.

Johnson, B. T. (1969). Mechanism for the degradation of 1,1,1-trichloro-2,2-bis(p-chlorophenyl)ethane by microorganisms. *Bacteriol. Proc.* A103:16.

Johnson, B. T. (1974). Aquatic food chain models for estimating bioaccumulation and biodegradation of xenobiotics. pp. IV—17—22 In De Freitas, A. S. W., D. J. Kushner and S. U. Quadri (eds). *Proceedings of the International Conference on Transport of Persistent Chemicals in Aquatic Ecosystems.* Nat. Res. Council Can.

Johnson, E. F., and U. Muller-Eberhard (1977). Resolution of two forms of cytochrome P—450 from liver microsomes of rabbits treated with 2,3,7,8-tetrachlorodibenzo-p-dioxin. *J. Biol. Chem.* 252:2839—2845.

Johnson, L. D., and J. C. Young (1983). Inhibition of anaerobic digestion by organic priority pollutants. *J. Water Pollut. Control Fed.* 55:1441—1449.

Johnson, L. M., and H. W. Talbot, Jr. (1983). Detoxification of pesticides by microbial enzymes. *Experientia* 39:1236—1246.

Johnston, J. B., and S. G. Robinson (1983). Genetic engineering and the development of new pollution control technolgies. Report No. UILU-ENG-83-0102, Advanced Environmental Control Technology Research Center, Univ. of Illionois, 131 p.

Jori, A., D. Calamari, F. Cattabeni, A. Di Domenico, C. L. Galli, E. Galli, A. Ramundo, and V. Silano (1982). Ecotoxicological profile of p-dichlorobenzene. *Ecotoxicol. Environ. Safety* 6:413—432.

Kachhy, A. N., and V. V. Modi (1976). Catechol metabolism in *Pseudomonas aeruginosa*: Regulation of *meta*-fission pathways. *Int. J. Exp. Biol.* 14:163—165.

Kaiser, J. P., and K. W. Hanselmann (1982). Fermentative mechanism of substituted monoaromatic compounds by a bacterial community from anaerobic sediments. *Arch. Microbiol.* 133:185—194.

Kaiser, K. L. E. (1983). A non-linear function for the approximation of octanol/water partition coefficients of aromatic compounds with multiple chlorine substitutions. .*Chemosphere* 12:1159—1167.

Kamp, P. F., and A. M. Chakrabarty (1979). Plasmids specifying
p-chlorobiphenyl degradation in enteric bacteria. pp. 275−285
In Timmis, K. N. and A. Puhler (eds.). *Plasmids of Medical,
Environmental and Commercial Importance.* Elsevier/North Holland Biomedical Press.

Kaneko, M., K. Morimoto, and S. Nambu (1976). The response of
activated sludge to a polychorinated biphenyl (KC-500). *Water
Res.* 10:157−163.

Karns, J. S., S. Duttagupta, and A. M. Chakrabarty (1983). Regulation of 2,4,5-trichlorophenoxyacetic acid and chlorophenol
metabolism in *Pseudomonas cepacia* AC1100. *Appl. Environ. Microbiol.* 46:1182−1186.

Katagiri, M., H. Maeno, S. Yammamoto, and O. Hayaishi (1965).
Salicylate hydroxylase, a monooxygenase requiring flavin adenine
dinucleotide. II. The mechanism of salicylate hydroxylation to
catechol. *J. Biol. Chem.* 240:3414−3417.

Kaufman, D. D. (1978). Degradation of pentachlorophenol in soil
and by soil microorganisms. pp. 27−39 In Rao, K. R. (ed).
Pentachlorophenol. Plenum Press, New York.

Kearnery, P. C. (1976). Biodegradable alternatives to persistent
pesticides. pp. 843−852 In Sharpley, J. M., and A. M. Kaplan
(eds.). *Proceedings of the Third International Biodegradation
Symposium.* Applied Science Publ., London.

Kearney, P. C., and D. D. Kaufman (1972). Microbial degradation
of some chlorinated pesticides. pp. 166−189 In *Degradation of
Synthetic Organic Molecules in the Bioshpere.* National Academy
of Sciences, Washington, D.C.

Kearney, P. C., J. R. Plimmer, and F. B. Guardia (1969). Mixed
chloroazobenzene formation in soil. *J. Agric. Food Chem.* 17:
1418−1419.

Kellog, S. T., D. K. Chatterjee, and A. M. Chakrabarty (1981).
Plasmid-assested molecular breeding: New technique for enhanced
biodegradation of persistent toxic chemicals. *Science* 214:1133−
1135.

Kenaga, E. E. (1974). Partitioning and uptake of pesticides in biological systems. pp. II−19−22 In De Freitas, A. S. W., D. J.
Kushner and S. U. Quadri (eds.). *Proceedings of the International Conference on Transport of Persistent Chemicals in
Aquatic Ecosystems.* Nat. Res. Council Can.

Kennedy, C. D. (1974). The absorption of benzoic acid and some
of its chlorine-substituted derivatives at an alkane/water inter-
face. *Pestic. Sci.* 5:675−690.

Khanna, S., and S. C. Fang (1966). Metabolism of ^{14}C-labeled 2,4-
dichlorophenoxyacetic acid in rats. *J. Agric. Food Chem.* 14:
500−503.

Kirk, T. K. (1984). Degradation of lignin. pp. 399–437 In Gibson,
    D. T. (ed.). *Microbial Degradation of Organic Compounds*. Mar-
    cel Dekker, Inc., New York.
Klecka, G. M., and D. T. Gibson (1981). Inhibition of catechol 2,3-
    dioxygenase from *Pseudomonas putida* by 3-chlorocatechol. *Appl.*
    Environ. Microbiol. 41:1159–1165.
Klibanov, A. (1983). Immobilized enzymes and cells as practical
    catalysts. *Science* 219:722–727.
Klibanov, A. M, B. N. Alberti, E. D. Morris, and L. M. Felshin
    (1980). Enzymatic removal of toxic phenols and anilines from
    wastewaters. *J. Appl. Biochem.* 2:414–421.
Klibanov, A. M., T-M. Tu, and K. P. Scott (1983). Peroxidase-
    catalyzed removal of phenols from coal conservation wastewaters.
    Science 221:259–261.
Kloskowski, R., I. Scheunert, W. Klein, and F. Korte (1981). Lab-
    oratory screening of distribution, conversion and mineralization
    of chemicals in the soil-plant-system and comparison to outdoor
    experimental data. *Chemosphere* 10:1089–1100.
Knackmuss, H. J., and W. Reineke (1973). Der einfluss von chlor-
    substituenten auf die oxygenierung von benzoat durch *Alcaligenes*
    eutrophus B9. *Chemosphere* 2:225–230.
Knackmuss, H. J., W. Beckmann, E. Dorn, and W. Reineke (1976).
    On the mechanism of the biological persistence of halogenated
    and sulfonated aromatic hydrocarbons. *Zbl. Bakt. Hyg., I.*
    Abt. Orig. B 162:127–137.
Knowlton, M. F., and J. M. Huckins (1983). Fate of radiolabeled
    sodium pentachlorophenate in littoral microcosms. *Bull. Environ.*
    Contam. Toxicol. 30:206–213.
Kobal, V. M., D. T. Gibson, R. E. Davis, and A. Garza (1973).
    X-ray determination of the absolute sterochemistry of the initial
    oxidation product formed from toluene by *Pseudomonas putida*
    39/D. *J. Am. Chem. Soc.* 95:4420–4421.
Kobayashi, H., and B. E. Rittman (1982). Microbial removal of ha-
    zardous organic compounds. *Environ. Sci. Technol.* 16:170A–
    183A.
Kobayashi, K. (1978). *Metabolism of Pentachlorophenol*. Plenum Press,
    New York.
Kozak, V. P., G. V. Simsiman, G. Chesters, D. Stensby, and J.
    Harkin (1979). Reviews of the environmental effects of pollutants:
    XI. Chlorophenols. U. S. Environmental Protection Agency. EPA-
    600/1–79–012. Cincinnati, Ohio.
Krulwich, T. A., and N. J. Pelliccione (1979). Catabolic pathways
    of coryneforms, nocardias, and mycobacteria. *Ann. Rev. Micro-*
    biol. 33:95–111.
Kuwatsuka, S. (1972). Degradation of several herbicides in soils
    under different conditions. pp. 385–400 in Matsumura, F., G.

M. Boush and T. Misato (eds.). *Environmental Toxicology of Pesticides.* Academic Press, New York.

Lackmann, R. K., W. J. Maier, and N. A. Shamat (1981). Removal of chlorinated organics by conventional biological waste treatment. Proc. 35th Ind. Waste Conf., Purdue Univ. Ann Arbor Sci. Publ., Inc., Woburn, Massachusetts. 502 p.

Lamberton, J. G., R. D. Inman, R. R. Claeys, W. A. Robson, and G. H. Arscott (1975). The metabolism of p,p'-DDE in laying Japanese quail and their incubated eggs. *Bull. Environ. Contam. Toxicol.* 14:657−664.

Larson, R. J., and A. G. Payne (1981). Fate of the benzene ring of linear alkylbenzene sulfonate in natural waters. *Appl. Environ. Microbiol.* 41:621−627.

Larway, P., and W. C. Evans (1965). Metabolism of quinol and resorcinol by soil pseudomonads. *Biochem. J.* 95:52.

Lay, M. M., and R. D. Ilnicki (1974). Peroxidase activity and propanil degradation in soil. *Weed Res.* 14:111−113.

Lech, J. J., A. H. Glickman, and C. N. Stratham (1978). Studies on the uptake, disposition, and metabolism of pentachlorophenol and pentachloroanisole in rainbow trout (*Salmon gairdneri*). pp. 107−113 In Rao, K. R. (ed). *Pentachlorophenol.* Plenum Press, New York.

Leemans, J. D. Inze, R. Villarroel, G. Engler, J. P. Hernalsteens, M. de Block, and M. van Montagu (1981). Plasmid mobilization as a tool for *in vivo* genetic engineering. pp. 401−409 In Levy, S. B., R. C. Clowes and E. L. Koenig (eds). *Molecular Biology, Pathogenicity, and Ecology of Bacterial Plasmids.* Plenum Press, New York.

Lehmicke, L. G., R. T. Williams, and R. L. Crawford (1979). ^{14}C-most-probable-number method for enumeration of active heterotrophic microorganisms in natural waters. *Appl. Environ. Microbiol.* 38:644−649.

Lehrbach, P. R., J. Zeyer, W. Reineke, H. J. Knackmuss, and K. N. Timmis (1984). Enzyme recruitment *in vitro*: Use of cloned genes to extend the range of haloaromatics degraded by *Pseudomonas* sp. strain B13. *J. Bacteriol.* 158:1025−1032.

Leigh, G. M. (1969). Degradation of selected chlorinated hydrocarbon insecticides. *J. Water Pollut. Control Fed.* 41:R450−R460.

Leisinger, T. (1983). Microorganisms and xenobiotic compounds. *Experientia* 39:1183−1191.

Lewis, D. L., H. P. Kollig, and T. L. Hall (1983). Predicting 2,4-dichlorophenoxyacetic acid ester transformation rates in periphyton-dominated ecosystems. *Appl. Environ. Microbiol.* 46:146−151.

Liem, H. H., U. Muller-Eberhard, and E. F. Johnson (1980). Differential induction by 2,3,7,8-tetrachlorodibenzo-p-dioxin of multiple forms of rabbit microsomal cytochrome P-450. Evidence for tissue specificity. *Mol. Pharmacol.* 18:565−570.

Liu, D., W. M. J. Strachan, K. Thomson, and K. Kwasniewska (1981). Determination of the biodegradability of organic compounds. *Environ. Sci. Technol.* 15:788−793.

Lovelock, J. E. (1975). Natural halocarbons in the air and in the sea. *Nature* 256:193−194.

Lu, P. Y., and R. L. Metcalf (1975). Environmental fate and biodegradability of benzene derivatives as studied in a model aquatic ecosystem. *Environ. Health Perspect.* 10:269−284.

Lu, P. Y., R. L. Metcalf, and L. K. Cole (1978). The environmental fate of ^{14}C-pentachlorophenol in laboratory model ecosystems. pp. 53−63 In Rao, K. R. (ed). *Pentachlorophenol.* Plenum Press, New York.

Ludzack, R. J., and M. B. Ettinger (1960). Chemical structures resistant to aerobic biochemical stabilization. *Purdue Univ. Eng. Bull. Ext. Ser.* 402−444.

Lyr, H. (1962). Detoxification of heartwood toxins and chlorophenols by higher fungi. *Nature* 195:289−290.

MacDonald, T. L. (1983). Chemical mechanisms of halocarbon metabolism. *CRC Crit. Rev. Toxicol.* 11:85−120.

Malaney, G. W., P. A. Lutin, J. J. Cibulka, and L. H. Hickerson (1967). Resistance of carcinogenic organic compounds to oxidation by activated sludge. *J. Water Pollut. Control Fed.* 39:2020−2028.

Marr, E. K., and R. W. Stone (1958). The bacterial oxidation of benzene. *Bacteriol. Proc.* P86:123.

Marr, E. K., and R. W. Stone (1961). Bacterial oxidation of benzene. *J. Bacteriol.* 81:425−430.

Martens, R. (1982). Concentrations and microbial mineralization of four to six ring polycyclic aromatic hydrocarbons in composted municipal waste. *Chemosphere* 11:761−770.

Mason, C. P., K. R. Edwards, R. E. Carlson, J. Pignatello, F. K. Gleason, and J. M. Wood (1982). Isolation of chlorine-containing antibiotic from the freshwater cyanobacterium *Scytonema hofmanni. Science* 215:400−402.

Mason, H. S., W. L. Fowlks, and E. Peterson (1955). Oxygen transfer and electron transport by the phenolase complex. *J. Am. Chem. Soc.* 77:2914−2915.

Matsumura, F. (1975). Environmental alteration of insecticide residues. pp. 325−354 In *Toxicology of Insecticides.* Plenum Press, New York.

Matsumura, F., and G. M. Boush (1967). Dieldrin: Degradation by soil microorganisms. *Science* 156:959−961.

Matsumura, F., and H. J. Benezet (1978). Microbial degradation of insecticides. pp. 623−667 In Hill, I. R. and S. J. L. Wright, (eds.). *Pesticide Microbiology.* Academic Press, New York.

McCarty, P. L., D. Argo, and M. Reinhard (1979). Operational experiences with activated carbon absorbers at water factory 21. *J. Am. Water Works Assoc.* 71:683–689.

McCarty, P. L., M. Reinhard, and B. E. Rittmann, (1981). Trace organics in groundwater. *Environ. Sci. Technol.* 14:40–51.

Menzie, C. M. (1978). *Metabolism of Pesticides*. Update II. U. S. Dept. Interior Fish Wildl. Serv. Spec. Sci. Rept. - Wildl. No. 212. Washington, D.C.

Menzie, C. M. (1980). *Metabolism of Pesticides*. Update III. U. S. Dept. Interior Fish Wildl. Serv. Spec. Sci. Rept. - Wildl. No. 232. Washington, D.C.

Metcalf, R. L., I. P. Kapoor, and A. S. Hirwe (1972). Development of persistent biodegradable insecticides related to DDT. pp. 244–259 In *Degradation of Synthetic Organic Molecules in the Biosphere*. National Academy of Sciences, Washington, D.C.

Metcalf, R. L., I. P. Kapoor, P. Y. Lu, C. K. Schuth, and P. Sherman (1973). Model ecosystem studies of the environmental fate of six organochlorine pesticides. *Environ. Health Perspect.* 4:35–44.

Metcalf, R. L., J. R. Sanborn, P. Y. Lu, and D. Nye (1975). Laboratory model ecosystem studies of the degradation and fate of radiolabeled tri-, tetra-, and pentachlorobiphenyl compared with DDE. *Arch. Environ. Contam. Toxicol.* 3:151–165.

Moore, S., and E. E. Staffeldt (1976). Enzymatic activity of soil fungi. pp. 711–718 In Sharpley, J. M. and A. M. Kaplan (eds.). *Proceedings of the Third International Biodegradation Symposium*. Applied Science Publ., London.

Motosugi, K., and K. Soda (1983). Microbial degradation of synthetic organochlorine compounds. *Experientia* 39:1214–1220.

Munnecke, D. M. (1981). The use of microbial enzymes for pesticide detoxification. pp. 251–269 In Leisinger, T., R. Hütter, A. M. Cook and J. Nuesch (eds.). *Microbial Degradation of Xenobiotics and Recalcitrant Compounds*. Academic Press, New York.

Murado, M. A., M. C. Tejedor, and G. Baluja (1976). Interactions between polychlorinated biphenyls (PCBs) and soil microfungi. Effects of Aroclor-1254 and other PCBs on *Aspergillus flavus* cultures. *Bull. Environ. Contam. Toxicol.* 15:768–774.

Murray, K., C. J. Duggleby, J. M. Sala-Trepat, and P. A. Williams (1972). The metabolism of benzoate and methylbenzoates via the *meta*-cleavage pathway by *Pseudomonas arvilla* MT-2. *Eur. J. Biochem.* 28:301–310.

Muster, C. J., L. A. Machattie, and J. A. Shapiro (1981). Transposition and rearrangements in plasmid evolution. pp. 349–358 In Levy, S. B., R. C. Clowes and E. L. Koenig (eds.). *Molecula Biology, Pathogenicity, and Ecology of Bacterial Plasmids*. Plenum Press, New York.

Neu, H. J., and K. Ballschmiter (1977). Abbau von chlorierten aromaten: Mikrobiologischer abbau der polychlorierten biphenyle (PCB). II. Biphenylole als metabolite der PCB. *Chemosphere* 6:419—423.

Neufield, R. D., and T. Valiknac (1979). Inhibition of phenol degradation by thiocyanate. *J. Water Pollut. Control Fed.* 51:2283—2291.

Neufeld, R. D., and T. Valiknac (1979). Inhibition of phenol degraphenol biokinetics. *J. Water Pollut. Control Fed.* 52:2367—2377.

Nozaki, M., Y. Kojima, T. Nakazawa, H. Fujisawa, K. Ono, S. Kotani, and O. Hayaishi (1966). Studies on the reaction mechanism of dioxygenases. pp. 347—367 In Block, K. and O. Hayaishi (eds). *Biological and Chemical Aspects of Oxygenases*. Maruzen Co. Ltd., Tokyo.

O'Conner, R. J., B. W. Weinrich, and W. A. Darlington (1964). Phenol and the microbial conversion of benzene to catechol. *Bacteriol. Proc.* P97:104—105.

O'Kelley, J. C., and T. R. Deason (1976). Degradation of pesticides by algae. EPA-600/3-76-022. U. S. Environmental Protection Agency, Office of Research and Development, Environmental Research Lab, Athens, Georgia. 41 pp.

Ohisa, N., and M. Yamaguchi (1979). *Clostridium* species and γ-BHC degradation in paddy soil. *Soil Biol. Biochem.* 11:645—649.

Ohisa, N., T. Kurihara, and M. Nakajima (1982). ATP synthesis associated with the conversion of hexachlorocyclohexane related compounds. *Arch. Microbiol.* 131:330—333.

Oloffs, P. C., and L. J. Albright (1974). Transport of some organochlorines in B.C. waters pp. I-89-92 In De Freitas, A. S. W., D. J. Kushner and S. U. Quadri (eds.). *Proceedings of the International Conference on Transport of Persistent Chemicals in Aquatic Ecosystems*. Nat. Res. Council Can.

Olsson, M. (1974). Time and space dependence of pollutant levels in aquatic biota, field studies. pp. III-40-60 In De Freitas, A. S. W. D. J. Kushner and S. U. Quadri (eds.). *Proceedings of the International Conference on Transport of Persistent Chemicals in Aquatic Ecosystems*. Nat. Res. Council Can.

Omori, T., and M. Alexander (1978). Bacterial and spontaneous dehalogenation of organic compounds. *Appl. Environ. Microbiol.* 35:512—516.

Ornston, L. N. (1971). Regulation of catabolic pathways in *Pseudomonas*. *Bacteriol. Rev.* 35:87—116.

Ornston, L. N., and D. Parke (1976). Evolution of catabolic pathways. *Biochem. Soc. Trans.* 4:468—473.

Ornston, L. N., and W. K. Yeh (1982). Recurring themes and repeated sequences in metabolic evolution. pp. 105—126 In A. M. Chakrabarty. (ed.). *Biodegradation and Detoxification of Environmental Pollutants*. CRC Press, Inc., Boca Raton, FL.

Oyler, A. R., R. J. Llukkonen, M. T. Lukasewycz, K. E. Heikkila,
    D. A. Cox, and R. M. Carlson (1983). Chlorine 'disinfection'
    chemistry of aromatic compounds. Polynuclear aromatic hydrocar-
    bons: Rates, products, and mechanisms. *Environ. Sci. Technol.*
    17:334—342.
Painter, H. A. (1974). Biodegradability. *Proc. R. Soc. Lond. B.*
    185:149—158.
Pal, D., J. B. Weber, and M. R. Overcash (1980). Fate of poly-
    chlorinated biphenyls (PCBs) in soil-plant systems. *Res. Rev.*
    74:45—98.
Papanastasiou, A. C., and W. J. Maier (1982). Dynamics of biode-
    gradation of 2,4-dichlorophenoxycacetate in the presence of glu-
    cose. *Biotechnol. Bioeng.* 25:2337—2346.
Papanastasiou, A. C., and W. J. Maier (1982). Kinetics of biode-
    gradation of 2,4-dichlorophenoxyacetate in the presence of glu-
    cose. *Biotechnol. Bioeng.* 24:2001—2011.
Pardini, R. S., J. C. Heidker, T. A. Baker, and B. Payne (1980).
    Toxicology of various pesticides and their decomposition products
    on mitochondrial electron transport. *Arch. Environ. Contam.*
    Toxicol. 9:87—97.
Paris, D. F., and D. L. Lewis (1973). Chemical and microbial de-
    gradation of ten selected pesticides in aquatic systems. *Res.*
    Rev. 45:95—124.
Paris, D. F., D. L. Lewis, J. T. Barnett, Jr., and G. L. Baughman
    (1975). Microbial degradation and accumulation of pesticides
    in aquatic systems. EPA-660/3-75-007. National Environmental
    Research Center, Office of Research and Development, USEPA,
    Corvallis, Oregon. 45 pp.
Paris, D. F., W. C. Steen, G. L. Baughman, and J. T. Barnett,
    Jr. (1981). Second-order model to predict microbial degradation
    of organic compounds in natural waters. *Appl. Environ. Micro-*
    biol. 41:603—609.
Paris, D. F., N. L. Wolfe, and W. C. Steen (1982). Structure-
    activity relationships in microbial transformation of phenols. *Appl.*
    Environ. Microbiol. 44:153—158.
Parke, D., and L. N. Ornston (1976). Constitutive synthesis of
    enzymes of the protocatechuate pathway and of the β-ketoadipate
    uptake system in mutant strains of *Pseudomonas putida*. *J.*
    Bacteriol. 126:272—281.
Parr, J. E., G. H. Willis, L. L. McDowell, C. E. Murphree, and
    S. Smith (1974). An automatic pumping sampler for evaluating
    the transport of pesticides in suspended sediment. *J. Environ.*
    Qual. 3:292—294.
Patel, T. R., and D. T. Gibson (1976). Bacterial *cis*-dihydrodiol
    dehydrogenases: Comparison of physiochemical and immunological
    properties. *J. Bacteriol.* 128:842—850.

Pavlou, S. P., R. N. Dexter, and J. R. Clayton, Jr. (1974). Chlo-
  inated hydrocarbons in coastal marine ecosystems. pp.II-31-35
  In De Freitas, A. S. W., D. J. Kushner and S. U. Quadri (eds.).
  *Proceedings of the International Conference on Transport of Per-
  sistent Chemicals in Aquatic Ecosystems.* Nat. Res. Council Can.

Pawlowsky, U., and J. A. Howell (1973). Mixed culture biooxidation
  of phenol. II. Steady state experiments in continuous culture.
  Biotechnol. Bioeng. 15:897−903.

Pawlowsky, U., J. A. Howell, and C. T. Chi (1973). Mixed culture
  biooxidation of phenol. III. Existence of multiple steady states
  in continuous culture with wall growth. *Biotechnol. Bioeng.*
  15:905−916.

Payne, W. J., W. J. Wiebe, and R. R. Christian (1970). Assays
  for biodegradability essential to unrestricted usage of organic
  compounds. *BioScience* 20:862−865.

Peakall, D. B., and J. L. Lincer (1970). Polychlorinated biphenyls,
  another long-life widespread chemical in the environment. *Bio-
  Science* 20:958−964.

Pemberton, J. M., and P. R. Fisher (1977). 2,4-D plasmids and
  persistence. *Nature* 268:732−733.

Peng, C. T., B. E. Gordon, W. R. Erwin, and R. M. Lemmon
  (1982). Dehalogenation and ring saturation by tritium atoms.
  Int. J. Appl. Radiat. Isot. 33:419−427.

Perry, J. J. (1979). Microbial cooxidations involving hydrocarbons.
  Microbiol. Rev. 43:59−72.

Petty, M. A. (1961). An introduction to the origin and biochemistry
  of microbial halometabolites. *Bacteriol. Rev.* 25:111−130.

Pfaender, F. K., and G. W. Bartholomew (1982). Measurement of
  aquatic biodegradation rates by determining heterotrophic uptake
  of radiolabeled pollutants. *Appl. Environ. Microbiol.* 44:159−164.

Pfister, R. M. (1972). Interactions of halogenated pesticides and
  microorganisms: A review. *CRC Crit. Rev. Microbiol.* 2:1−33.

Pitter, P. (1976). Determination of biological degradability of organic
  substances. *Water Res.* 10:231−235.

Plimmer, J. R. (1970). The photochemistry of halogenated herbicides.
  Res. Rev. 33:47−74.

Plimmer, J. R. (1972). Principles of photodecomposition of pesticides.
  pp 279−290 In *Degradation of Synthetic Organic Molecules in
  the Biosphere.* National Academy of Sciences, Washington, DC.

Plimmer, J. R., P. C. Kearney, and D. W. von Endt (1967). Mechanism of conversion of DDT to DDD by *Aerobacter aerogenes.*
  Bacteriol. Proc. A43:8.

Poiger, J., J-R. Buser, H. Weber, U. Zweifel, and C. Schlatter
  (1982). Structure elucidation of mammalian TCDD-metabolites.
  Experientia 38:484−486.

Priest, B., and R. J. Stephens (1975). Studies on the breakdown of p-chlorophenyl methylcarbamate. I. In soil. *Pestic. Sci.* 6: 53—59.

Pritchard, R. H., and N. B. Grover (1981). Control of plasmid replication and its relationship to incompatibility. pp. 271—278 In Levy, S. B., R. C. Clowes and E. L. Koenig (eds). *Molecular Biology Pathogenicity, and Ecology of Bacterial Plasmids.* Plenum Press, New York.

Radding, S. B., D. H. Liu, H. L. Johnson, and T. Mill (1977). Review of the environmental fate of selected chemicals. EPA-560/5-77-003. 147 p.

Raymond, R. L., and v. W. Jamison (1971). Biochemical activites of *Nocardia. Adv. Appl. Microbiol.* 14:93—122.

Reichardt, P. B., B. L. Chadwick, M. A. Cole, B. R. Robertson, and D. K. Button (1981). Kinetic study of the biodegradation of biphenyl and its monochlorinated analogues by a mixed marine microbial community. *Environ. Sci. Technol.* 15:75—79.

Reineke, W. (1984). Microbial metabolism of halogenated aromatic compounds. pp. 319—360 In Gibson, D. T. (ed.). *Microbial Degradation of Organic Compounds.* Marcel Dekker, Inc., New York.

Reineke, W., and H-J. Knackmuss (1979). Construction of haloaromatics utilizing bacteria. *Nature* 277:385—386.

Reineke, W., D. J. Jeenes, P. A. Williams, and H-J. Knackmuss (1982). TOL plasmid pWWO in constructed halobenzoate-degrading *Pseudomonas* strains: Prevention of *meta* pathway. *J. Bacteriol.* 150:195—201.

Reiner, A. M. (1971). Metabolism of benzoic acid by bacteria: 3,5-Cyclohexadiene-1,2-diol-1-carboxylic acid is an intermediate in the formation of catechol. *J. Bacteriol.* 108:89—94.

Reiner, A. M., and G. D. Hegeman (1971). Metabolism of benzoic acid by bacteria. Accumulation of (-)-3,5-cyclohexadiene-1,2-diol-1-carboxylic acid by a mutant strain of *Alcaligenes eutrophus. Biochemistry* 10:2530—2536.

Renner, G., E. Richter, and K. P. Schuster (1978). Synthesis of hexachlorobenzene metabolites. *Chemosphere* 8:669—674.

Ribbons, D. W. (1970). Specificity of monohydric phenol oxidations by *meta* cleavage pathways in *Pseudomonas aeruginosa* T1. *Arch. Mikrobiol.* 74:103—115.

Ribbons, D. W., and R. W. Eaton (1982). Chemical transformations of aromatic hydrocarbons that support the growth of microorganisms. pp. 59—84 in A. M. Chakrabarty (ed.). *Biodegradation and Detoxification of Environmental Pollutants.* CRC Press, Inc., Boca Raton FL.

Ribbons, D. W., P. Keyser, D. A. Kunz, B. F. Taylor, R. W.
    Eaton, and B. N. Anderson (1984). Microbial degradation of
    phthalates. pp. 371–397 in Gibson, D. T. (ed.). *Microbial De-
    gradation of Organic Compounds*, Marcel Dekker, Inc., New
    York.
Rich, S., and J. G. Horsfall (1954). Relation of polyphenol oxidases
    to fungitoxicity. *Proc. Natl. Acad. Sci. USA* 40:139–145.
Rittman, B. E., E. J. Bouwer, J. E. Schreiner, and P. L. McCarty
    (1980). Biodegradation of trace organic compounds in ground
    water systems. Technical Report No. 255, Stanford University
    Dept. Civil Eng., 48 p.
Roberts, J. L., Jr., and D. T. Sawyer (1981). Facile degradation
    by superoxide ion of carbon tetrachloride, chloroform. methy-
    lene chloride, and p,p'-DDT in aprotic media. *J. Am. Chem.
    Soc.* 103:712–714.
Roberts, P. V., J. Schreiner, and G. D. Hopkins (1982). Field
    study of organic water quality changes during ground water
    recharge in the Palo Alto baylands. *Water Res.* 16:1025–1035.
Rogoff, M. H. (1958). Dissimilation of methylnaphthalenes by *Pseu-
    domonas* spp. *Bacteriol. Proc.* P87:123–124.
Rogoff, M. H. (1961). Oxidation of aromatic compounds by bacteria.
    Adv. Appl. Microbiol. 3:193–221.
Rozich, A. F., A. F. Gaudy, Jr., and P. D. D'Adamo (1983). Pre-
    dictive model for treatment of phenolic wastes by activated
    sludge. *Water Res.* 10:1453–1466.
Rubin, H. E., and M. Alexander (1983). Effect of nutrients on the
    rates of mineralization of trace concentrations of phenol and p-
    nitrophenol. *Environ. Sci. Technol.* 17:104–107.
Rubin, H. E., R. V. Subba-Rao, and M. Alexander (1982). Rates
    of mineralization of trace concentrations of aromatic compounds
    in lake water and sewage samples. *Appl. Environ. Microbiol.*
    43:1133–1138.
Russell, L. L., C. B. Cain, and D. I. Jenkins (1983). Impact of
    priority pollutants on publicly owned treated works processes:
    A literature review. Proc. 37th Ind. Waste Conf., Purdue Univ.,
    Ann Arbor Sci. Publ., Inc., Ann Arbor, Michigan. 871 p.
Saeger, V. W., and G. E. Thompson (1980). Biodegradabolity of
    halogen-substituted diphenylmethanes. *Environ. Sci. Technol.*
    14:705–709.
Safe, S. H. (1984). Microbial degradation of polychlorinated bi-
    phenyls. pp. 361–369 in Gibson, D. T. (ed.). *Microbial Degra-
    dation of Organic Compounds*. Marcel Dekker, Inc., New York.
Safe, S., C. Wyndham. A. Crawford, and J. Kohli (1978). Metabolism:
    Detoxification or toxification. pp. 299–307 In Hutzinger, O., I.
    H. van Lelyveld and B. C. J. Zoeteman (eds.). *Aquatic Pollu-
    tants: Transformation and Biological Effects*. Pergamon Press,
    New York.

Saleh, F. Y., G. F. Lee, and H. W. Wolf (1982). Selected organic pesticides, behavior, and removal from domestic wastewater by chemical and physical processes. *Water Res.* 16:479—488.

Schafer-Ridder, M., U. Brocker, and E. Vogel (1976). Naphthalene 1,4-endoperoxide. *Angew. Chem.* 15:228—229.

Schauerte, W., J. P. Lay, W. Klein, and F. Korte (1982). Influence of 2,4,6-trichlorophenol and pentachlorophenol on the biota of aquatic systems. *Chemosphere* 11:71—79.

Schauerte, W., J. P. Lay, W. Klein, and F. Korte (1982). Long-term fate of organochlorine xenobiotics in aquatic ecosystems. *Ecotoxicol. Environ. Safety* 6:560—569.

Schink, B., and N. Pfennig (1982). Fermentation of trihydroxyben-zenes by *Pelobacter acidigallici* gen. nov. sp. nov., a new strictly anaerobic, non-sporeforming bacterium. *Arch. Microbiol.* 133:195—201.

Schmidt, E., M. Hellwig, and H-J. Knackmuss (1983). Degradation of chlorophenols by a defined mixed microbial community. *Appl. Environ. Microbiol.* 46:1038—1044.

Schultz, M. E., and O. C. Burnside (1980). Absorption, translo-cation, and metabolism of 2,4-D and glyphosate in hemp dog-bane (*Apocynum cannabinum*). *Weed Sci.* 28:13—20.

Schultz, M. E., and O. C. Burnside (1980). Effect of lanolin or lanolin + starch rings on absorption and translocation of 2,4-D or glyphosate in hemp dogbane (*Apocynum cannabinum*). *Weed Sci.* 28:149—152.

Seidman, M. M., A. Toms, and J. M. Wood (1969). Influence of side-chain substituents on the position of cleavage of the benzene ring by *Pseudomonas fluorescens*. *J. Bacteriol.* 97:1192—1197.

Seiler, J. P. (1978). The genetic toxicology of phenoxy acids other than 2,4,5-T. *Mutat. Res.* 55:197—226.

Sethunathan, N. (1973). Microbial degradation of insecticides in flooded soil and in anaerobic culture. *Res. Rev.* 47:143—165.

Shamsuzzaman, K. M., and E. A. Barnsley (1974). The regulation of naphthalene metabolism in *Pseudomonas*. *Biochem. Biophys. Res. Commun.* 60:582—589.

Shamsuzzaman, K. M., and E. A. Barnsley (1974). The regulation of naphthalene oxygenase in *Pseudomonas*. *J. Gen. Microbiol.* 83:165—170.

Shelton, D. R., and J. M. Tiedje (1984). General method for deter-mining anaerobic biodegradation potential. *Appl. Environ. Micro-biol.* 47:850—857.

Shiaris, M. P., and J. J. Cooney (1983). Replica plating method for estimating phenanthrene-utilizing and phenanthrene-come-tabolizing microorganisms. *Appl. Environ. Microbiol.* 45:706—710.

Silver, S., and T. G. Kinscherf (1982). Genetic and biochemical basis for microbial transformations and detoxification of mercury

and mercurial compounds. pp. 85–103 In A. M. Chakrabarty
(ed). *Biodegradation and Detoxification on Environmental Pollutants*, CRC Press, Inc., Boca Raton, FL.

Simkins, S., and M. Alexander (1984). Models for mineralization
kinetics with the variables of substrate concentration and population density. *Appl. Environ. Microbiol.* 47:1299–1306.

Siuda, J. F., and J. F. DeBernardis (1973). Naturally occurring
halogenated organic compounds. *Lloydia* 36:107–143.

Slater, J. H., and A. T. Bull (1982). Environmental microbiology:
Degradation. *Phil. Trans. R. Soc. Lond.* B 297:575–597.

Sleeper, B. P., and R. Y. Stanier (1950). The bacterial oxidation
of aromatic compounds. I. Adaptive patterns with respect to
polyphenolic compounds. *J. Bacteriol.* 59:117–127.

Sleeper, B. P., M. Tsuchida, and R. Y. Stanier (1950). The bacterial oxidation of aromatic compounds. II. The preparation of
enzymatically active dried cells and the influence thereon of
prior patterns of adaptation. *J. Bacteriol.* 59:129–133.

Sloane, N. H., C. Crane, and R. L. Mayer (1951). Studies on the
metabolism of p-aminobenzoic acid by *Mycobacterium smegmatis*.
J. Biol. Chem. 193:452–458.

Sloane, N. H., M. Samuels, and R. L. Mayer (1954). Factors affecting the hydroxylation of aniline by *Mycobacterium smegmatis*.
J. Biol. Chem. 206:751–755.

Smith, A., and R. B. Cain (1965). Utilization of halogenated aromatic compounds by *Nocardia erythropolis*. *J. Gen. Microbiol.*
41:xvi.

Smith, A., E. K. Tranter, and R. B. Cain (1968). The utilization
of some halogenated aromatic acids by *Nocardia*. Effects on
growth and enzyme induction. *Biochem. J.* 106:203–209.

Smith, B. S. W., J. D. Jones, and W. C. Evans (1952). The aromatic oxidative metabolism of certain benzene ring compounds
by soil bacteria. *Biochem. J.* 50:xxviii.

Soulas, G. (1982). Mathematical model for microbial degradation of
pesticides in the soil. *Soil Biol. Biochem.* 14:107–115.

Spain, J. C., and P. A. van Veld (1983). Adaptation of natural
microbial communities to degradation of xenobiotic compounds:
Effects of concentration, exposure time, inoculum, and chemical
structure. *Appl. Environ. Microbiol.* 45:428–435.

Spokes, J. R., and N. Walker (1974). A novel pathway of benzoate
metabolism in *Bacillus* species. *Ann. Microbiol. Enzymol.* 24:
307–315.

Stanier, R. Y., and J. L. Ingraham (1954). Protocatechuic acid
oxidase. *J. Biol. Chem.* 210:799–808.

Steen, W. C., and S. W. Karichkhoff (1981). Biosorption of hydrophobic organic pollutants by mixed microbial populations. *Chemosphere* 10:27–32.

Stotzky, G., and V. N. Krasovsky (1981). Ecological factors that affect the survival, establishment, growth, and genetic recombination of microbes in natural habitats. pp 31—42 In Levy, S. B., R. C. Clowes and E. L. Koenig (eds). *Molecular Biology, Pathogenicity, and Ecology of Bacterial Plasmids*, Plenum Press, New York.

Subba-Rao, R. V., and M. Alexander (1977). Effect of chemical structure on the biodegradability of 1,1,1-trichloro-1,1-bis(*p*-chlorophenyl)ethane (DDT). *J. Agric. Food Chem.* 25:327—329.

Subba-Rao, R. V., and M. Alexander (1982). Effect of sorption on mineralization of low concentrations of aromatic compounds in lake water samples. *Appl. Environ. Microbiol.* 44:659—668.

Subba-Rao, R. V., H. E. Rubin, and M. Alexander (1982). Kinetics and extent of mineralization of organic chemicals at trace levels in fresh water and sewage. *Appl. Environ. Microbiol.* 43:1139—1150.

Suett, D. L. (1975). Persistence and degradation of chlorfenvinphos, chlormephos, disulfoton, phorate, and pirimphos-ethyl following spring and late-summer soil application. *Pestic. Sci.* 6:385—393.

Sundstrom. G., O. Hutzinger, and S. Safe (1976). The metabolism of chlorobiphenyls—A review. *Chemosphere* 5:267—298.

Szczepanik-Van Leeuwen, P. A., and W. R. Penrose (1983). Functional properties of a microcosm of the freshwater benthic zone and the effects of 2,4-dichlorophenol. *Arch. Environ. Contam. Toxicol.* 12:427—437.

Szetela, R. W., and T. Z. Winnicki (1981). A novel method for determining the parameters of microbial kinetics. *Biotechnol. Bioeng.* 23:1485—1490.

Tabak, H. H., and E. F. Barth (1978). Biodegradability of benzidine in aerobic suspended growth reactors. *J. Water Pollut. Control Fed.* 50:552—558.

Tabak, H. H., C. W. Chambers, and P. W. Kabler (1964). Microbial metabolism of aromatic compounds. I. Decomposition of phenolic compounds and aromatic hydrocarbons by phenol-adapted bacteria. *J. Bacteriol.* 87:910—919.

Tabak, H. H., S. A. Quave, C. I. Mashni, and E. F. Barth (1981). Biodegradability studies with organic priority pollutant compounds. *J. Water Pollut. Control Fed.* 53:1503—1518.

Thom, N. S., and A. R. Agg (1975). The breakdown of synthetic organic compounds in biological processes. *Proc. R. Soc. Lond. B.* 189:347—357.

Treccani, V. (1962). Microbial degradation of hydrocarbons. *Prog. Ind. Microbiol.* 4:1—33.

Treccani, V. (1965). Microbial degradation of aliphatic and aromatic hydrocarbons. *Z. Allg. Mikrobiol.* 5:332—341.

Treccani, V. (1974). Microbial degradation of aromatic compounds:
    Influence of methyl and alkyl substituents. pp. 533–547 In
    Spencer, B. (ed.), *Industrial Aspects of Biochemistry*. Federa-
    tion of European Chemical Societies.
Treccani, V. (1976). Biodegradation of surface-active agents. pp.
    457–466 In Paoletti, R., R. Jacini and R. Porcellati (eds.).
    Lipids, Vol.2-*Technology*. Raven Press, New York.
Trevors, J. T. (1982). Effect of temperature on the degradation
    of pentachlorophenol by *Pseudomonas* species. *Chemosphere* 11:
    471–475.
Trudgill, P. W. (1984). Microbial degradation of the alicyclic ring.
    pp. 131–180 In Gibson, D. T. (ed.). *Microbial Degradation of
    Organic Compounds*. Marcel Dekker, Inc., New York.
Trudgill, P. W. (1984). The microbial metabolism of furans. pp 295–
    308 In Gibson, E. T. (ed.). *Microbial Degradation of Organic
    Compounds*. Marcel Dekker, Inc., New York.
Tucker, E. S., V. W. Saeger, and O. Hicks (1975). Activated sludge
    primary biodegradation of polychlorinated biphenyls. *Bull. En-
    viron. Contam. Toxicol.* 14:705–713.
Tulp, M. Th. M., and O. Hutzinger (1978). Rat metabolism of poly-
    chlorinated dibenzo-p-dioxins. *Chemosphere* 9:761–768.
Tyler, J. E., and R. K. Finn (1974). Growth rates of a pseudo-
    monad on 2,4-dichlorophenoxyacetic acid and 2,4-dichlorophenol.
    Appl. Microbiol. 28:181–184.
Vandenbergh, P. A., R. H. Olsen, and J. F. Colaruotolo (1981).
    Isolation and genetic characterization of bacteria that degrade
    chloroaromatic compounds. *Appl. Environ. Microbiol.* 42:737–739.
Van Engers, L. (1978). Mineralization of organic matter in the sub-
    soil of a waste disposal site: A laboratory experiment. *Soil Sci.*
    126:22–28.
Van Oss, C. J. (1978). Phagocytosis as a surface phenomenon.
    Annu. Rev. Microbiol. 32:19–39.
Varma, M. M., L. W. Wan, and C. Prasad (1976). Acclimation of
    wastewater bacteria by induction or mutation selection. *J. Water
    Pollut. Control Fed.* 48:832–834.
Veber, K., J. Zahradnik, and I. Breyl (1980). Efficiency and rate
    of elimination of polychlorinated biphenyls from wastewaters by
    means of algae. *Bull. Environ. Contam. Toxicol.* 25:841–845.
Veerkamp, W., R. Pel, and O. Hutzinger (1983). Transformation
    of chlorobenzoic acids by a *Pseudomonas* sp.: Comparison of
    batch and chemostat cultures. *Chemosphere* 12:1337–1343.
Vind, H. P. (1976). The role of microorganisms in the transport
    of chlorinated insecticides. pp. 793–797 In Sharpley, J. M.
    and A. M. Kaplan (eds.), *Proceedings of the Third International
    Biodegradation Symposium*. Applied Science Publ., London.

Virtanen, M. T., and M. L. Hattula (1982). The fate of 2,4,6-tri-chlorophenol in an aquatic continuous-flow system. *Chemosphere* 11:641−649.

Virtanen, M. T. A. Roos, A. U. Arstila, and M. L. Hattula (1980). An evaluation of a model ecosystem with DDT. *Arch. Environ. Contam. Toxicol.* 9:491−504.

Vogel, E., H. H. Klug, and M. Schafer-Ridder (1976). Syn- and anti-naphthalene 1,2:3,4-dioxide. *Angew. Chem.* 15:229−230.

Waler, J. D., and R. R. Colwell (1974). Microbial petroleum degra-dation: Use of mixed hydrocarbon substrates. *Appl. Microbiol.* 27:1053−1060.

Walker, J. D., and R. R. Colwell (1974). Microbial petroleum degra-rates of components of petroleum. *Can. J. Microbiol.* 22:1209−1213.

Wallnofer, P. R., S. Safe, and O. Hutzinger (1971). Metabolism of the systemic fungicides 2-methylbenzanilide and 2-chlorobenza-nilide by *Rhizopus japonicus*. *Pestic. Biochem. Physiol.* 1:458−463.

Walsh, G. E., K. A. Ainsworth, and L. Faas (1977). Effects and uptake of chlorinated naphthalenes in marine unicellular algae. *Bull. Environ. Contam. Toxicol.* 18:297−302.

Wang, Y. S., R. V. Subba-Rao, and M. Alexander (1984). Effect of substrate concentration and organic and inorganic compounds on the occurrence and rate of mineralization and cometabolism. *Appl. Microbiol.* 47:1195−1200.

Ware, G. W., and C. C. Roan (1970). Interaction of pesticides with aquatic microorganisms and plankton. *Res. Rev.* 33:15−45.

Weber, H., H. Poiger, and C. Schlatter (1982). Acute oral toxicity of TCDD-metabolites in male guinea pigs. *Toxicol.¯Lett.* 14:117−122.

Weber, W. J., Jr., N. H. Corfis, and B. E. Jones (1983). Removal of priority pollutants in integrated activated sludge-activated carbon treatment systems. *J. Water Pollut. Control Fed.* 55:369−376.

Webley, D. M., R. B. Duff, and V. C. Farmer (1959). Effect of substitution in the side-chain on beta-oxidation of aryloxy-alkylcarboxylic acids by *Nocardia opaca*. *Nature* 183:748−749.

Wigmore, G. J., and D. W. Ribbons (1980). p-Cymene pathway in *Pseudomonas putida*: Selective enrichment of defective mutants by using halogenated substrate analogs. *J. Bacteriol.* 143:816−824.

Williams, J. H. (1975). Persistence of chlorfenvinphos in soils. *Pestic. Sci.* 6:501−509.

Williams, P. A. (1978). Microbial genetics relating to hydrocarbon degradation. pp 135−164 In Watkinson, R. J. (ed.). *Develop-ments in Biodegradation of Hydrocarbons-1*. Applied Science Pub-lishers Ltd., London.

Williams, P. A., and K. Murray (1974). Metabolism of benzoate and
    the methylbenzoates by *Pseudomonas putida (arvilla)* MT-2:
    Evidence for the existence of a TOL plasmid. *J. Bacteriol.* 120:
    416—423.
Williams, P. A., and M. J. Worsey (1976). Plasmids and catabolism.
    Biochem. Soc. Trans. 4:466—468.
Williams, P. A., F. A. Catterall, and K. Murray (1975). Metabolism
    of naphthalene, 2-methylnaphthalene, salicylate, and benzoate
    by *Pseudomonas* PG: Regulation of tangential pathways. *J. Bac-
    teriol.* 124:679—685.
Williams, P. P. (1977). Metabolism of synthetic organic pesticides
    by anaerobic microorganisms. *Res. Rev.* 66:63—135.
Wiseman, A., J. A. Gondal, and P. Sims (1975). 4'-Hydroxylation
    of biphenyl by yeast containing cytochrome P-450: radiation and
    thermal stability, comparisons with liver enzyme (oxidized and
    reduced forms). *Biochem. Soc. Trans.* 3:278—281.
Wodzinski, R. S., and D. Bertolini (1972). Physical state in which
    naphthalene and bibenzyl are utilized by bacteria. *Appl. Micro-
    biol.* 23:1077—1081.
Wolfe, N. L., D. F. Paris, W. C. Steen, and G. L. Baughman
    (1980). Correlation of microbial degradation rates with chemical
    structure. *Environ. Sci. Technol.* 14:1143—1144.
Wolfe, N. L., R. G. Zepp, P. Schlotzhauer, and M. Sink (1982).
    Transformation pathways of hexachlorocyclopentadiene in the
    aquatic environment. *Chemosphere* 11:91—101.
Wong, P. T. S., and K. L. E. Kaiser (1974). Bacterial degradation
    of polychlorinated biphenyls. II. Rate studies. *Bull. Environ.
    Contam. Toxicol.* 13:249—256.
Wood, J. M. (1982). Chlorinated hydrocarbons: Oxidation in the
    biosphere. *Environ. Sci. Technol.* 16:291A—296A.
Woodcock, D. (1978). Microbial degradation of fungicides, fumigants,
    and nematocides. pp. 731—780 in Hill, I. R. and S. J. L.
    Wright (eds.). *Pesticide Microbiology.* Academic Press, New York.
Worsey, M. J., F. C. H. Franklin, and P. A. Williams (1978).
    Regulation of the degradative pathway enzymes coded for by
    the TOL plasmid (pWWO) from *Pseudomonas putida* MT-2. *J.
    Bacteriol.* 134:757—764.
Wright, S. J. L. (1978). Interactions of pesticides with micro-algae.
    pp 535—602 in Hill, I. R. and S. J. L. Wright (eds.). *Pesticide
    Microbiology.* Academic Press, New York.
Wyrill, J. B., III, and O. C. Burnside (1976). Absorption, trans-
    location, and metabolism of 2,4-D and glyphosate in common
    milkweed and hemp dogbane. *Weed Sci.* 24:557—566.
Yagi, O., and R. Sudo (1980). Degradation of polychlorinated bi-
    phenyls by microorganisms. *J. Water Pollut. Control Fed.* 52:
    1035—1043.

Yamamoto, S., M. Katagiri, H. Maeno, and O. Hayaishi (1965). Salicylate hydroxylase, a monooxygenase requiring flavin adenine dinucleotide. I. Purification and general properties. *J. Biol. Chem.* 240:3408–3413.

Yano, K., and K. Arima (1968). Metabolism of aromatic compounds by bacteria II. *m*-Hydroxybenzoic acid hydroxylase A and B; 5-dihydroshikimic acid, a precursor of protocatechuic acid, a new pathway from salicylic acid to gentisic acid. *J. Gen. Appl. Microbiol.* 4:241–258.

Yu, C-A., and I. C. Gunsalus (1970). Crystalline cytochrome P-450$_{CAM}$. *Biochem. Biophys. Res. Commun.* 40:1431–1436.

Ziffer, H., D. M. Jerina, D. T. Gibson, and V. M. Kobal (1973). Absolute stereochemistry of the (+)-cis-1,2-dihydroxy-3-methyl-cyclohexa-3,5-diene produced from toluene by *Pseudomonas putida*. *J. Am. Chem. Soc.* 95:4048–4049.

Ziffer, H., K. Kabuto, D. T. Gibson, V. M. Kobal, and D. M. Jerina (1977). The absolute stereochemistry of several *cis*-dihydrodiols microbially produced from substituted benzenes. *Tetrahedron* 33:2491–2496.

Zitko, V. (1974). Trends of PCB and DDT in fish and aquatic birds. pp III-61-64 In De Freitas, A. S. W., D. J. Kushner and S. U. Quadri (eds.). *Proceedings of the International Conference on Transport of Persistent Chemicals in Aquatic Ecosystems.* Nat. Res. Council Can.

Zitko, V. (1983). "Shorthand" numbering of chlorobiphenyls. *Chemosphere* 12:835–836.

Zitko, V. (1984). Methods for chemical characterization of biodegradation. pp 29–42 In Gibson, D. T. (ed.), *Microbial Degradation of Organic Compounds.* Marcel Dekker, Inc., New York.

Zoulalian, V., F. Bessou, A. Tessier, P. G. Campbell, S. A. Visser, and J. P. Villeneuve (1974). Dynamique de degradation du phenol dans le fleuve Saint-Laurent. pp II-53-58 In De Freitas, A. W. S., D. J. Kushner and S. U. Quadri (eds.). *Proceedings of the International Conference on Transport of Persistent Chemicals in Aquatic Ecosystems.* Nat. Res. Council Can.

Appendix

This alphabetical illustrated list of compounds provides a ready reference for the structures of the major compounds noted in this book. Primary substrate compounds as well as intermediate metabolites are shown, to better assist the reader in following the discussions of compound biotransformation.

I.

$$CH_3$$
$$HC=O$$

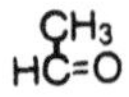

ACETALDEHYDE

2.

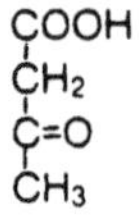

$$COOH$$
$$CH_2$$
$$C=O$$
$$CH_3$$

ACETOACETIC ACID

3.

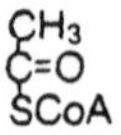

$$CH_3$$
$$C=O$$
$$SCoA$$

ACETYL-CoA

4.

$$COOH$$
$$CH_2$$
$$C-COOH$$
$$CH$$
$$COOH$$

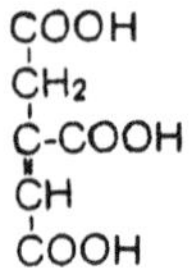

ACONITIC ACID

5.

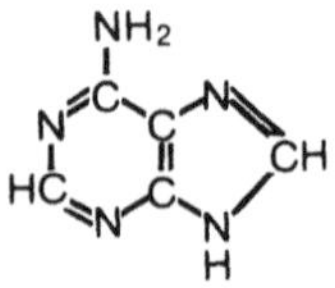

ADENINE

6.

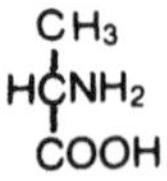

$$CH_3$$
$$HCNH_2$$
$$COOH$$

ALANINE

7.

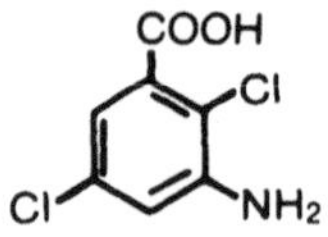

AMIBEN (chloramben,
3-amino-2,5-dichlorobenzoic acid)

8.

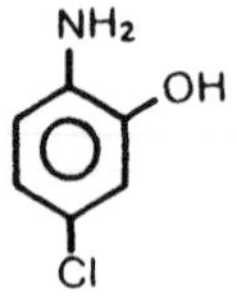

2-AMINO-5-
CHLOROPHENOL

9.

ANILINE

10.

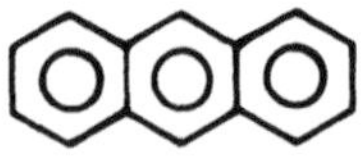

ANTHRACENE

II.

$$NHCOOCH_2C\equiv CCH_2Cl$$

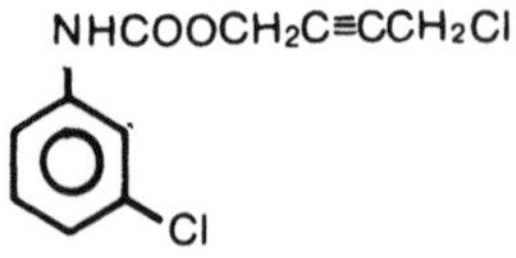

BARBAN [(3-chlorophenyl)-
carbamic acid 4-chloro-
2-butynyl ester]

12.

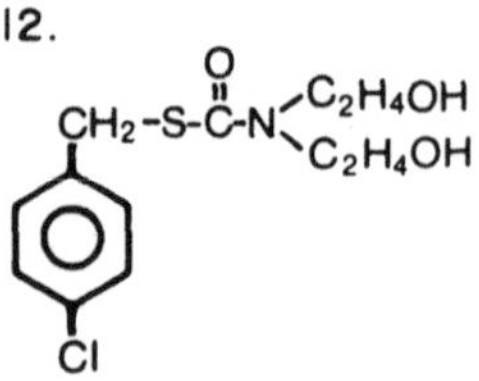

BENTHIOCARB (S-4-
chlorobenzyl-N,N-
diethylthiolcarbamate)

13.

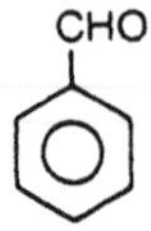

BENZALDEHYDE

14.

BENZENE

15.

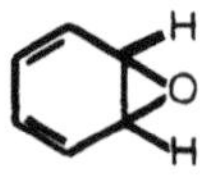

BENZENE 1, 2-OXIDE

6.

BENZENESULFONIC
ACID

21.

BIPHENYL

26.

4-CARBOXYMETHYLENEBUT-
2-EN-4-OLIDE

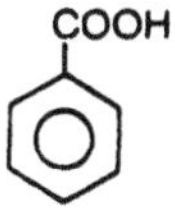

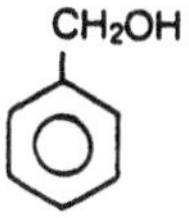

7.

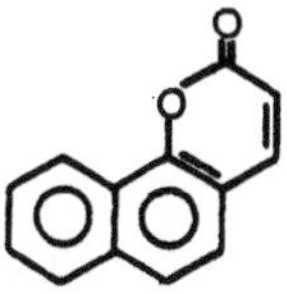

7, 8-BENZOCOUMARIN

22.

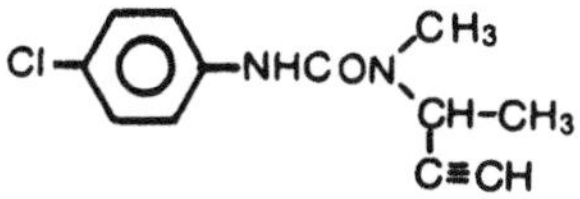

BUTURON [3-(4-chlorophenyl)-
1-isobutynyl-1-methylurea]

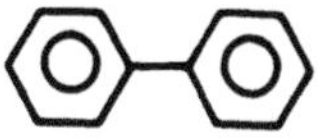

27.

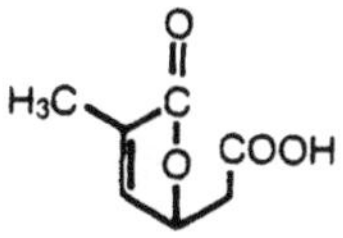

4-CARBOXYMETHYLENE-2-
METHYL-Δ^a-BUTENOLIDE

8.

BENZOIC
ACID

23.

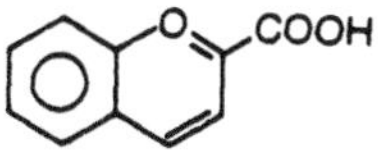

2-CARBOXYBENZO-
PYRILIUM

28.

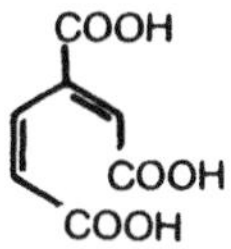

3-CARBOXY-
cis, cis-MUCONIC
ACID

9.

BENZYL
ALCOHOL

24.

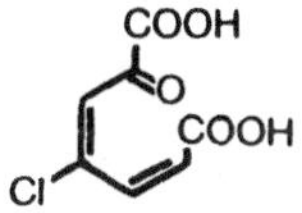

7-CARBOXY-4-CHLORO-
2-KETO-HEPT-3,5-
DIENOIC ACID

29.

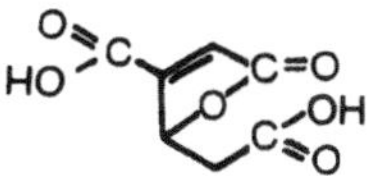

3-CARBOXYMUCONOLACTONE

10.

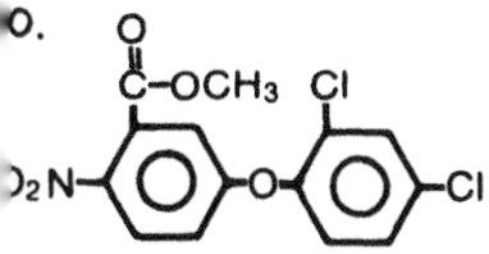

FENOX [methyl 5-(2, 4-dichloro-
phenoxy)-2-nitrobenzoate]

25.

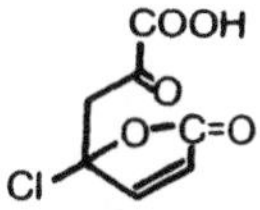

7-CARBOXY-4-CHLORO-
2-KETO-HEPT-
4, 7-LACTONE

30.

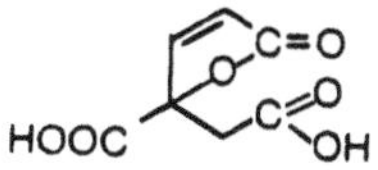

4-CARBOXYMUCONOLACTONE

31.

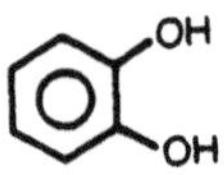

CATECHOL

32.

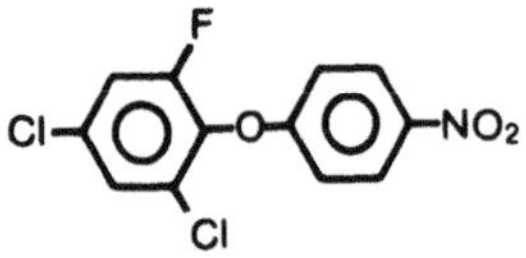

CFNP (2, 4-dichloro-6-fluorophenyl-
4'-nitrophenyl ether)

33.

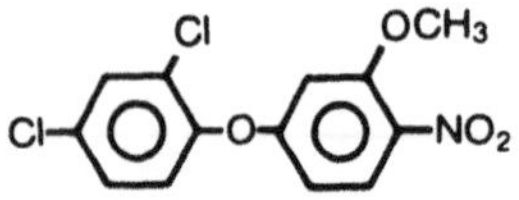

CHLOMETHOXYNIL (2, 4-dichloro-
phenyl-3'-methoxy-4'-nitrophenyl
ether)

34.

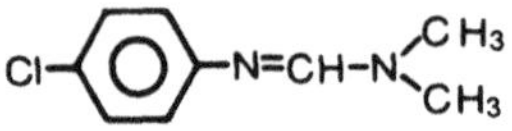

CHLORDIMEFORM [N-(4-chloro-
-o-tolyl)-N',N'-dimethylformamidine]

35.

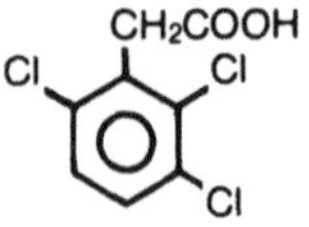

CHLORFENAC (fenac,
2,3,6-trichlorophenylacetic acid)

36.

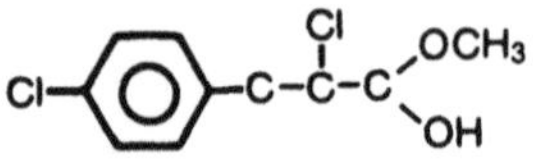

CHLORFENPROP METHYL
[methyl 2-chloro-3-
(4-chlorophenyl) propionate]

37.

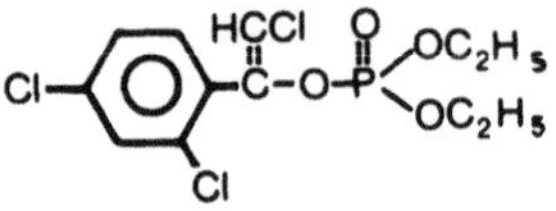

CHLORFENVINPHOS
2-chloro-1-(2, 4-dichlorophenyl)-
vinyl diethyl phosphate]

38.

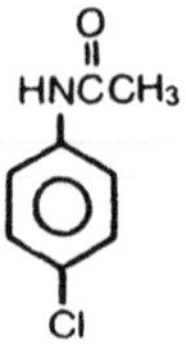

4-CHLOROACETANILIDE

39.

4-CHLOROANILINE

40.

CHLOROBENZENE

41.

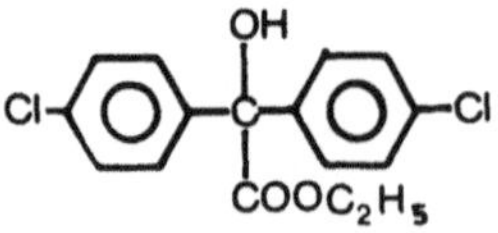

CHLOROBENZILATE
(ethyl 4,4'-dichlorobenzilate)

42.

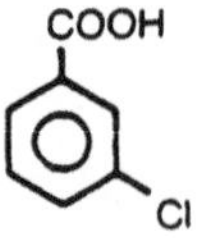

3-CHLORO-
BENZOIC
ACID

43.

4-CHLOROBENZOIC ACID

44.

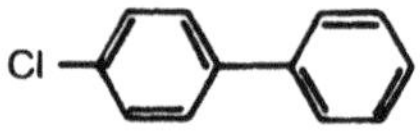

4-CHLOROBIPHENYL

45.

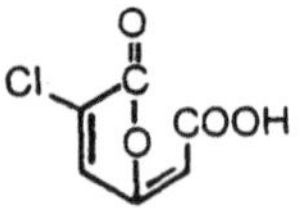

2-CHLORO-4-CARBOXY-
METHYLENE-BUT-2-ENOLIDE

46

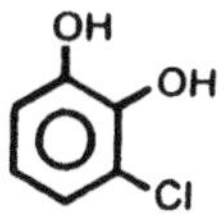

OH
OH
Cl

3-CHLOROCATECHOL

51.

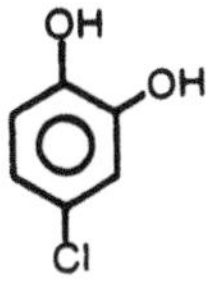

COOH
OH
OH
H
Cl

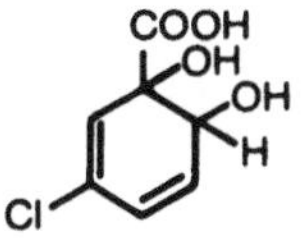

5, CHLORO-3, 5-CYCLOHEXADIENE
1, 2-DIOL- 1-CARBOXYLIC ACID

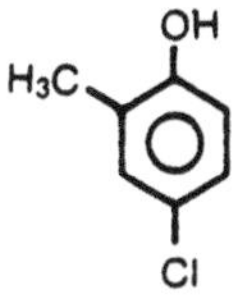

56.

O
‖
HNCCH₃
OH
Cl

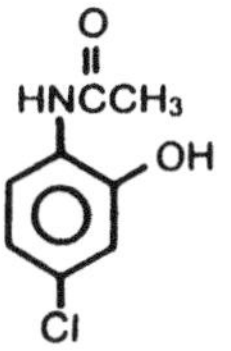

4-CHLORO-2-HYDROXY-
ACETANILIDE

47.

OH
OH
Cl

4-CHLOROCATECHOL

52.

H OH
OH
H
Cl

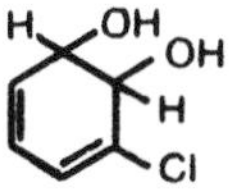

3-CHLORO-1, 2-DIHYDROXY-
CYCLOHEXA-3, 5-DIENE

57.

OCH₂COOH
H₃C
OH
Cl

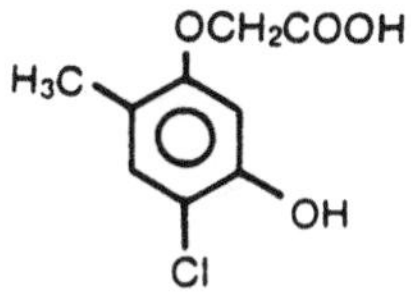

(4-CHLORO -5- HYDROXY-
2- METHYL PHENOXY)-
ACETIC ACID

48.

OH
H₃C
Cl

5-CHLORO-o-CRESOL

53.

Cl
Cl
Cl H OH OH
H
Cl

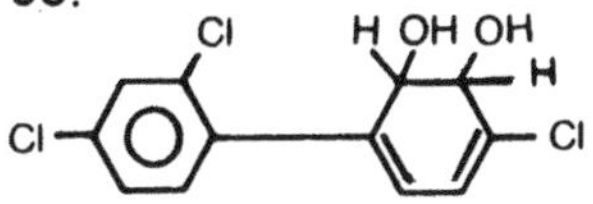

1-CHLORO-2, 3-DIHYDROXY-
4-(2, 4-DICHLOROPHENYL)-
HEXA-4, 6-DIENE

58.

HO₂C
O
OHC
Cl

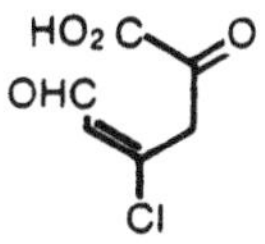

4-CHLORO-2-HYDROXY-
MUCONIC SEMIALDEHYDE

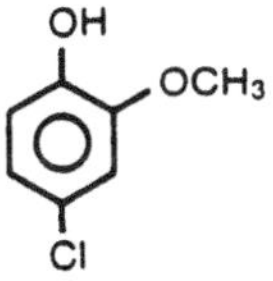

49.

COOH
OH
OH
H
Cl

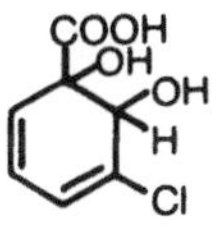

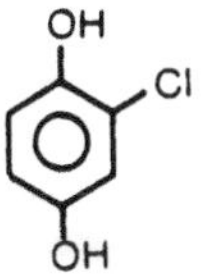

3-CHLORO-3, 5-CYCLOHEXADIENE
1, 2-DIOL- 1-CARBOXYLIC ACID

54.

OH
OCH₃
Cl

4-CHLOROGUAIACOL

59.

Cl
Cl
O OH OH
O
Cl

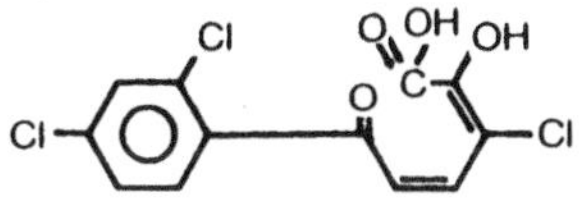

3-CHLORO-2-HYDROXY-6-(2,
4-DICHLOROPHENYL) HEXA-2,
4-DIENOIC ACID

50.

COOH
OH
OH
H
Cl

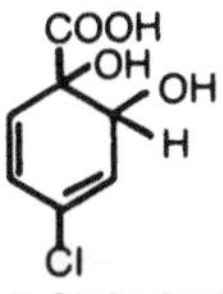

4-CHLORO-3, 5-CYCLOHEXADIENE
1, 2-DIOL- 1-CARBOXYLIC ACID

55.

OH
Cl
OH

CHLOROHYDROQUINONE

60.

OCH₂COOH
OH
Cl

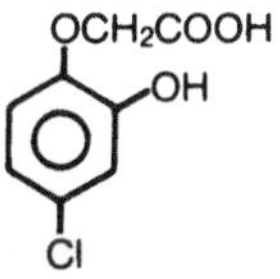

4-CHLORO-2-HYDROXY-
PHENOXY ACETIC ACID

61.

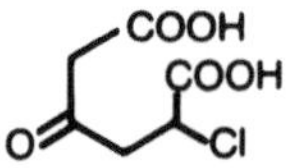

2-CHLORO-4-KETOADIPIC ACID

66.

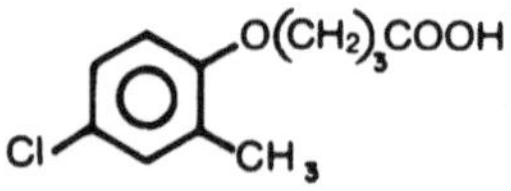

MCPB [4-(4-CHLORO-
2-METHYLPHENOXY)
BUTYRIC ACID]

71.

3-CHLOROMUCONIC ACID

62.

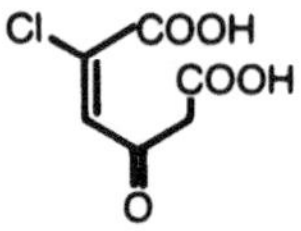

2-CHLOROMALEYLACETIC ACID

67.

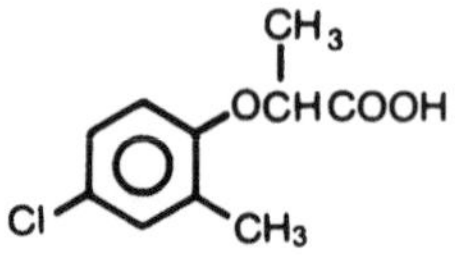

2-(4-CHLORO-2-METHYL
PHENOXY)PROPIONIC ACID

72.

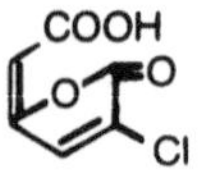

2-CHLOROMUCONOLACTONE

63.

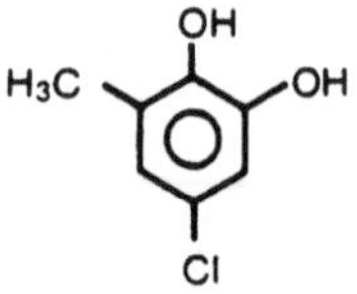

5-CHLORO-
3-METHYLCATECHOL

68.

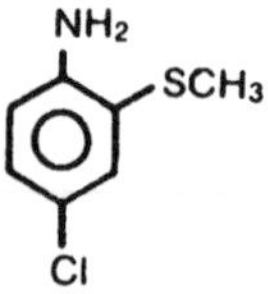

4-CHLORO-2-METHYL-
THIOBENZENEAMINE

73.

4-CHLORONITROBENZENE

64.

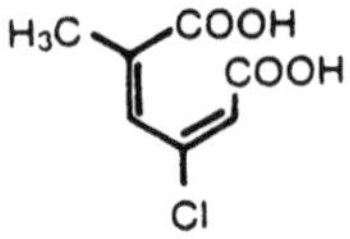

cis, cis-4-CHLORO-
2-METHYL-
MUCONIC ACID

69.

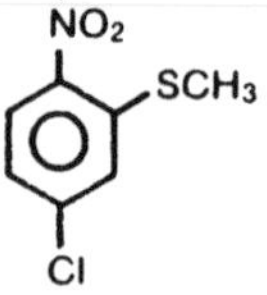

4-CHLORO-2-METHYLTHIO-
1-NITROBENZENE

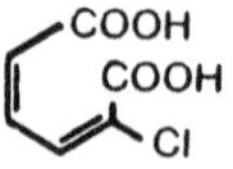

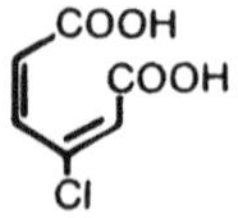

74.

4-CHLORONITROSO-
BENZENE

65.

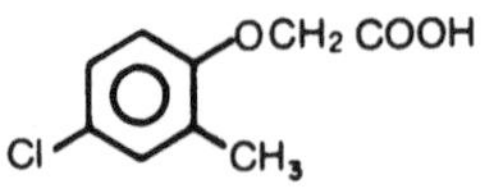

MCPA
(4-CHLORO-2-METHYL
PHENOXYACETIC ACID)

70.

2-CHLOROMUCONIC ACID

75.

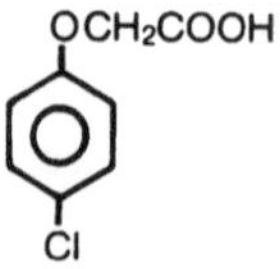

4-CHLOROPHENOXY-
ACETIC ACID

76.

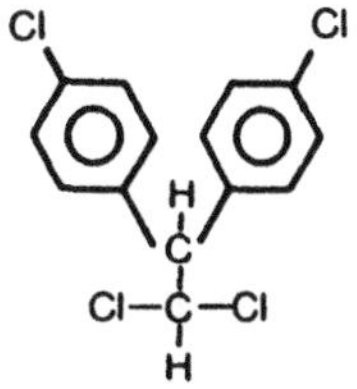

1, 1-bis (p-CHLORO-
PHENYLETHANE)

77.

4-CHLOROPHENYL-
HYDROXYLAMINE

78.

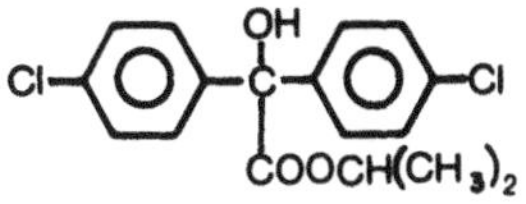

CHLOROPROPYLATE
(isopropyl 4, 4'-dichlorobenzilate)

79.

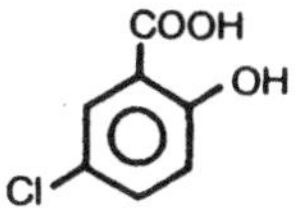

5-CHLOROSALICYLIC ACID

80.

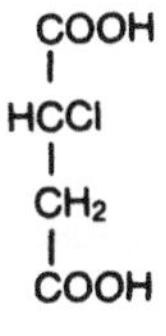

COOH
|
HCCl
|
CH₂
|
COOH

CHLOROSUCCINIC ACID

81.

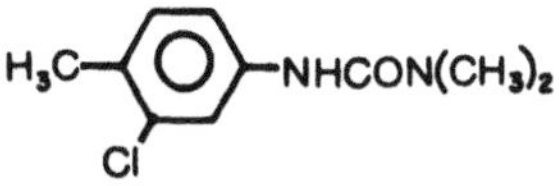

CHLOROTOLURON
(3-(3-chloro-4-methylphenyl)-
1,1-dimethylurea)

82.

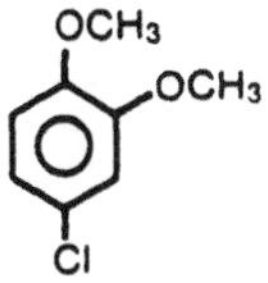

4-CHLOROVERATROLE

83.

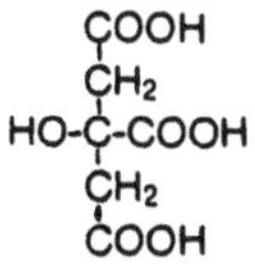

COOH
|
CH₂
|
HO-C-COOH
|
CH₂
|
COOH

CITRIC ACID

84.

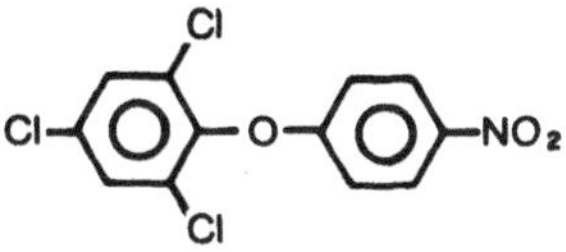

CNP
(2,4,6-TRICHLOROPHENYL-
4'-NITROPHENYL ETHER)

85.

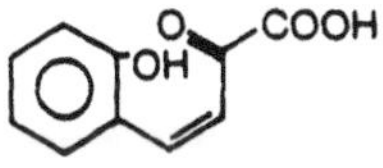

COUMARALDEHYDE

86.

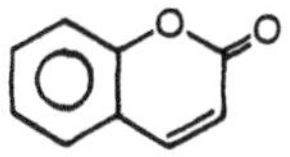

COUMARIN

87.

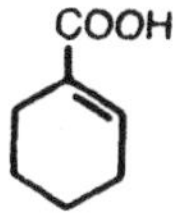

CYCLOHEX-1-ENE-
1-CARBOXYLIC ACID

88.

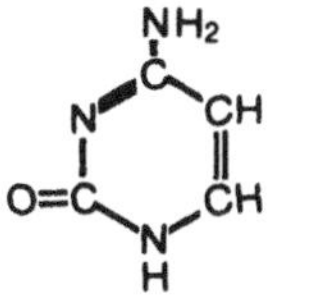

CYTOSINE

89.

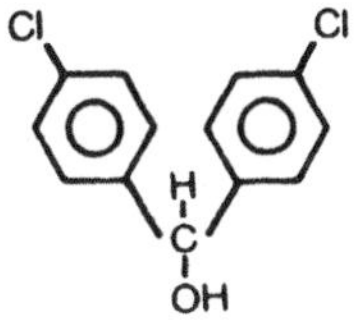

DBH
(DICHLOROBENZHYDROL)

90.

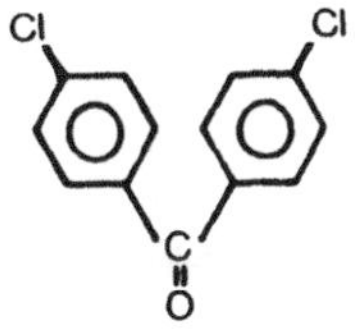

DBP
(DICHLOROBENZOPHENONE)

91.

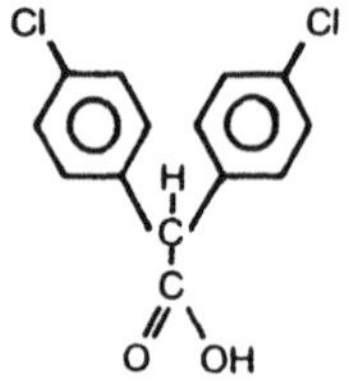

DDA
[2,2-bis(p-CHLOROPHENYL)
ACETIC ACID]

92.

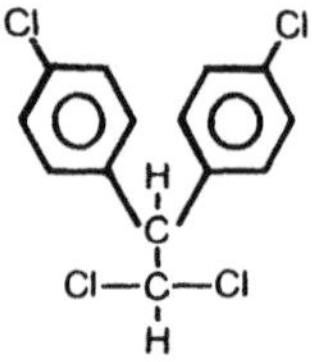

DDD
[1,1-DICHLORO-2,2-bis
(p-CHLOROPHENYL)ETHANE]

93.

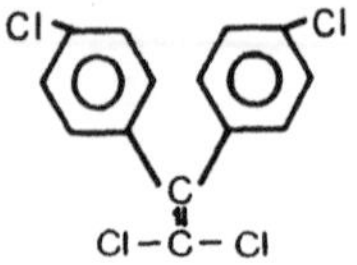

DDE
[2,2-bis(p-CHLOROPHENYL)-
1,1-DICHLOROETHYLENE]

94.

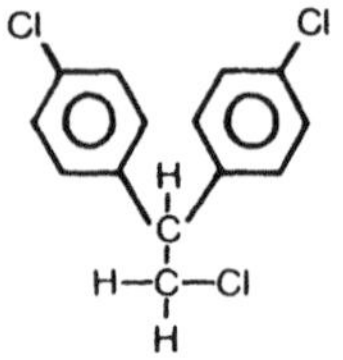

DDMS
[1,1-bis(p-CHLOROPHENYL-
2-CHLOROETHANE)]

95.

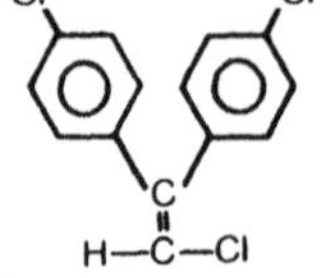

DDMU
[1-CHLORO-2,2-bis(p-CHLORO-
PHENYL)ETHYLENE]

96.

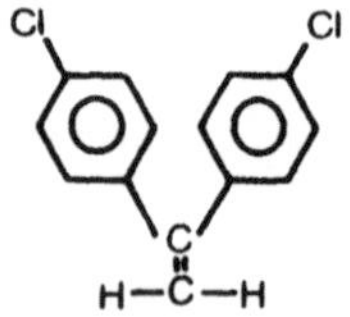

DDNU
[unsym-bis(p-CHLOROPHENYL)
ETHYLENE]

97.

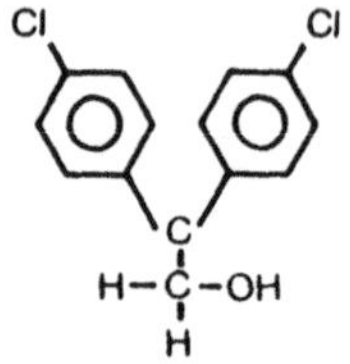

DDOH
[2,2-bis(p-CHLOROPHENYL)-
ETHANOL]

98.

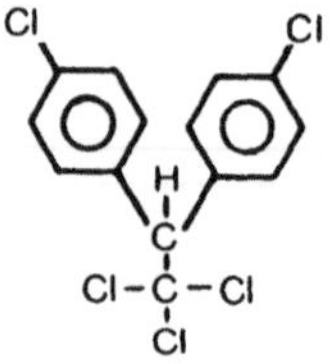

DDT
[1,1,1 TRICHLORO-2,2-bis-
(p-CHLOROPHENYL)ETHANE]

99.

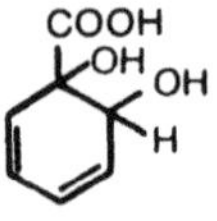

DHB
(3,5-CYCLOHEXADIENE-
1,2-DIOL-1-
CARBOXYLIC ACID)

100.

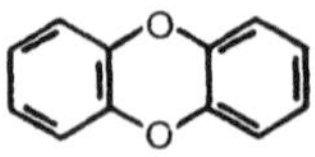

DIBENZO-p-DIOXIN

101.

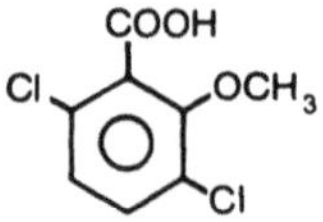

DICAMBA (3,6-DICHLORO-
o-ANISIC ACID)

102.

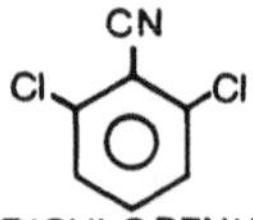

DICHLOBENIL
(2,6-DICHLOROBENZONITRILE)

103.

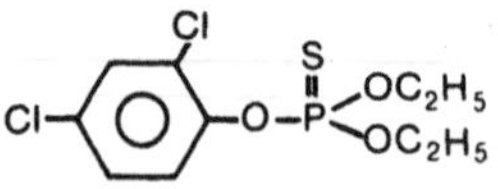

DICHLOFENTHION
[o,o-DIETHYL-o-
(2,4-DICHLOROPHENYL)-
PHOSPHOROTHIOATE]

104.

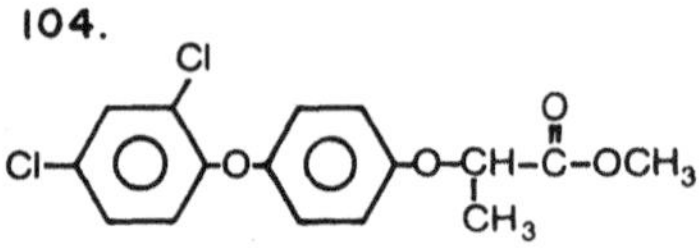

DICHLORFOP METHYL
[(±)-METHYL 2-[4-(2,4-DI-
CHLOROPHENOXY)PHENOXY]-
PROPIONATE]

105.

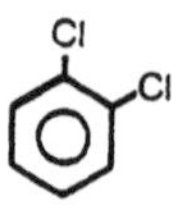

1,2-DICHLOROBENZENE

106.

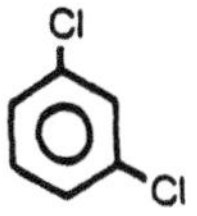

Cl

1, 3-DICHLOROBENZENE

111.

Cl   Cl

OH

OH

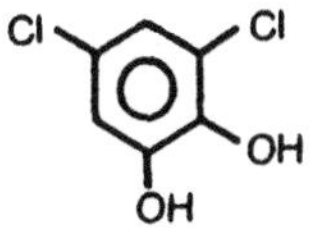

3, 5-DICHLOROCATECHOL

116.

OCH₂COOH

Cl

OH

Cl

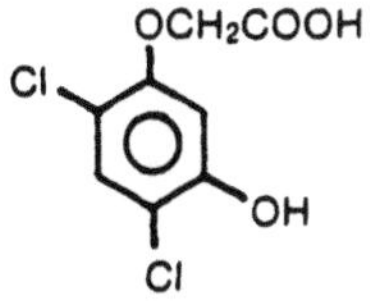

(2, 4-DICHLORO- 5-
HYDROXYPHENOXY)-
ACETIC ACID

107.

Cl

Cl

1, 4- DICHLOROBENZENE

112.

COOH
OH
OH
H

Cl   Cl

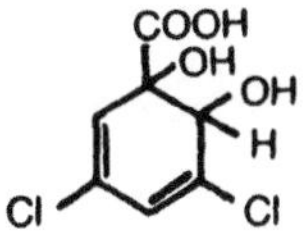

3,5-DICHLORO-3, 5-CYCLO-
HEXADIENE - 1, 2-DIOL -1 -
CARBOXYLIC ACID

117.

OCH₂COOH

Cl

Cl

OH

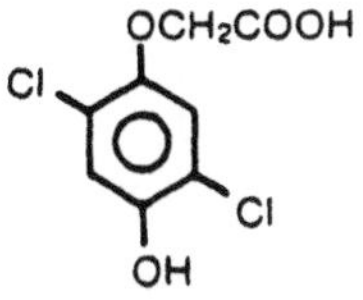

(2, 5 -DICHLORO -4-
HYDROXYPHENOXY)-
ACETIC ACID

108.

NH₂
Cl

Cl

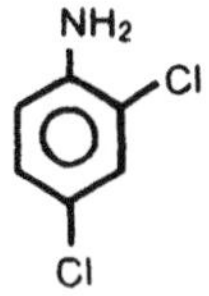

2, 4-DICHLORO-
BENZENEAMINE

113.

OH
OCH₃

Cl

Cl

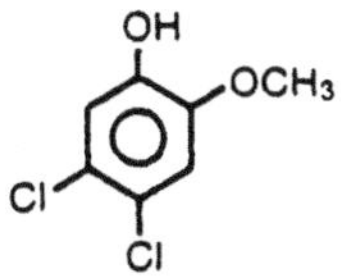

4, 5-DICHLOROGUAIACOL

118.

COOH
COOH

Cl   Cl

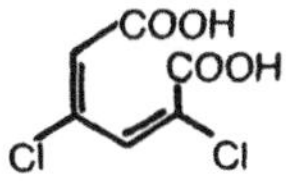

DICHLOROMUCONIC   ACID

109.

Cl

Cl   COOH

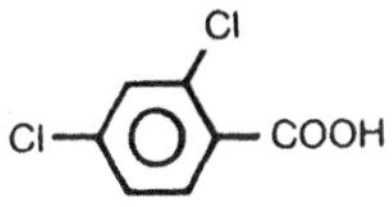

2, 4-DICHLOROBENZOIC ACID

114.

OH
Cl   Cl

OH

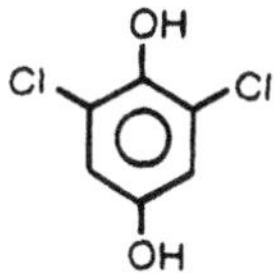

DICHLOROHYDROQUINONE

119.

NO₂
Cl

Cl

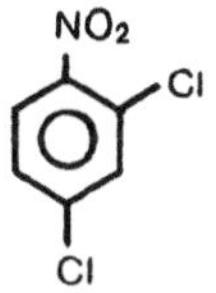

2, 4-DICHLORO-
1-NITROBENZENE

110.

COOH

Cl   Cl

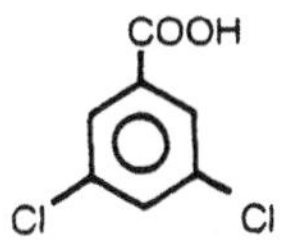

3, 5-DICHLORO-
BENZOIC ACID

115.

HO₂C   O
OHC
Cl   Cl

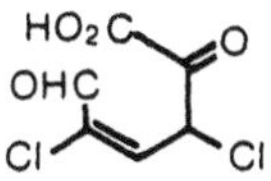

3, 5-DICHLORO-2-
HYDROXYMUCONIC
SEMIALDEHYDE

120.

OH
Cl

Cl

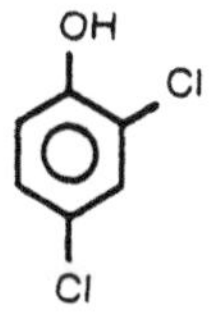

2, 4-DICHLOROPHENOL
(2, 4-DCP)

121.

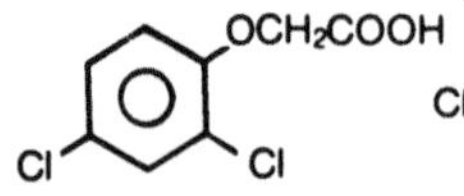

(2,4-DICHLORO-
PHENOXYACETIC ACID)

126.

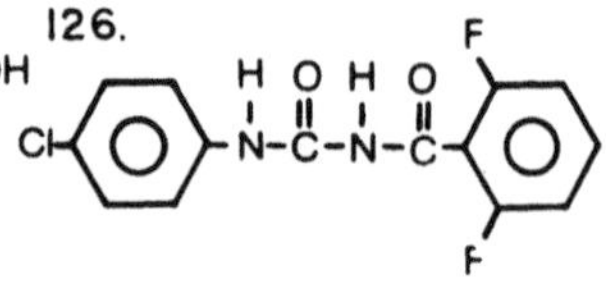

DIFLUBENZURON
[1-(4-CHLOROPHENYL)-
3-(2,6-DIFLUOROBENZOYL)UREA]

131.

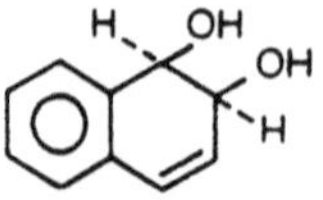

cis-1, 2-DIHYDRO-1, 2-
DIHYDROXYNAPHTHALENE

122.

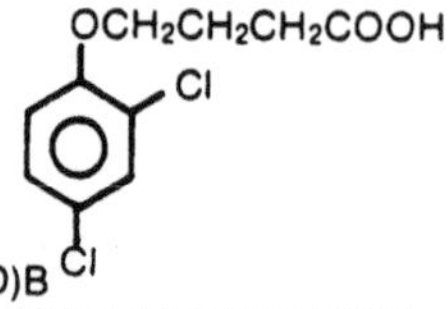

4-(2,4-D)B
4-(2, 4-DICHLOROPHENOXY)
BUTYRIC ACID

127.

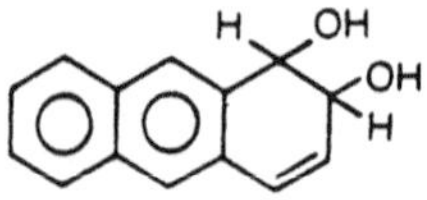

cis 1, 2- DIHYDRO-1, 2-
DIHYDROXYANTHRACENE

132.

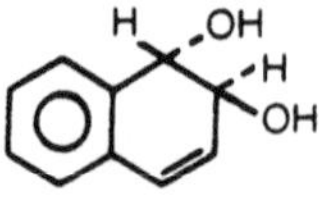

trans-1, 2-DIHYDRO-
1, 2-DIHYDROXYNAPHTHALENE

123.

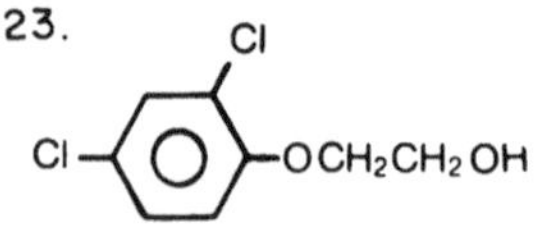

(2, 4-DICHLOROPHENOXY)
ETHANOL

128.

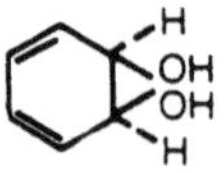

cis-1, 2-DIHYDRO-
1, 2-DIHYDROXYBENZENE

133.

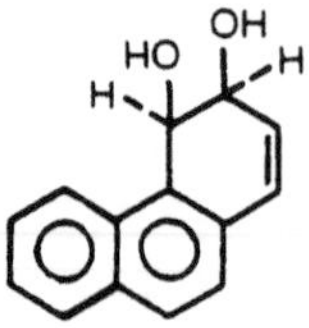

cis-3, 4-DIHYDRO-3, 4-
DIHYDROXYPHENANTHRENE

124.

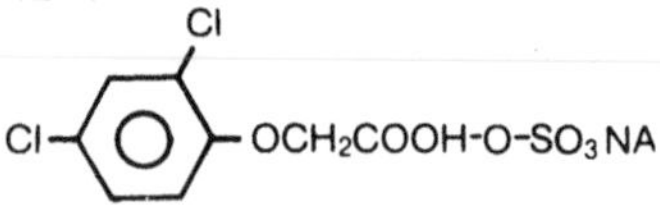

2-(2,4-DICHLOROPHENOXY)
ETHYL SULFATE,
SODIUM SALT

129.

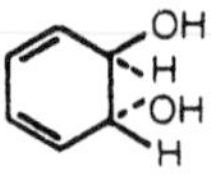

trans-1, 2-DIHYDRO-1,2
DIHYDROXYBENZENE

134.

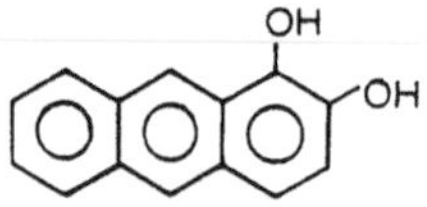

1, 2- DIHYDROXY-
ANTHRACENE

125.

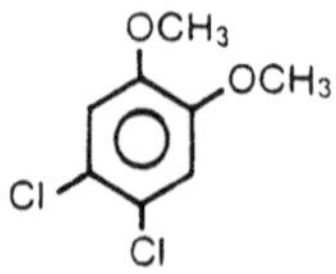

4,5-DICHLOROVERATROLE

130.

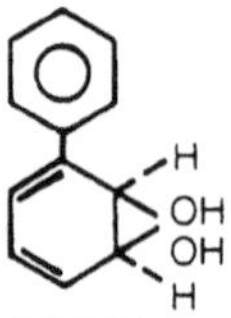

cis-2, 3-DIHYDRO-
2, 3-DIHYDROXYBIPHENYL

135.

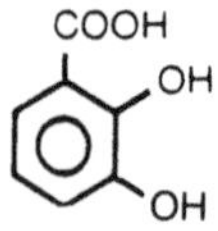

2,3-DIHYDROXYBENZOIC ACID

136.

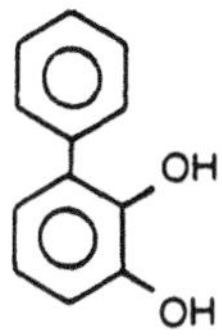

2, 3-DIHYDROXYBIPHENYL

141.

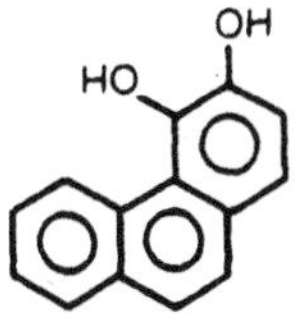

3, 4-DIHYDROXYPHENANTHRENE

146.

GENTISIC ACID

137.

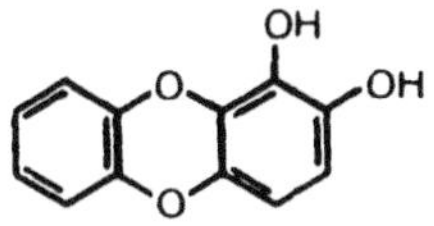

1, 2-DIHYDROXY-
DIBENZO-p-DIOXIN

142. DPM
DICHLORODIPHENYL METHANE

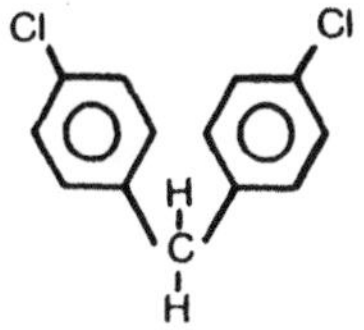

147.

CHO
|
COOH

GLYOXYLIC ACID

138.

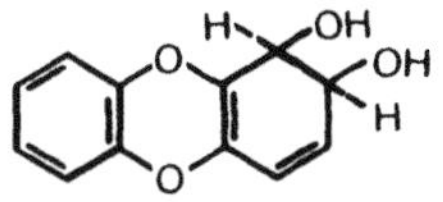

1, 2-DIHYDROXY-
1, 2-DIHYDRODIBENZO-p-DIOXIN

143.

COOH
|
CH
‖
CH
|
COOH

FUMARIC ACID

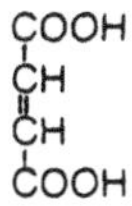

148.

GUANINE

139.

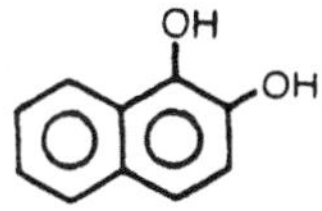

1, 2-DIHYDROXYNAPHTHALENE

144.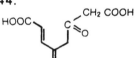

FUMARYLACETOACETIC
ACID

149.

HEXACHLOROBENZENE

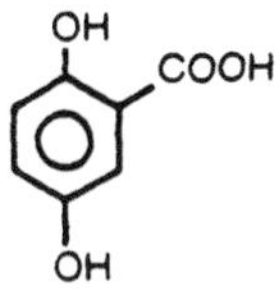

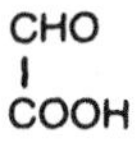

140.

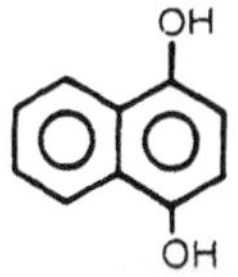

1, 4-DIHYDROXYNAPHTHALENE

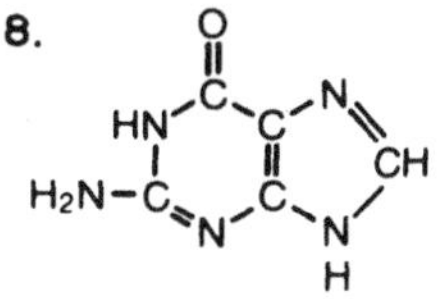

145.

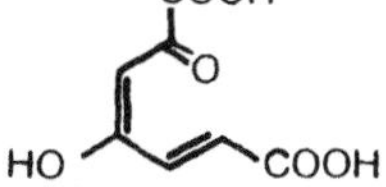

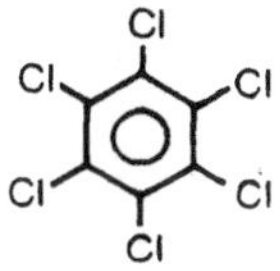

FUMARYLPYRUVIC ACID

150.

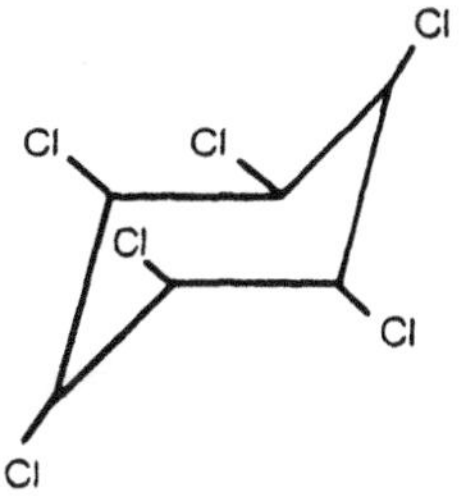

HEXACHLOROCYCLOHEXANE

151.

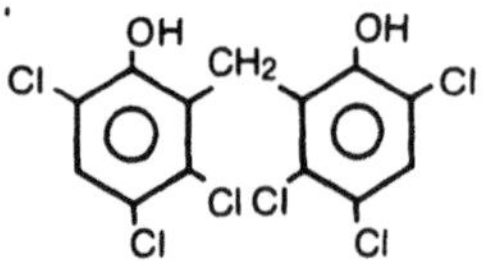

HEXACHLOROPHENE
[2, 2'-METHYLENE bis –
(3, 4, 6-TRICHLOROPHENOL)]

152.

HOMOGENTISIC ACID

153.

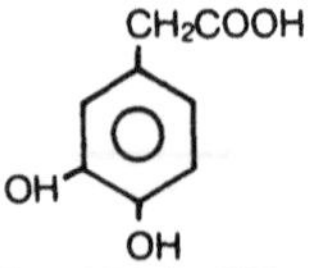

HOMOPROTOCATECHUIC
ACID

154.

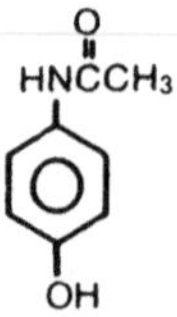

4-HYDROXYACETANILIDE

155.

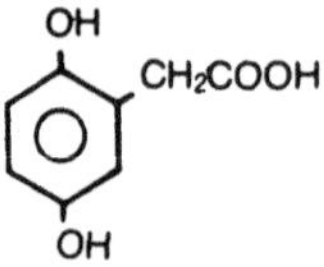

4-HYDROXYANILINE

156.

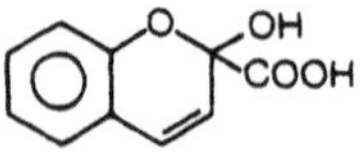

o-HYDROXYBENZALPYRUVIC
ACID.

157.

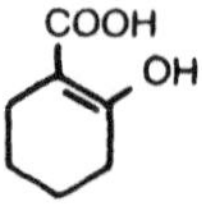

o-HYDROXYBENZOIC
ACID

158.

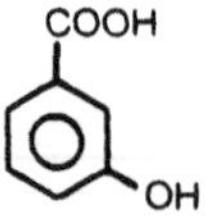

m-HYDROXYBENZOIC
ACID

159.

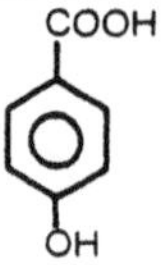

p-HYDROXYBENZOIC ACID

160.

2-HYDROXY-4-CARBOXY-
MUCONIC SEMIALDEHYDE

161.

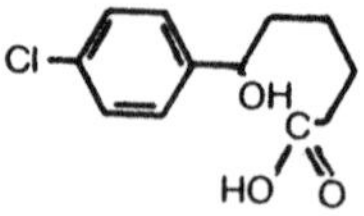

6-HYDROXY-6-(4'CHLOROPHENYL)–
HEXANOIC ACID

162.

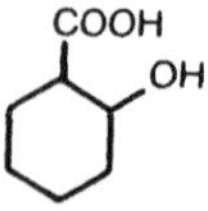

2-HYDROXY-
CYCLOHEXANE–
CARBOXYLIC ACID

163.

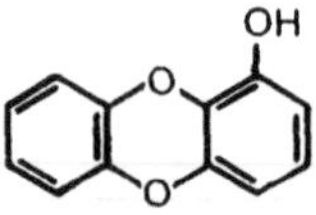

1-HYDROXY-
DIBENZO-p-
DIOXIN

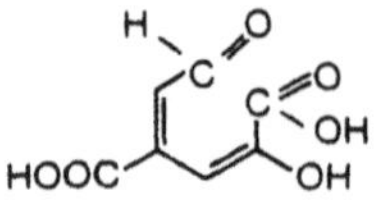

164.

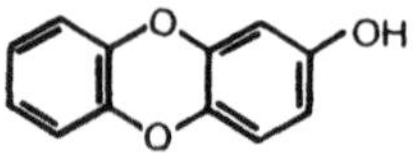

2-HYDROXY-
DIBENZO-p-
DIOXIN

165.

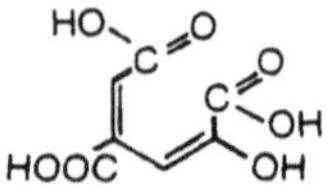

2-HYDROXY-4-CARBOXY-
MUCONIC ACID

166.

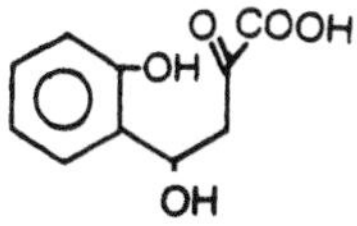

4-HYDROXY-4-ọ-
HYDROXYPHENYL-
2-OXOBUTYRIC ACID

167.

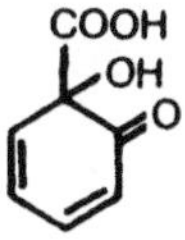

∝ -HYDROXY-
β-KETOCARBOXYLIC
ACID

168.

HYDROXYMALONIC
SEMIALDEHYDE

169.

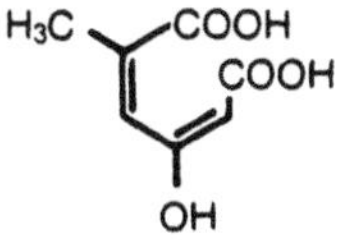

4-HYDROXY-2-METHYL-
MUCONIC ACID

170.

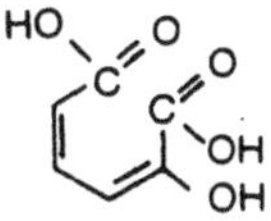

2-HYDROXYMUCONIC ACID

171.

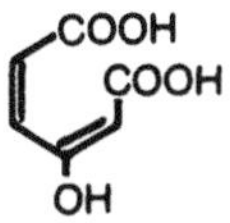

3-HYDROXYMUCONIC ACID

172.

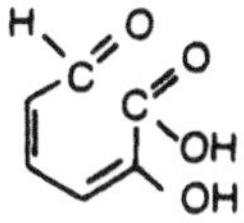

2-HYDROXYMUCONIC
SEMIALDEHYDE

173.

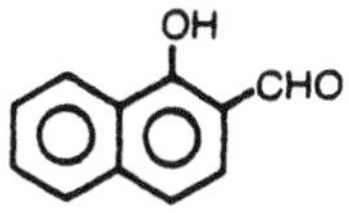

1-HYDROXY-2-
NAPHTHALDEHYDE

174.

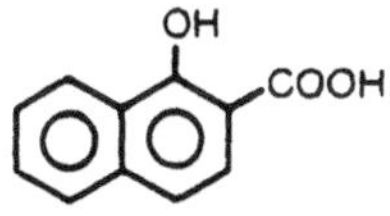

1-HYDROXY-2-
NAPHTHOIC ACID

175.

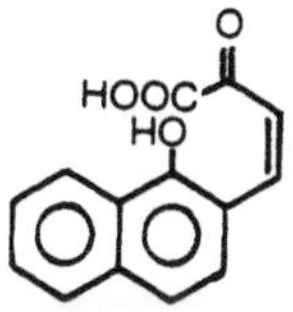

cis-4-(1-HYDROXY-
NAPHTH-2-YL)-2-OXOBUT-
3-ENOIC ACID

176.

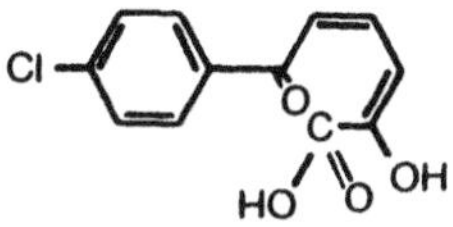

2-HYDROXY-6-OXO-6-(4'-
CHLOROPHENYL)HEXA-2,
4-DIENOIC ACID

177.

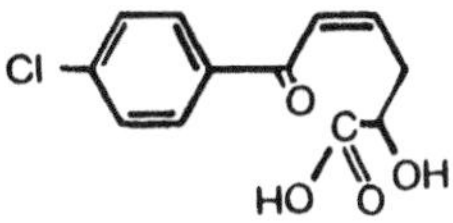

2-HYDROXY-6-OXO-6-(4'-
CHLOROPHENYL)-4-HEXENOIC
ACID

178.

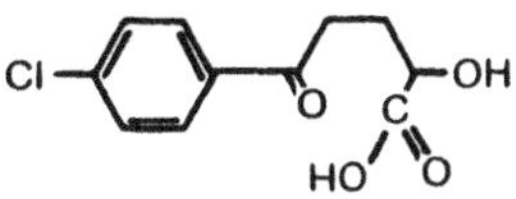

2-HYDROXY-5-OXO-5-(4'-
CHLOROPHENYL-
PENTANOIC ACID

179.

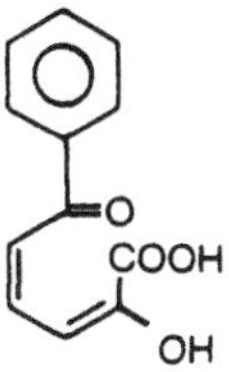

2-HYDROXY-6-OXO-6-
PHENYLHEXA-2,4-DIENOIC ACID

180.

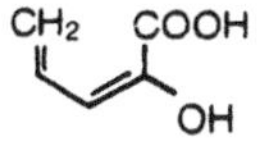

2-HYDROXYPENTA-2,4-DIENOIC ACID

181.

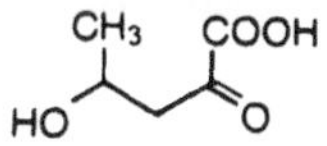

4-HYDROXY-2-OXOVALERIC ACID

182.

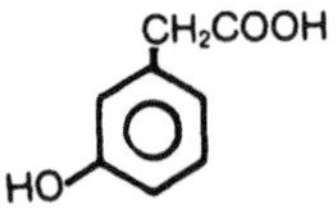

CH₂COOH

3-HYDROXYPHENYLACETIC ACID

183.

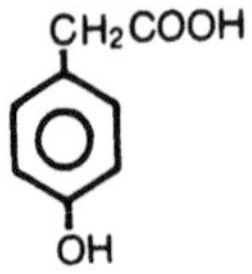

CH₂COOH

OH

4-HYDROXYPHENYLACETIC ACID

184.

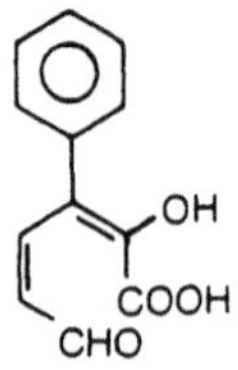

OH

COOH

CHO

2-HYDROXY-3-PHENYLMUCONIC
SEMIALDEHYDE

185.

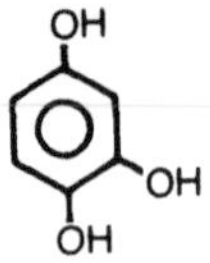

OH

OH

OH

HYDROXYQUINOL

186.

H   OH

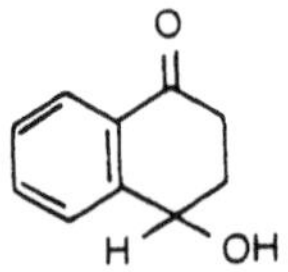

4-HYDROXY-1-TETRALONE

187.

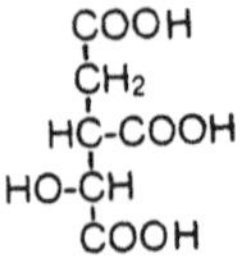

COOH
|
CH₂
|
HC-COOH
|
HO-CH
|
COOH

ISOCITRIC ACID

188.

C=O
O
C
OH

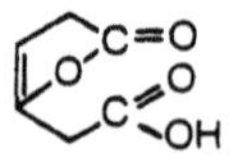

3-KETOADIPATE ENOL-LACTONE

189.

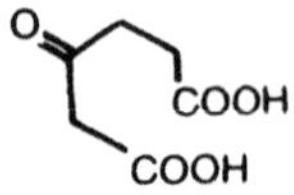

O

COOH

COOH

β-KETOADIPIC ACID

190.

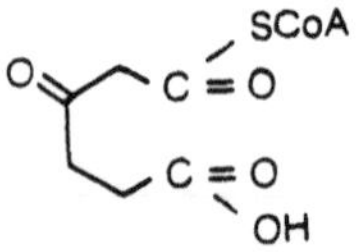

SCoA
C = O
O
C = O
OH

3-KETOADIPYL CoA

191.

COOH
O

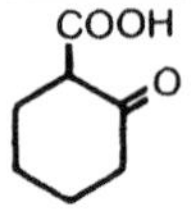

2-KETO-
CYCLOHEXANE-
CARBOXYLIC ACID

192.

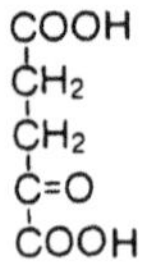

COOH
|
CH₂
|
CH₂
|
C=O
|
COOH

2-KETOGLUTARIC ACID

193.

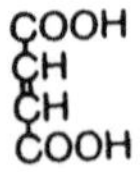

COOH
|
CH
‖
CH
|
COOH

MALEIC ACID

194.

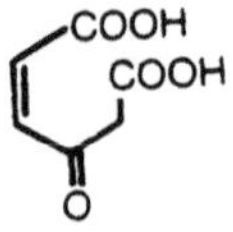

COOH
COOH

O

MALEYL-
ACETIC·ACID

195.

COOH CH₂COOH
C
O

O

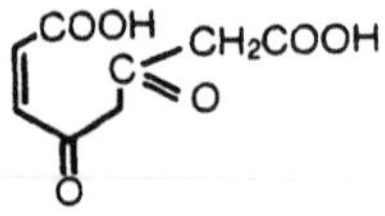

MALEYLACETOACETIC ACID

196.

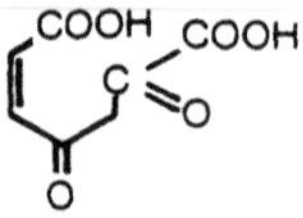

COOH COOH
C
O

O

MALEYLPYRUVIC ACID

197.

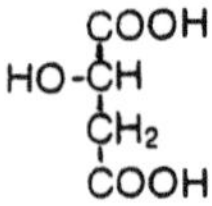

COOH
|
HO-CH
|
CH₂
|
COOH

MALIC ACID

198.

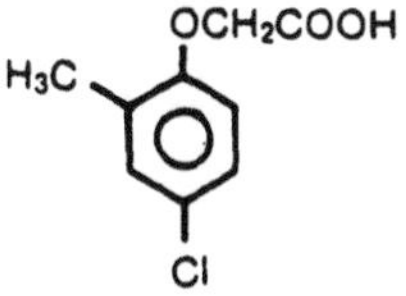

MCPA
(4-CHLORO-2-METHYL-
PHENOXYACETIC ACID)

199.

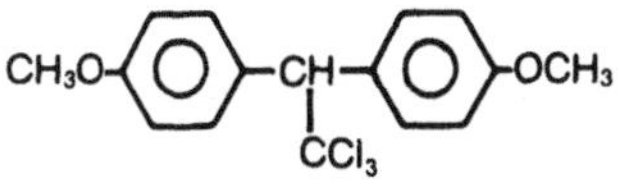

METHOXYCHLOR
[2,2-bis (p-methoxyphenyl)-1,1,1-
trichloroethane]

200.

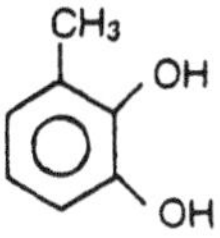

3-METHYLCATECHOL

201.

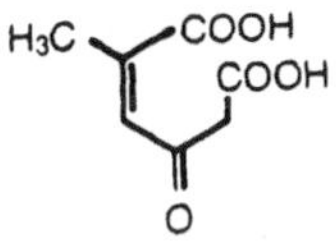

3-METHYLMALEYL ACETATE

202.

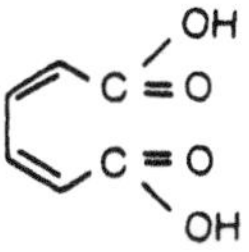

MUCONIC ACID

203.

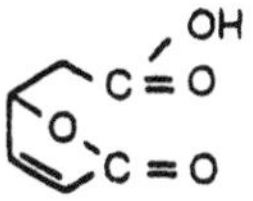

MUCONOLACTONE

204.

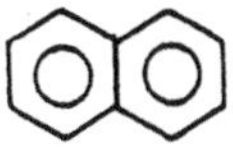

NAPHTHALENE

205.

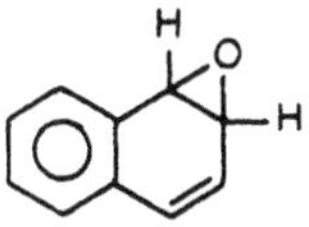

NAPHTHALENE 1, 2-OXIDE

206.

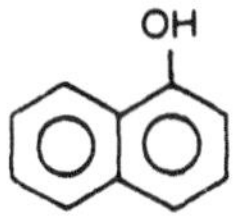

1-NAPHTHOL

207.

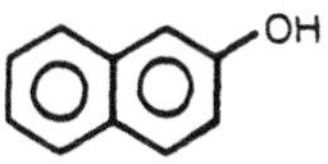

2-NAPHTHOL

208.

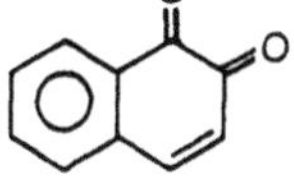

1, 2-NAPHTHOQUINONE

209.

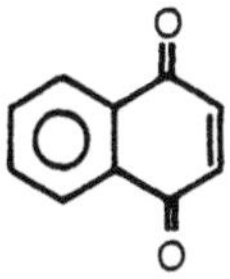

1, 4-NAPHTHOQUINONE

210.

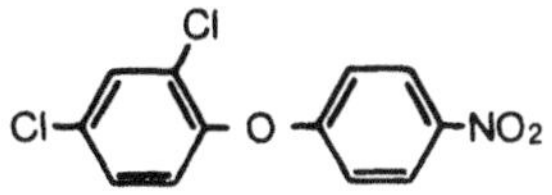

NITROFEN
(2, 4 - DICHLOROPHENYL -
4'-NITROPHENYL ETHER)

211.

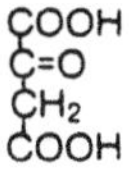

OXALOACETIC ACID

212.

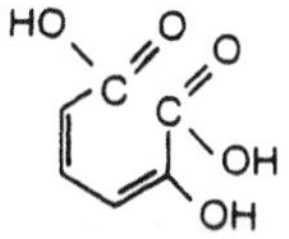

4-OXALOCROTONIC ACID

213.

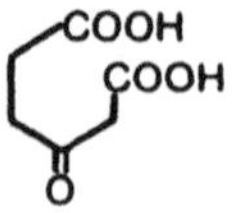

3-OXOADIPIC ACID

214.

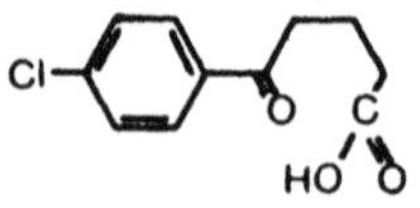

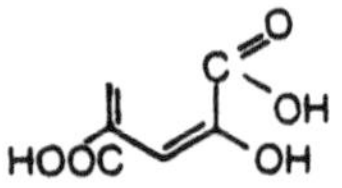

2-OXO-4-CARBOXYPENT-
4-ENOATE

215.

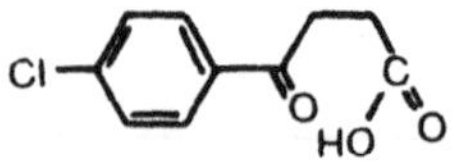

4-OXO-4-(4'-CHLOROPHENYL)-
BUTANOIC ACID

216.

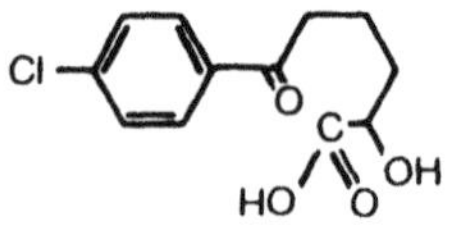

6-OXO-6-(4'-CHLOROPHENYL)-
2-HYDROXYHEXANOIC ACID

217.

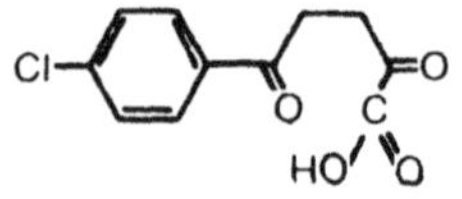

2, 5-OXO-5-(4'-CHLOROPHENYL)
PENTANOIC ACID

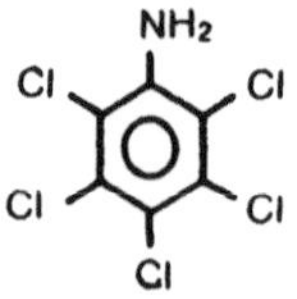

218.

5-OXO-5-(4'-CHLOROPHENYL)
PENTANOIC ACID

219.

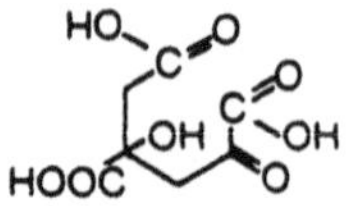

2-OXO-4-HYDROXY-4-
CARBOXYMUCONIC ACID

220.

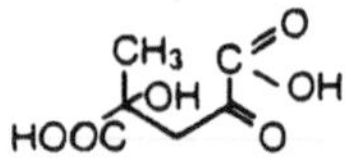

2-OXO-4-HYDROXY-
CARBOXYPENTANOIC ACID

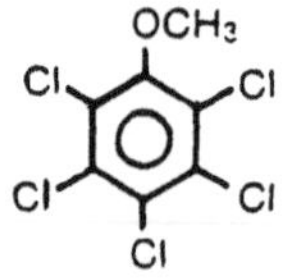

221.

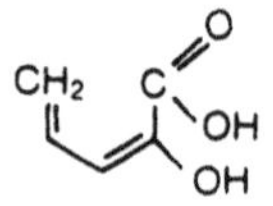

2-OXOPENT-4-ENOIC ACID

222.

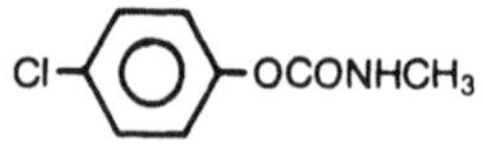

PCMC
(4-CHLOROPHENYL
N-METHYLCARBAMATE)

223.

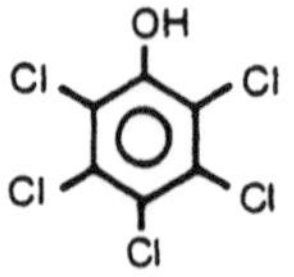

PCP
(PENTACHLOROPHENOL)

224.

PENTACHLOROANILINE

225.

PENTACHLOROANISOLE

226.

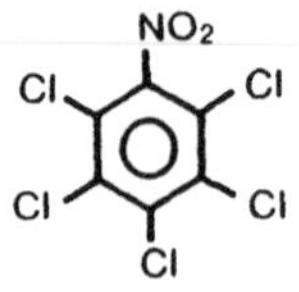

PENTACHLORONITROBENZENE

227.

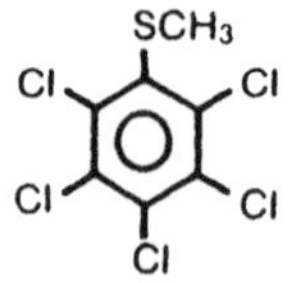

PENTACHLOROTHIOANISOLE

228.

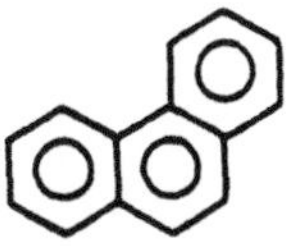

PHENANTHRENE

233.

COOH

OH

OH

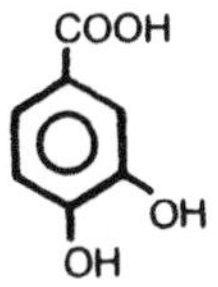

PROTOCATECHUIC ACID

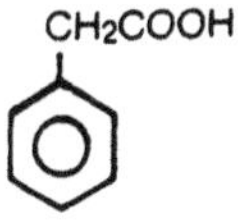

238.

COOH
CH₂
CH₂
COOH

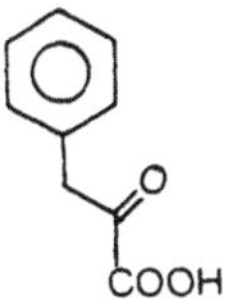

SUCCINIC ACID

229.

OH

PHENOL

234.

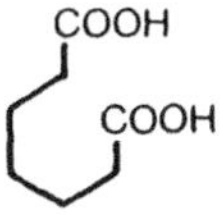

CH₃
|
C = O
|
COOH

PYRUVIC ACID

239.

COOH
CH₂
CH₂
C=O
SCoA

SUCCINYL-CoA

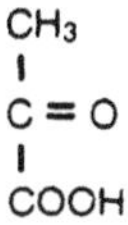

230.

CH₂COOH

PHENYLACETIC ACID

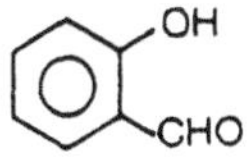

235.

OH

CHO

SALICYLALDEHYDE

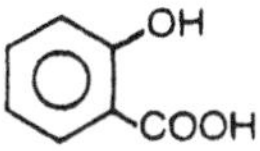

240.

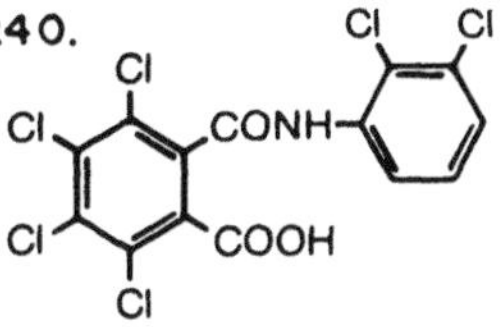

TECHLOFTHALAM
[N-(2, 3-dichlorophenyl)-3, 4,
5, 6-tetrachlorophthalamic acid]

231.

COOH

PHENYLPYRUVIC ACID

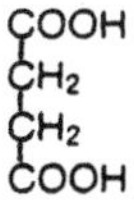

236.

OH

COOH

SALICYLIC ACID

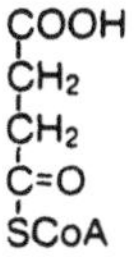

241.

Cl   Cl

Cl   Cl

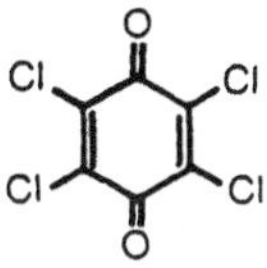

TETRACHLOROBENZOQUINONE

232.

COOH

COOH

PIMELIC ACID

237.

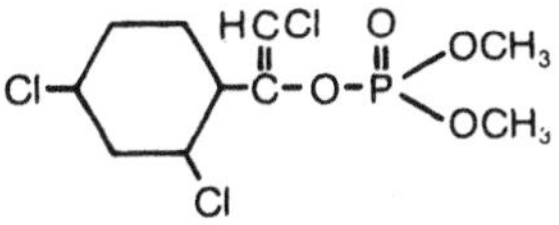

SD 8280
[2-chloro-1-(2, 4-dichlorophenyl)-
vinyl dimethyl phosphate]

242.

OH
Cl   OH

Cl   Cl
Cl

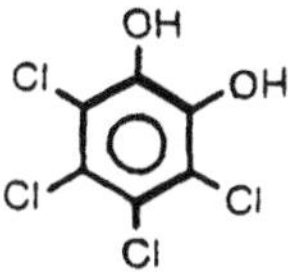

TETRACHLOROCATECHOL

243.

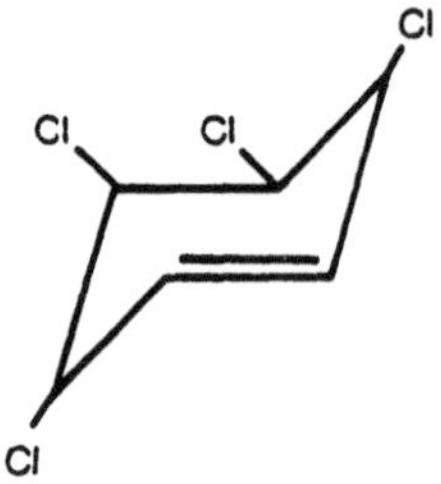

TETRACHLOROCYCLOHEXENE

244.

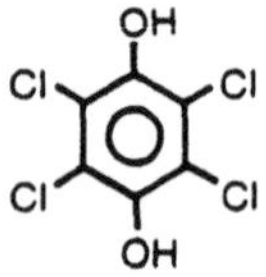

TETRACHLOROHYDROQUINONE

245.

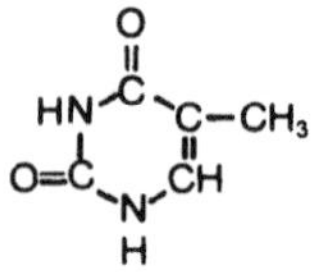

THYMINE

246.

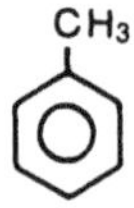

TOLUENE

247.

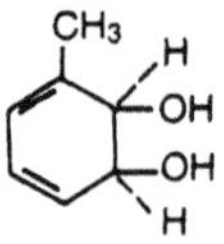

cis-TOLUENE
DIHYDRODIOL

248.

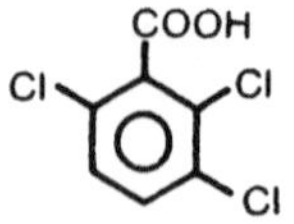

2, 3, 6-TRICHLORO-
BENZOIC ACID

249.

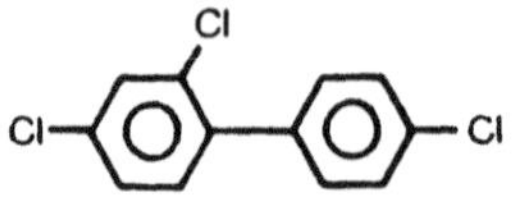

2, 4, 4'-TRICHLOROBIPHENYL

250.

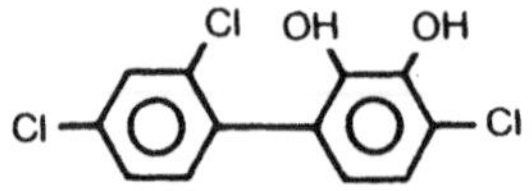

2, 4, 4'-TRICHLORO-2',
3'-DIHYDROXYBIPHENYL

251.

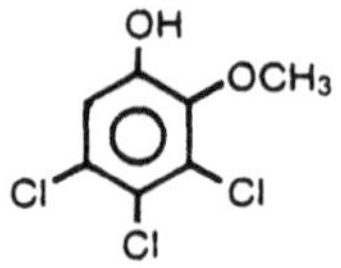

3, 4, 5-TRICHLOROGUAIACOL

252.

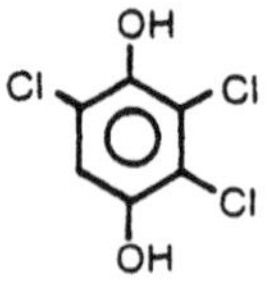

TRICHLOROHYDROQUINONE

253.

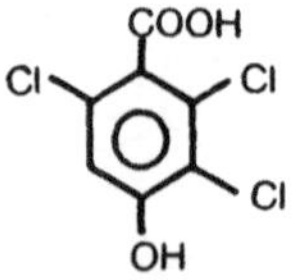

2, 3, 6-TRICHLORO-
4-HYDROXYBENZOIC ACID

254.

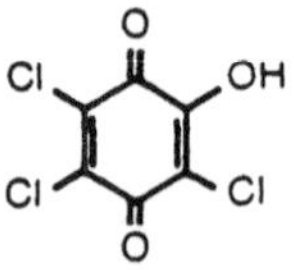

TRICHLOROHYDROXY-
BENZOQUINONE

255.

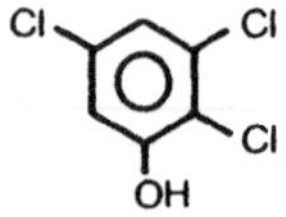

2, 3, 5-TRICHLORO-
PHENOL

256.

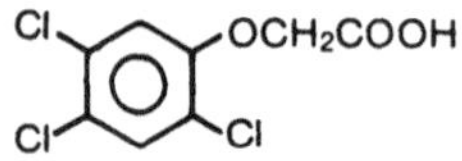

2, 4, 5-T
(2, 4, 5-TRICHLORO-
PHENOXYACETIC ACID)

257.

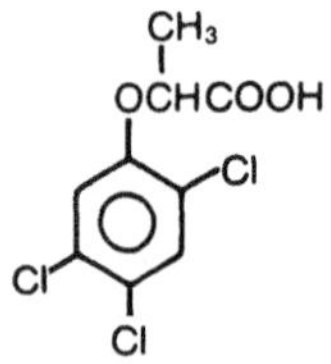

2-(2, 4, 5-TRICHLOROPHENOXY)PRO-
PIONIC ACID
(SILVEX)

258.

3, 4, 5-TRICHLOROSYRINGOL

259.

3, 4, 5-TRICHLOROVERATROLE

260.

URACIL

Index

About the Authors

Melissa L. Rochkind-Dubinsky is Senior Microbiologist for Toxicology and Risk Management Services at International Technology Corporation in Knoxville, Tennessee, where she designs and manages sampling programs for hazardous waste site characterization and conducts environmental endangerment assessments. She also evaluates applications of bioremediation to hazardous contamination of the evironment. Prior to this she was a postdoctoral research associate at the University of Tennessee in Knoxville and taught at the University of New Hampshire in Durham. Her research focuses on the fate and effects of toxic compounds with respect to microorganisms and aquatic ecology studies. She is a member of the American Society for Microbiology and the Society for Risk Analysis. Dr. Rochkind-Dubinsky received the B.S. degree (1975) in biological sciences from the University of Maryland in College Park, M.S. degree (1979) in marine science from the University of South Florida in Tampa, and Ph.D. degree (1983) in microbiology from the University of New Hampshire in Durham.

Gary S. Sayler is Professor of Microbiology and Ecology at the University of Tennessee in Knoxville, where he has taught since 1976. He also holds appointments in the Environmental Toxicology and Biotechnology Programs. The author or coauthor of over 60 articles and book chapters, he concentrates his research on biodegradation, plasmid evolution, and the ecological and toxicological impacts of environmental contaminants on the structure and function of microbial communities. An editorial board member of the *Journal of Microbiological Methods, Industrial Microbiology*, and *Applied and Environmental Microbiology*, he is a member of the American Society for Microbiology, American Association for the Advancement of Science, American Chemical Society, Society for Environmental Toxicology and Chemistry, and Society for Industrial Microbiology. Dr. Sayler received the B.S. degree (1971) in bacteriology from North Dakota State University in Fargo, Ph.D. degree (1974) in bacteriology and biochemistry from the University of Idaho in Moscow, and conducted postdoctoral research at the University of Maryland (1975).

James W. Blackburn is an Adjunct Research Associate in the University of Tennessee at Knoxville's Energy, Environment, and Resources Center. A biochemical engineer currently involved in the dynamics of catabolic genotypes in mixed-culture biological treatment systems, he has published over 40 reports, papers, and presentations in environmental science and biotechnology. He was recently the Manager of International Technology Corporation's biotechnology development effort focusing on technology applications to hazardous waste control using microbial processes. A registered Professional Engineer in Ohio, he is a member of the American Institute of Chemical Engineers, American Society for Microbiology, and American Association for the Advancement of Science. Mr. Blackburn received the B.S. degree (1973) in chemical engineering from the University of Cincinnati and pursued graduate studies in chemical engineering and microbiology at the University of Michigan and the University of Tennessee.